MACROCOGNITION IN TEAMS

Macrocognition in Teams

Theories and Methodologies

Edited by

MICHAEL P. LETSKY
Office of Naval Research, Arlington, USA,

NORMAN W. WARNER
Naval Air Systems Command, USA

STEPHEN M. FIORE
University of Central Florida, USA

C.A.P. SMITH
Colorado State University, USA

ASHGATE

Published by
Ashgate Publishing Limited
Gower House
Croft Road
Aldershot
Hampshire GU11 3HR
England

Ashgate Publishing Company
Suite 420
101 Cherry Street
Burlington, VT 05401-4405
USA

www.ashgate.com

British Library Cataloguing in Publication Data
Macrocognition in teams : theories and methodologies. -
(Human factors in defence)
1. Teams in the workplace 2. Cognition
I. Letsky, Michael P.
658.4'022

ISBN: 978-0-7546-7325-5

Library of Congress Cataloging-in-Publication Data
Macrocognition in teams : theories and methodologies / by Michael P. Letsky ... [et al.].
p. cm. -- (Human factors in defence)
Includes bibliographical references and index.
ISBN 978-0-7546-7325-5
1. Teams in the workplace. 2. Cognition. I. Letsky, Michael P.

HD66.M335 2008
658.4'022--dc22

2008005948

Printed and bound in Great Britain by
MPG Books Ltd, Bodmin, Cornwall.

Contents

List of Figures		*xi*
List of Tables		*xv*
Preface		*xiii*
Chapter 1	**Macrocognition in Teams**	**1**
	Michael Letsky and Norman W. Warner	
	Macrocognition in teams	1
	The Program	1
	Program objectives	2
	Limitations and scope	2
	Military perspective	3
	CKI program thrusts	4
	Research questions	5
	Model of collaborative team activity	6
	The macrocognition construct	7
	Theories of human cognition	7
	Characteristics of macrocognition	9
	Need for empirical research	10
	Summary	11
	Note	11
	References	11
Chapter 2	**Empirical Model of Team Collaboration Focus on Macrognition**	15
	Norman W. Warner and Michael P. Letsky	
	Introduction	15
	Empirical model of team collaboration	16
	Model components	19
	Empirical model of team collaboration example	24
	Inputs to the team	25
	Problem area characteristics	25
	Collaboration stages and cognitive processes	26
	Empirical validation of model	29
	Conclusions	30
	References	31

Chapter 3 Shared Mental Models and Their Convergence 35
Sara A. McComb
Shared mental models 36
Mental model convergence 38
Measuring shared mental models 41
Shared mental models and macrocognition 44
Implications 45
References 46

Chapter 4 Communication as Team-level Cognitive Processing 51
Nancy J. Cooke, Jamie C. Gorman and Preston A. Kiekel
Team communication and its relation to team cognition 53
Automating communication analysis 54
Applications of communication analysis 59
Conclusions 62
Acknowledgments 63
References 63

Chapter 5 Collaboration, Training, and Pattern Recognition 65
Steven Haynes and C.A.P. Smith
Introduction 65
Prior research 66
Method 71
Decision task and experimental procedure 72
Manipulation 73
Incentives 73
Data analysis and results 73
Outcome quality 74
Time to decision 74
Sharing 74
Bumping 77
Discussion 78
Implications for practice 80
Conclusion 81
Future research 81
Acknowledgments 81
References 82

Chapter 6 Toward a Conceptual Model of Common Ground in Teamwork 87
John M. Carroll, Gregorio Convertino, Mary Beth Rosson and Craig H. Ganoe
Introduction 87
A conceptual model of common ground in teamwork 87
Model implementation: example manipulations and effects 96
Discussion: toward a generative model 101

	Model productivity	102
	References	103
Chapter 7	**Agents as Collaborating Team Members**	**107**
	Abhijit V. Deshmukh, Sara A. McComb and Christian Wernz	
	Agent-based cognitive augmentation	108
	Designing effective human–agent teams	112
	Augmenting team macrocognitive processes	117
	Conclusions	121
	References	122
Chapter 8	**Transferring Meaning and Developing Cognitive Similarity in Decision-making Teams: Collaboration and Meaning Analysis Process**	**127**
	Joan R. Rentsch, Lisa A. Delise and Scott Hutchison	
	Introduction	127
	Distributed teams	128
	Cognitive similarity in teams	130
	Information-sharing among team members	131
	Transfer of expert knowledge to increase knowledge interoperability in teams	132
	Externalizing tacit knowledge	134
	Collaboration and meaning analysis process	137
	Acknowledgment	140
	Note	140
	References	140
Chapter 9	**Processes in Complex Team Problem-solving: Parsing and Defining the Theoretical Problem Space**	**143**
	Stephen M. Fiore, Michael Rosen, Eduardo Salas, Shawn Burke and Florian Jentsch	
	Overview of approach	143
	Individual and team knowledge-building	145
	UR about information – processes subservient to individual knowledge-building	147
	UR about teammates – processes subservient to team knowledge-building	148
	Developing shared problem conceptualization	150
	Option generation and team consensus development	153
	Outcome appraisal	156
	Processes occurring across functions	156
	Conclusions	157
	Acknowledgments	159
	References	159

Chapter 10 Augmenting Video to Share Situation Awareness More Effectively in a Distributed Team **165**
David Kirsh
Passing the bubble of situational awareness 167
Current technology and science 168
Slideshows are better than video's for presenting to non-experts 168
Animation vs stills: A brief review 171
Conjectures, testbeds and methods 173
Experiment one 174
Results: Experiment one 178
Experiment two 181
Results: Experiment two 183
Conclusions 184
Acknowledgments 185
References 185

Chapter 11 EWall: A Computational System for Investigating and Supporting Cognitive and Collaborative Sense-making Processes **187**
Paul Keel, William Porter, Mathew Sither and Patrick Winston
Introduction 187
EWall user interface 189
EWall agent system 194
Discussion 203
Conclusion and future work 205
Acknowledgments 205
References 206

Chapter 12 DCODE: A Tool for Knowledge Transfer, Conflict Resolution and Consensus-building in Teams **209**
Robert A. Fleming
Introduction 209
Basic cognitive assessment tags (CATs) 212
Knowledge elicitation approach 214
Display of CATs 215
Consensus 218
Summary 220
References 220

Chapter 13 Modeling Cultural and Organizational Factors of Multinational Teams **223**
Holly A.H. Handley and Nancy J. Heacox
Introduction 223
Support for team collaboration 223
Culture and work processes 224
I-DecS decision support tool 226

Usability and utility assessments 229
Integration with decision support system for coalition operations 231
Conclusion 231
Acknowledgment 232
References 232
Appendix: Cultural Dimensions and Organization Impacts 235
References for Appendix 236

Chapter 14 CENTER: Critical Thinking in Team Decision-making 239
Kathleen P. Hess, Jared Freeman and Michael D. Coovert
Introduction 239
Collaborative critical thinking theory 240
CENTER 245
CENTER validation study 249
Fielding CENTER 254
Conclusion 255
Acknowledgments 255
References 255

Chapter 15 Measuring Situation Awareness through Automated Communication Analysis 259
Peter W. Foltz, Cheryl A. Bolstad, Haydee M. Cuevas, Marita Franzke, Mark Rosenstein and Anthony M. Costello
Introduction 259
Situation awareness and team performance 260
TeamPrints: LSA-based analysis of real-time communication 262
Exploratory analyses: The NEO mission scenario data set 264
C3Fire simulation experiment 266
Conclusions and future directions 272
Acknowledgments 273
References 273

Chapter 16 Converging Approaches to Automated Communications-based Assessment of Team Situation Awareness 277
Shawn A. Weil, Pacey Foster, Jared Freeman, Kathleen Carley, Jana Diesner, Terrill Franz, Nancy J. Cooke, Steve Shope and Jamie C. Gorman
Introduction 277
Situation awareness and macrocognition 278
Multiple communications assessment approaches 280
Assessing situation awareness in large organizations: the Enron example 284
Methods 286
Results: Semantic and social network analyses of the Enron e-mail corpus 287
Results: Communication flow processes in the Enron corpus 291

Discussion and conclusions 295
Acknowledgments 300
References 300

Chapter 17 Shared Lightweight Annotation TEchnology (SLATE) for Special Operations Forces 305
Mark St. John and Harvey S. Smallman
Introduction 305
SLATE's two key innovations 306
SOF characteristics and requirements 306
SLATE: Integration of innovative collaboration concepts 309
Summary 318
Acknowledgments 319
References 319

Chapter 18 JIGSAW – Joint Intelligence Graphical Situation Awareness Web for Collaborative Intelligence Analysis 321
Harvey S. Smallman
Overview 321
JIGSAW: Joint Intelligence Graphical Situation Awareness Web 326
JIGSAW's support for team collaboration: A macrocognitive perspective 331
Summary 334
Acknowledgments 335
References 335

Chapter 19 The Collaboration Advizor Tool: A Tool to Diagnose and Fix Team Cognitive Problems 339
David Noble
Introduction 339
CAT cognitive framework 340
CAT functions 344
Summary 349
References 349

Chapter 20 Collaborative Operational and Research Environment (CORE): A Collaborative Testbed and Tool Suite for Asynchronous Collaboration 351
Elizabeth M. Wroblewski and Norman W. Warner
Collaboration in today's military 352
An empirical model of team collaboration 352
Naturalistic decision-making versus laboratory research 353
The team collaboration research spectrum 354
CORE testbed 358
CORE tool suite 360

Conclusion 361
References 361

Chapter 21 Plug-and-Play Testbed for Collaboration in the Global Information Grid 365
Alex Bordetsky and Susan Hutchins
Global information grid and collaborative technology 365
Plug-and-play testbed with reachback to GIG 366
Maritime interdiction operation experiments 367
Testbed networking layer: Self-forming wireless tactical network topology 368
Collaborative network tools 369
MIO scenario 370
Cognitive complexity of scenarios 373
Conceptual model of team collaboration 374
Types of problem-solving situations 374
Approach to measuring team collaboration 375
Coding process 375
Results 376
Significant macrocognitive processes 376
Firefighters from 9-11 379
New coding categories 382
Discussion 382
Challenges 383
Conclusions 383
References 383

Chapter 22 Naturalistic Decision-making Based Collaboration Scenarios 385
Elizabeth Wroblewski and Norman W. Warner
Introduction 385
Naturalistic decision-making scenarios 386
Collaboration scenarios 388
Generating an effective collaboration scenario 391
Common pitfalls 393
Conclusion 393
References 394

Chapter 23 Macrocognition Research: Challenges and Opportunities on the Road Ahead 397
Stephen M. Fiore, C.A.P. Smith and Michael P. Letsky
Summarizing theoretical and methodological developments in macrocognition research 397
How findings can make a difference 401
Research gaps and challenges for the future 404

Measuring macrocognition in the 'head' and in the 'world' 406
Conclusions 412
Acknowledgments 413
References 413

Index *417*

List of Figures

Figure 1.1	The current empirical model of team collaboration	6
Figure 2.1	Model evolution of team collaboration	17
Figure 2.2	Current empirical model of team collaboration	18
Figure 3.1	The mental model convergence process	39
Figure 4.1	The Input-Process-Output framework of team information processing; examples of input, process, and output variables are shown	52
Figure 5.1	Model of individual and team collaboration	68
Figure 5.2	Multiplayer collaborative game	72
Figure 6.1	Group process model	90
Figure 6.2	Activity system model	92
Figure 6.3	Integrated conceptual model	94
Figure 6.4	Empirical model: variables manipulated, measured, and controlled in our study of common ground in teamwork	97
Figure 7.1	Conceptual framework for designing effective human–agent teams	112
Figure 7.2	Examples of agent support	119
Figure 8.1	Components and cognitive processes of the C-MAP	138
Figure 9.1	Individual and team knowledge-building	146
Figure 9.2	Development of shared problem conceptualization	151
Figure 9.3	Consensus development process	153
Figure 9.4	Macrocognitive processes occurring across stages	157
Figure 10.1	Factorial design of experiment one	174
Figure 10.2	An annotated Starcraft game	176
Figure 10.3	Typical growth curves for players before and after receiving a bubble, and for bubble-free controls	177
Figure 10.4	Bubble recipients perform better	179
Figure 10.5	Live versus canned. Despite the long hours spent creating canned presentations, live presentations delivery more	179

Figure 10.6 Well chosen stills are better than video 180
Figure 10.7 Attributes found in different forms of live presentation 182
Figure 10.8 Venue 1 – presenter and venue 2 – recipient 183

Figure 11.1 Card layout and functions 190
Figure 11.2 Workspace view 192
Figure 11.3 Agent system 194
Figure 11.4 Design criteria 196

Figure 12.1 Elicitation and encoding of Cognitive Assessment Tags (CATS) 211
Figure 12.2 Basic IOB structure, and the CAT template 216
Figure 12.3 Two completed IOBs 216
Figure 12.4 A sample individual workspace and summary chart 217
Figure 12.5 Summary option recommendations from three team participants 218
Figure 12.6 Display of IOBs with maximum disparity between individual CATs 219

Figure 13.1 I-DecS simulation output 228
Figure 13.2 I-DecS 'compatibility of ways of doing business' 228

Figure 14.1 A structural model of collaboration (Warner, Letsky, and Cowen, in press) to which collaborative critical thinking serves as a global resource 243
Figure 14.2 Collaborative critical thinking consists of four types of team member interactions in a cyclical and adaptive process 245
Figure 14.3 CENTER probes are simple, unobtrusive and rapidly addressed 248
Figure 14.4 Responses to CENTER probes are summarized as means and range 248
Figure 14.5 CENTER's instant message capability 250

Figure 15.1 Total lost cells and lost houses by team and scenario at SAGAT stop 270
Figure 15.2 Total lost cells by scenario and team number at end of scenarios 270

Figure 16.1 Network representations of e-mail data 289
Figure 16.2 Similarity of flow patterns between critical Enron events and non-critical controls 292
Figure 16.3 Graph of C-values for the critical versus the August 1st baseline and one non-critical event 293
Figure 16.4 Chain difference between non-critical and baseline and chain difference between critical time periods and the August 1st baseline 294

Figure 17.1 The laptop/desktop version allows users to visually track and interact with multiple mission threads for command centers and reach-back assets 307
Figure 17.2 The palmtop version maximizes usability for the reduced screen real estate available on a mobile palmtop display and shows a single mission thread for dismounted, fielded users 307
Figure 17.3 Sequence of steps for receiving and viewing an infob, and responding by making a new infob 317

Figure 18.1 The intelligence cycle: the generic business process underlying analysis (modified from Heuer 1999) 323
Figure 18.2 Illustration of JIGSAW's three main modules 327
Figure 18.3 I-Scapes and their five main features 330
Figure 18.4 Illustration of JIGSAW's support for the four macrocognitive stages of the Warner, Letsky and Cowen (2005) model 333

Figure 19.1 Representation of knowledge risks, importance multiplier, and behavioural symptoms in CAT 344
Figure 19.2 Overall team diagnosis 345
Figure 19.3 Team responses to team questions about goals 346
Figure 19.4 Key team issues 347
Figure 19.5 Recommendations 348

Figure 20.1 Iterative team collaboration research spectrum 355
Figure 20.2 CORE tool suite and testbed 358

Figure 21.1 NPS testbed network with unmanned vehicles and sensors in San Francisco Bay 368
Figure 21.2 Area included in the testbed broadband wireless link to unmanned systems 369
Figure 21.3 A team of teams works within the MIO collaborative network 370

List of Tables

Table 2.1	Significant macro cognitive processes within each collaboration stage (experiments I and II)	29
Table 4.1	Correlations between LSA density component and other content metrics	58
Table 5.1	ANOVA results for outcome quality descriptive statistics	75
Table 5.2	Time to decision in seconds per trial	75
Table 5.3a	Sharing tool usage counts per trial	76
Table 5.3b	Correct sharing tool usage counts per trial	76
Table 5.4	Average number of bumps per trial	77
Table 5.5a	Average number of bumps per trial in core locations	78
Table 5.5b	Average number of bumps per trial in slot locations	78
Table 6.1	Examples of manipulations and measures by model component	98
Table 14.1	Collaborative critical thinking behaviors defined and applied to two objects of analysis: team products and team process	244
Table 14.2	CENTER measures critical thinking activities applied to team products and processes	247
Table 14.3	Seven experimental conditions	250
Table 14.4	Summary of means and sample size by condition	253
Table 15.1	Illustrative example of team communications during a C3Fire scenario	269
Table 16.1	Critical events at Enron from August – December 2001	287
Table 16.2.	Executive group	291
Table 16.3	Dates and descriptions of critical or non-critical events	293
Table 21.1	Maritime interdiction operations scenario participants and roles	372
Table 21.2	Cognitive process coding frequencies for air warfare, MIO and 9-11 firefighters scenarios	377–378
Table 21.3	Excerpt from MIO scenario: communications coding for developing solution alternatives	380

Table 21.4	MIO scenario: knowledge interoperability development and agreement on a final plan	381
Table 23.1	Conceptualizing macrocognition measures	406

Preface

This volume is about the study of high-level cognitive processing in teamwork. It represents the first forum upon which we hope the science of macrocognition will grow and prosper. We hope this volume motivates a deeper and richer dialogue amongst those interested in team performance in one-of-a-kind scenarios. Although the topic addressed here is focused on the study of high-level mental processing in complex problem-solving teams, the research products have the potential to be a major factor in our evolution toward global and international problem resolution. More specifically, the domain this volume considers includes teams engaging in brainstorming, decision-making, intelligence analysis, disaster relief, mission planning and many other activities that share the properties of short-fused, high stakes, one-of-a-kind, complex problem-solving. Such teams also often include multidisciplinary or multicultural members, uncertainty of source information and participants that are often geographically distributed and engaging in teamwork asynchronously. These characteristics broadly define the domain of interest considered here.

The evolving military strategy of 'Defense Transformation' dictates the use of small, autonomous, quick-response teams rather than use of massive forces with large weapons platforms. So too, the commercial world seems to be wrestling with these same issues as geometrically advancing sociotechnical systems produce floods of data and international partnerships create challenges in the rapid transfer of knowledge. Teams that work in this environment will need different sets of tools and new capabilities that might be described as advanced knowledge processing and the ability to rapidly exchange meaning among distributed team members.

In this book we have focused upon 'macrocognition' as the term descriptive of the cognitive processes to be studied. The term has been used previously to describe the study of cognition in real-to-life situations, but here we expand the concept through a reductionism approach that proposes that these fundamental concepts can be studied both in a laboratory environment and in the field setting, with the results being extensible for use in other applications. A more detailed definition of macrocognition and a conceptual model is offered in the introductory chapter. The purpose of this preliminary model is to serve as a conceptual framework to understand macrocognition in teams along with focusing research efforts. Since the study of macrocognition is truly interdisciplinary including contributions from psychology, philosophy, computer science, mathematics and business management many of the terms used that have the same label have different meanings across the contributing disciplines. This has been a significant problem both in terms of communication of findings as well as determining where best to publish the research findings. To date the psychology literature base has been the main target of publications generated by the program. As the program continues to grow, efforts will be made to expand this presence into other areas. In terms of selected readings in the various chapters, it may be worthwhile to study the definitions of terms as provided in Chapter 2 to provide a contextual reference for non-sequential topics throughout the text.

As previously mentioned, macrocognition is one component that comprises team performance. For the purpose of defining our scope of effort, we recognize that training, human behavior, team formation, expertise and many other factors contribute to team performance. However, for study of our central topic these factors are assumed to be held constant. It is also recognized that many enabling technologies contribute to team performance including sociotechnical systems, human factors in design issues and human performance modeling of perception, attention and reaction. These technologies are regarded as necessary and supporting tools in the study of team performance but are not considered as an area to be examined within the scope of this text.

The topic of study described herein had its inception in the fall of 1999 with the establishment of the Collaboration and Knowledge Interoperability Program at the Office of Naval Research. The nature of the program is 'Discovery and Invention', ONR's term for fundamental research. The program is supplemented by Small Business Innovative Research (SBIR) projects which serve as a source of funding for applications work. A Multidisciplinary University Research Initiative (MURI) project was started in 2005, adding a significant additional increment of basic research effort. The structure and strategy of the CKI program was initiated then from these funding sources in the form of a basic research investment, supplemented by two testbeds for technology demonstration purposes and a series of applications projects to develop tools to support team problem-solving and decision-making. The structure of this volume also follows from this alignment of projects. The first section reports on the contributions of research grants, both at various universities and Navy labs. The second section reports the results of the various applications (SBIR) projects that are part of the program and the final section discusses the structure and employment of the two research product testbeds that are part of the program.

In summary, this book is intended to serve both academic and applications interests. Products could be employed in both military and commercial settings as is the objectives of the SBIR projects. The contribution of this volume and the program is aimed at improving team performance, specifically quick response, *ad-hoc* teams faced with complex problem-solving tasks. This objective is undertaken with the expectation that these kinds of teams with these characteristics and domains will become ubiquitous in the future of team cognition.

Michael P. Letsky

Chapter 1

Macrocognition in Teams

Michael Letsky and Norman W. Warner

Macrocognition in teams

The concept of *Macrocognition in Teams* began as the research kernel of the larger issue of how to understand and facilitate complex collaborative team activity – specifically collaboration in quick-response *ad hoc* teams. This topic was driven by interest from both commercial and military communities as the forces of evolving sociotechnical systems, globalization and pervasive information accessibility changed the nature and dynamics of team activity. It is becoming increasingly evident that there are internalized, non-quantifiable, mental processes at work as teams collect, filter, process and share information for problem-solving purposes. More specifically, teams are engaging in the process of knowledge building including creating new 'team-level' knowledge that is not the mental property of any one individual. It is this knowledge building process that macrocognition seeks to investigate. What are the supporting components, are they measurable, are they consistently present, are they subject to lab experimentation, are they extensible and how do they interact and affect team performance? This was the motivation for the initiation of the Collaboration and Knowledge Interoperability (CKI) Program that was established at the Office of Naval Research (ONR) in 1999.

The Program

The core of the CKI Program is focused on basic research, in the terms of the ONR, 'Discovery and Invention'. The program is principally an academic grants program with a multidisciplinary perspective and a core in cognitive science. The largest basic research project is in the form of a Multidisciplinary University Research Initiative (MURI) which is managed by the University of Central Florida. About half of the CKI program consists of a number of small business innovative research (SBIR) initiatives, which are 2–4 year projects with the specific objective of producing usable products from the underlying CKI technology that can be transitioned to various user communities. The program therefore maintains a research focus on basic science phenomena and also produces short-term products as individual components of the technology mature. Customers are typically military-based but the SBIR program also strongly encourages commercialization.

Program objectives

The objective of the CKI program is to respond to emerging needs in both the military and business environments to better understand and improve the effectiveness of team decision-making in complex, data-rich situations. Specifically, there is a need to better understand cognitive processes employed when collaborating to solve problems that are characterized as one-of-a-kind, time compressed, high-stake problems, supported by uncertain, open source data. Advances in communications technology and the need to involve a broader community in team decision-making further characterize the decision-making environment as distributed and asynchronous in nature and involving interdisciplinary and multicultural participants. In order to simplify the description of the domain of interest, the term 'Transformational Teams' (TT) is used to represent characteristics of interest. TT is defined by the following characteristics:

- Unstructured, agile teams
- Distributed and asynchronous team member relationships
- Heterogeneous members (multidisciplinary, multicultural, etc.)
- *Ad-hoc*, naturalistic decision-making (NDM) and problem-solving tasks
- Short duration, high-stress problems
- Uncertainty in source information
- Dynamic information
- Rotating team members.

The science basis of the program is in understanding team cognition and knowledge building/sharing in support of team decision-making. Specific issues include individual and team knowledge construction, development of knowledge interoperability, development of shared awareness and achieving team consensus. Principal military drivers are derived from the need to support new joint and coalition decision-making demands in a defense transformation environment that includes asymmetric warfare challenges (Jensen 2002). In short, the program attempts to integrate the human component into the Defense Department's Network-Centric Operations (NCO) concept. This will be accomplished by delivering the tools to make human connectivity an integral component in the platforms and sensors that comprise a Network-Centric battlespace. Products of the program include tools that provide the cognitive and computational support for manipulation of knowledge often used in visual displays for decision-making, along with knowledge used to represent and transfer meaning for improved situational awareness for improved team performance in strategic-level decision-making. Examples of operational uses include open source intelligence analysis, Special Operations support, in-route mission analysis and crisis action team support.

Limitations and scope

The focus of the inquiry in this area is from a cognitive perspective. It is recognized, however, that the area is interdisciplinary in nature and in very close alignment with information science as an enabling technology. Further, the program does not focus on classic social and human behavior issues in view of the distributed and asynchronous nature of the problem area. Other associated cognitive-based technologies, including computational human performance models such as ACT-R (Anderson and Matessa 1998) and SOAR (Kennedy and Trafton 2006) are recognized as important tools but are also considered enabling/supporting technologies to address the central problem of team cognition and problem-solving.

Military perspective

From a military perspective, National Defense Policy Transformation from platform-based, large-scale battles to quick reaction, mobile force operations in discrete events has resulted in new challenges in team decision making. Network-Centric Operations policies have been refined, from a Navy perspective, into the FORCEnet concept which requires fully netted situational awareness in a state of the art naval enterprise data network – fully interoperable and seamlessly integrated – to provide information superiority in direct support of the warfighter. Timely responsiveness in this new environment will require a renewed focus on leveraging the ability of human operators to perform complex and collaborative analysis and decision making under severe time constraints (Joint Vision 2010). Characteristics of this new environment include:

- Asymmetric warfare
- Asynchronous and distributed command-level decision-making
- Operations with joint, coalition, non-government and volunteer organization partners
- Dealing with open-source (uncertain, conflicting, partial, non-official) data
- More focus on humanitarian relief, disaster aid and politically-charged operations
- Rapidly changing team members and associated organizational structures
- Culturally diverse partners
- Short turnaround, high stakes, crisis-driven decision-making
- More human interface with agents/automated systems.

Given that these characteristics represent the military environment of the future, the following capabilities will be required:

- Knowledge interoperability among heterogeneous participants
- Conceptual and executable analytical tools to support collaboration
- Language independent, situation-at-a-glance visualization of meaning
- Development of team-shared situational awareness
- Establish teamwork knowledge (common ground and trust)
- Agent interfaces that can react to human variation
- Flexible team/organizational response to workload variation
- explicitly
- Models/computational methods for team naturalistic decision-making
- Cognitive-friendly operator interface with supporting agents.

CKI program thrusts

The program objective is to develop cognitive science-based tools, models, computational methods, and human–agent interfaces to help attain common situational awareness among heterogeneous, distributed team members engaged in asynchronous, quick response collaboration for issue resolution, course-of-action selection or decision-making. The functions are performed in a collaborative team environment with *collaborative* defined as the cognitive aspects of joint analysis or problem-solving for the purpose of attaining shared understanding sufficient to achieve situational awareness for decision-making or creation of a product. A key component of collaboration is *knowledge building*, the identification, collection, fusion and presentation of data for the purpose of attaining comprehension of a situation.

The CKI program consists of four principal thrusts:

1. Conversion of data into knowledge – individual and team knowledge development
2. Develop knowledge interoperability – standardized representation of meaning
3. Team shared understanding – mental model convergence
4. Team consensus development – selection and validation of team solution

Individual and team knowledge development

Individual knowledge development is the process of individual team member conversion of data into knowledge about the situation at hand. The process includes collecting, fusing, filtering, representing and displaying (visualizing) knowledge produced in a format that can be shared. Products of this process are often knowledge objects or iconic representations of accumulated knowledge sets. Team knowledge development is defined as all team members participating to clarify information for building common team knowledge. The process includes each team member collecting and internally synthesizing the new team knowledge into their current knowledge structures.

Develop knowledge interoperability

A capability to make knowledge 'interoperable', that is, convertible into a format that can be understood by all, including multidisciplinary and multicultural team members. Supporting technology should provide a translation capability to deal with individual personal ontologies and perceptual anchors or, in essence, concept translators. The supporting technology should take advantage of common perceptual sorting processes such as size, color, shape, virtual position, etc.

Develop team shared understanding

The process of reaching team agreement about the characteristics and meaning of the situation at hand within the bounds of the objectives of the collaborative activity (decision-making, intelligence analysis, COA selection, etc.). Not all team members will have equal depth and breadth in all supporting knowledge but each has sufficient knowledge to complete his role as a team member. Key to achieving this capability is the availability of measures and metrics necessary to quantify levels of shared understanding.

Team consensus development

The process of selection and validation of a team solution. Problem solving and solution generation is a key component of team consensus development, which includes the process of creating new ideas and solution alternatives. Possible support tools would help individuals to observe and identify trends and to identify new relationships. Support tools include capabilities for critical thinking, negotiating, brainstorming, creative thinking, sorting and displaying knowledge. During validation of a team solution, analytical and computational tools may be needed to sort and compare aspects of various complex alternatives. Measures of mental model convergence would be required along with advanced voting mechanisms. Tools are also needed to 'set boundaries' or to keep deliberation in line with overall objectives and/or constraints of the solution space. Support may also be needed for a 'what-if' analysis capability and mental simulation of alternatives/combinations.

Research questions

From the above thrusts arise the following research questions:

- How does asynchronous and distributed activity affect team collaboration – what are useful support tools?
- Can knowledge structures and/or mental models be described/classified/taxonomized?
- What methods are there to develop convention-free representation of meaning/knowledge?
- How are joint conventions of meaning established?
- Do mental models converge in the collaborative process? If so, what is the mechanism?
- Model the process of developing shared understanding/collaborative knowledge.
- What are the methods for collecting and storing expertise?
- How does team trust/understanding individual goals/identification of differences affect collaboration?
- What is the role of critical thinking in collaboration?

- How is knowledge developed/distributed/stored in a collaborative group, role of transactive memory?
- What are the non-language/text factors in collaboration?
- What is the perceptual basis of shared understanding?
- What role should agents play in aiding collaboration?
- What role should computational models play in supporting collaboration?
- What metrics can be developed to measure collaboration performance?

Model of collaborative team activity

Given the described problem domain, an initial model of team collaboration was developed (see Chapter 2, this volume) as a framework to understand team collaboration during one-of-a-kind, time-critical, asynchronous, distributed collaborative problem-solving. Figure 1.1 illustrates the various components of the team collaboration model, which are described in detail in Chapter 2.

The objectives of the proposed team collaboration model are to:

- Identify and understand the macrocognitive processes consistently present during collaborative team activity.
- Conceptually integrate and visualize the relationships among identified macro cognitive processes.
- Provide a model-based approach for experimentation and hypothesis testing
- Provide a framework to guide development of computational models of macro cognitive processes and relationships.
- Develop metrics and measures to quantify the impact of macro cognitive processes on team collaboration performance.

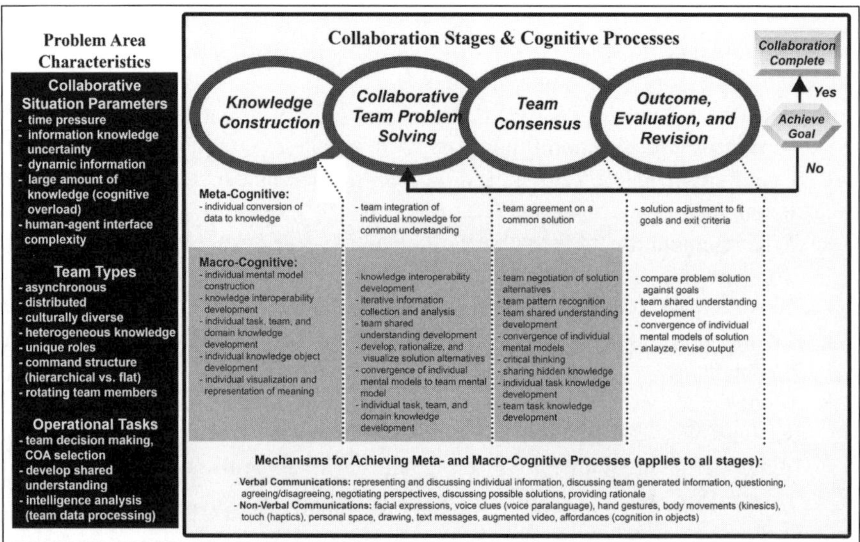

Figure 1.1 The current empirical model of team collaboration

- Provide design guidelines for tool development supporting team collaborative activity.
- Identify and prioritize future research topics/issues in team collaboration.

As further discussed in Chapter 2, the cognitive processes in this model are described at a *macro level* (e.g., knowledge interoperability development, team shared understanding development, convergence of individual mental models to team mental model) rather than at the *micro level* (e.g., information processing, neural–cognitive). The lower-level cognitive activities such as perception of new information (Marr and Poggio 1979), storage into working and long term memory (Card, Moran and Newell 1983), memory chucking (Newell and Simon 1972) are considered micro cognitive activities and not part of the model. This investigation of the cognitive processes at the macro level was undertaken because of the limitations in using microcognitive processes to explain higher-order decision-making mechanisms.

The macrocognition construct

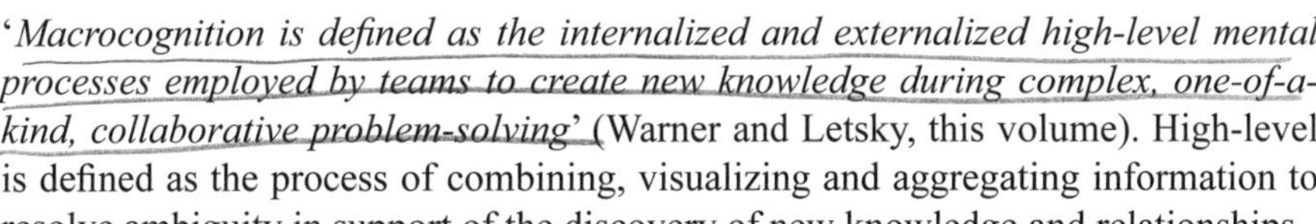

'*Macrocognition is defined as the internalized and externalized high-level mental processes employed by teams to create new knowledge during complex, one-of-a-kind, collaborative problem-solving*' (Warner and Letsky, this volume). High-level is defined as the process of combining, visualizing and aggregating information to resolve ambiguity in support of the discovery of new knowledge and relationships.

Internalized processes are those higher-level mental (cognitive) processes that occur at the individual or the team level, and are not expressed externally (e.g., writing, speaking, gesture), and can only be measured indirectly via qualitative metrics (e.g., cognitive mapping, think aloud protocols, multidimensional scaling, etc.), or surrogate quantitative metrics (e.g., pupil size, galvanic skin response, functional magnetic resonance imaging [fMRI], etc.). These processes can become either fully or partially externalized when they are expressed in a form that relates to other individual's reference/interpretation systems (e.g., language, icons, gestures, boundary objects).

Externalized processes are those higher-level mental (cognitive) processes that occur at the individual or the team level, and which are associated only with actions that are observable and measurable in a consistent, reliable, repeatable manner or through the conventions of the subject domain have standardized meanings.

To fully understand the meaning and scope of the term macrocognition, it is necessary to compare and contrast this concept with the various theories of human cognition.

Theories of human cognition

Early philosophers such as Descartes (1641) and Kant (1781) have examined basic questions of the origin of knowledge and human thought. However, it was not until after the inability of behaviorism (Skinner 1985, 1989) to explain internal representations that cognitive psychology emerged. A multitude of theories of human cognition began to develop. Several theories (Weiner 1948; Shannon 1949; Simon 1979, 1989; Miller 1988; Wickens 1992) explain human cognition in terms of *an information-processing* model, which focuses on information representation, processing and computation. Other theories (Chomsky 1957; Cohen and Schaff 1973) believe that the development of knowledge and the processes of thought are achieved through human *language*. Piaget (1975) focused on *developmental biology*, which described not only the different components of human cognition but also the developmental stages of cognition. As computer science matured, several theories developed (Newell and Simon 1956; Feigenbaum and Feldman 1963; Minsky 1981; Anderson 1993) that explain human cognition in terms of a *computer computational model*. These computational models varied in how they explained cognition ranging from computational logic, production rules to frames. Other theories (Rumelhart 1990; Churchland 1989; Rosenberg 1988) emphasize physiology in understanding human cognition. These theories use *physiological neural networks represented in computational models* to explain cognition and its processes. Davidson, Deuser and Sternberg (1994) proposed a theory of *metacognition*, which is knowledge of one's own cognitive processes, in explaining how human cognitive processes are used in problem-solving. These various theories of human cognition all concentrate on the *individual* as the unit of analysis and focus on the *internalized mental representations of the individual*. Recent theories of human cognition show a shift away from the individual and internalized mental representations to a concentration on *groups*, and *externalized representations* (e.g., artifacts, communities). Such theories include:

- **Mediated cognition**: Vygotsky (1934/1986) proposed a different view of cognition as social and collaboratively mediated.
- **Distributed cognition**: Suchman (1987); Winograd (1996); Pea (1993) and Hutchins (1996) emphasize not viewing the mind as isolated from artifacts and other people.
- **Situated learning**: Lave's (1988, 1991) approach of applying situated perspective to learning showed how learning is a community process.
- **Knowledge building**: Scardamalia and Bereiter (1996) developed the concept of community learning with a model of collaborative knowledge building in computer-supported environments.
- **Meaning making**: Koschmann (1999) presented knowledge building as meaning making, which draws upon theories of conversation analysis and ethnomethodology.
- **Activity Theory**: Nadri and Kaptelinin (1996) characterize activity theory as an attempt to integrate three perspectives: (1) the objective, (2) the ecological, and (3) the sociocultural. Activity theory analyzes human beings in their *natural* environment and takes into account cultural factors and developmental aspects of human cognition. Mental processes manifest themselves in external actions performed by a person, so they can be verified and corrected.
- **Macrocognition**: Klein, Moon, Klein, Hoffman and Hollnagel (2003) use the term macrocognition to describe the cognitive functions that are performed by individuals, teams and technology in *natural* versus artificial laboratory decision-making settings. Macrocognition often involves ill-defined goals,

whereas micro cognitive tasks usually have well-defined goals. Macrocognition comprises the mental activities that must be successfully accomplished to perform a task or achieve a goal. *Microcognition* is defined as the building blocks of cognition, the processes that are invariant and serve as the basis for all kinds of thinking and perceiving.

- **Group cognition**: Stahl (2006) and Cooke (2005) focus on examining how the group constructs intersubjective knowledge that appears in the group discourse rather than theorizing about the internal mental process of the individual. Stahl found that meaning is created across the utterances of different people. Therefore, meaning is not the product of the mind but a characteristic of group dialog. These discoveries required a theoretical paradigm shift in how teams build knowledge and collaborate. Under this new theoretical paradigm, team members need to exercise *high-level cognitive activities* (e.g., team knowledge development about the problem; producing knowledge artifacts for problem clarification; negotiating to reach team agreement).

Characteristics of macrocognition

The term macrocognition (Warner and Letsky, this volume) consists of some of the same characteristics embodied in the recent theories of human cognition together with several unique characteristics (highlighted in **bold**). Macrocognition in collaborative teams consists of the following characteristics:

- The unit of analysis includes both the individual team member and the whole team because of the unique macrocognitive processes operating at the individual and team level.
- Cognitive activities are analyzed at a high level because of the limitations in using microcognitive processes to explain higher order decision-making mechanisms
- **Focus on both internalized and externalized mental processes employed by team members during complex, one-of-a-kind, collaborative problem-solving**
- **Macrocognitive processes can be empirically studied in the lab and in operational field settings given domain-rich collaborative problem-solving scenarios**
- Macrocognitive processes (i.e., **internalized and externalized**) occur during team member interaction (i.e., socially and collaboratively mediated) and are influenced by the artifacts in the environment
- Macrocognitive processes develop and change over time
- Macrocognitive processes are domain dependent and collaboration environment dependent (e.g., face-to-face versus asynchronous, distributed collaboration tools).

Need for empirical research

Macrocognition in team collaboration is a relatively new area of human cognition. As such extensive empirical research is needed to address the previously outlined research questions. This empirical research, which is being conducted by the various CKI principal investigators, will provide a deeper understanding of macrocognition and how this construct can be supported by various collaboration tools to enhance team collaboration. A critical component to conducting empirical research in macrocognition is the use of *sensitive measures* of the various macrocognitve processes. Also, effective *metrics* for measuring macrocognition has been a problem in past research. Current techniques used to measure macrocognition have included verbal protocol analysis, discourse analysis, concept mapping, multidimensional scaling, subjective questionnaires, latent semantic analysis along with measuring outcome measures such as quality of decisions/solutions, and time to develop a solution. Some recent advances in macrocognition measures include automated analysis of team communication data (e.g., flow techniques and dynamic modeling) developed by Cooke, Gorman, and Kiekel (Chapter 4), structured questionnaires for measuring mental model convergence (McComb, Chapter 3), and a collaboration and meaning analysis process (C-MAP) tool to measure externalization of knowledge (Rentsch, Chapter 8). However, additional measures are needed that are sensitive to measuring the various *internalized* and *externalized* macrocognitive processes, which are outlined in the team collaboration model and described in Chapter 2. As part of the Multidisciplinary University Research Initiative (MURI) at the University of Central Florida, one major technical thrust is the development of new sensitive measures and metrics for measuring internalized and externalized macrocognitive processes. These new measures and metrics should provide researchers with the ability to effectively quantify macrocognitive processes resulting in a deeper scientific understanding of macrocognition.

In addition to the empirical project research conducted by the CKI investigators, the CKI program has developed two major testbeds for simulation of operational conditions for testing of maturing products and concepts. The **C**ollaborative **O**perational and **R**esearch **E**nvironment (CORE) testbed located at the Naval Air Systems Command, Patuxent River, Maryland has the capability to evaluate mature CKI products (e.g., CAT tool, Chapter 19; I-DecS, Chapter 13; SLATE, Chapter 17) using collaborative teams to solve realistic military scenarios (e.g., Non-Combatant Evacuation Scenarios and Intelligence Analysis and Mission Planning Scenarios). These scenarios also serve as an experimental tool to validate the macrocognitive processes. In addition, these scenarios are provided to CKI investigators, which provides realistic context for their research.

Also, this testbed has the capability as an empirical research environment for examining specific macrocognitive processes during complex collaborative team problem-solving. Chapters 20 and 22 describe the CORE testbed and the scenarios respectively. The second testbed is the Testbed for Collaboration in the Global Information Grid located at the Naval Postgraduate School (NPS), Monterey, California. The NPS testbed focuses on understanding team collaboration and the associated macrocognitive processes in a high-fidelity, complex, decision-making environment, specifically the Global Information Grid. Chapter 21 describes the NPS testbed in detail.

Summary

The chapters of this book are centered on three focus areas: (1) understanding the theoretical constructs of macrocognition in collaborative teams, (2) development of collaboration tools to support macrocognitive processes during team collaboration, and (3) development and use of collaborative testbeds and realistic problem-solving scenarios for CKI product evaluation and transition. Chapter 2 provides our initial thoughts on how teams collaborate given one-of-a-kind, time compressed, high-stakes problems along with the macrocognitive processes that support that type of team collaboration. Our initial thoughts are represented in the form of a conceptual model, which is based on two major empirical studies. We envision that this model will be refined as more empirical research is conducted on the specific questions identified earlier in this chapter. Chapters 3 through 10 present our latest research on specific areas of macrocognition such as convergence of mental model content and structure, communication in team cognition, collaborative team pattern recognition, agent support of macrocognitive processes, constructing common ground, measuring knowledge structures for meaning transfer along with a taxonomy of macrocognitive processes. Chapters 11 through 19 describe various team collaboration tools developed under the CKI program, which are designed to support various macrocognitive processes during team collaboration. Chapters 20 through 22 offer a description and discussion of the CKI testbeds along with the realistic scenarios used in these testbeds. The book concludes with Chapter 23, which describes the lessons learned in understanding macrocognition in collaborative teams together with a roadmap of where we need to go in the future.

Note

The views, opinions, and findings contained in this article are the authors and should not be construed as official or as reflecting the views of the Department of Navy.

References

Anderson, J.R. (1993). *Rules of the Mind.* (Hillsdale, NJ: Lawrence Erlbaum Associates).

Anderson, J.R. and Matessa, M. (1998). 'The Rational Analysis of Categorization and the ACT-R Architecture', in M. Oaksford and N. Chater (eds) *Rational Models of Cognition,* pp. 197–217. (Oxford: Oxford University Press).

Card, S., Moran, T. and Newell, A. (1983). *The Psychology of Human–Computer Interaction*. (Hillsdale, NJ: Erlbaum).

Chomsky, N. (1957). *Syntactic Structures*. (The Hague: Mouton).

Churchland, P.M. (1989). *A Neurocomputational Perspective: The Nature of Mind and the Structure of Science*. (Cambridge, MA: Bradford/MIT Press).

Cohen, R. and Schaff, A. (1973). *Language and Cognition*. (New York, NY: McGraw-Hill).

Cooke, N.J. (2005). 'Communication in Team Cognition', *Office of Naval Research Collaboration and Knowledge Management Workshop Proceedings*, University of San Diego, San Diego, CA. Available at http://www.onr.navy.mil/sci_tech/34/341/cki/.

Davidson, J.E., Deuser, R. and Sternberg, R.J. (1994). 'The Role of Metacognition in Problem Solving', in J. Metcalfe and A.P. Shimamura, (eds), *Metacognition*. (Cambridge, MA: MIT Press).

Descartes, R. (1641). *Descartes' Meditations*. Translated and edited by D. Manley and C. Taylor. Retrieved June 2, 2003. http://philos.wright.edu/descartes/meditations.html.

Feigenbaum, E.A. and Feldman, J. (eds) (1963). *Computers and Thought*. (New York: McGraw-Hill).

Hutchins, E. (1996). *Cognition in the Wild*. (Cambridge, MA: MIT Press).

Jensen, J.A. (2002). 'Joint Tactics, Techniques, and Procedures for Virtual Teams'. Assistant Deputy for Crisis Operations, USCINCPAC (J30-OPT), Camp H. M. Smith.

Joint Vision 2010, Chairman of the Joint Chiefs of Staff, Pentagon, Washington, DC, 2002.

Kant, I. (1781). *The Critique of Pure Reason*.Translated by Norman Kemp Smith. (New York: St. Martin's Press, 1965).

Kennedy, W.G., and Trafton, J.G. (2006). 'Long-term Learning in Soar and ACT-R', *Proceedings of the Seventh International Conference on Cognitive Modeling*, pp. 162–168). (Trieste, Italy: Edizioni Goliardiche).

Klein, G., Ross, K., Moon, B., Klein, D., Hoffman, R. and Hollnagel, E. (2003). *Macrocognition*. Published by the IEEE Computer Society Intelligent Systems, Volume 18(3), 81–85.

Koschmann, T. (1999). 'Meaning Making: Special Issue', *Discourse Processes*, 27(2), 1–10.

Lave, J. (1988). *Cognition in Practice: Mind, Mathematics and Culture in Everyday Life*. (Cambridge: Cambridge University Press).

Lave, J. (1991). 'Situating Learning in Communities of Practice', in L. Resnick, J. Levine and S. Teasley (eds), *Perspectives on Socially Shared Cognition*, pp. 63–83. (Washington, DC: APA).

Marr, D. and Poggio, T. (1979). 'A Computational Theory of Human Stereo Vision', *Proceedings of Royal Society of London* Vol. B 204, 301–328.

Miller, B.B. (1988). 'Managing Information as a Resource', in J. Rabin and E.M. Jackowski (eds), *Handbook of Information Resource Management*, pp. 3–33. (New York, NY: Dekker).

Minsky, M. (1981). A Framework for Representing Knowledge', in J. Haugeland (ed.), *Mind Design*, pp. 95–128. (Cambridge, MA: MIT Press).

Nardi, B. (ed.) (1996). *Context and Consciousness: Activity Theory and Human-Computer Interaction*. (Cambridge, MA: MIT Press).

Newell, A. and Simon, H.A. (1956). The Logic Theory Machine: A Complex Information Processing System', *IRE Trans. Inf. Theory* IT-2, 61–79.

Newell, A. and Simon, H.A. (1972). *Human Problem Solving*. (Englewood Cliffs, NJ: Prentice Hall).

Pea, R. (1993). 'Practices of Distributed Intelligence and Designs for Education', in G. Salomon (ed.), *Distributed Cognitions: Psychological and Educational Considerations*, pp. 47–87. (Cambridge: Cambridge University Press).

Piaget, J. (1975). 'Piaget's Developmental Psychology', in Hilgard, E. and Bower, G. (eds), *Theories of Learning*, pp. 318–346. The Century Psychology Series. (Englewood Cliffs, NJ: Prentice-Hall, Inc.).

Rosenberg, A. (1988). *Philosophy of Social Science*. (Boulder, CO: Westview).

Rumelhart, D.E. (1990). 'Brain Style Computation: Learning and Generalization', in S.F. Zornetzer, J.L. Davis, and C. Lau (eds), *An Introduction to Neural and Electronic Networks*, pp. 405–420. (San Diego, CA: Academic Press).

Scardamalia, M., and Bereiter, C. (1996). 'Computer Support for Knowledge-Building Communities', in T. Koschmann (ed.), *CSCL: Theory and Practice of an Emerging Paradigm*, pp. 249–268. (Hillsdale, NJ: Lawrence Erlbaum Associates).

Simon, Herbert A. (1979/1989). *Models of Thought*. (New Haven, CT: Yale University Press).

Shannon, C.E. (1949). 'A Mathematical Theory of Communication', *Bell System Technical Journal* 27, 379–423 and 623–656.

Skinner, B.F. (1989). 'The Origins of Cognitive Thought', *American Psychologist* 44, 13–18.

Skinner, B.F. (1985). 'Cognitive Science and Behaviourism', *British Journal of Psychology* 76, 291–301.

Stahl, G. (2006) *Group Cognition: Computer Support for Building Collaborative Knowledge*. (Cambridge, MA: MIT Press).

Suchman, L. (1987). *Plans and Situated Actions: The Problem of Human-Machine Communication*. (Cambridge: Cambridge University Press).

Vygotsky, L. (1934/1986). *Thought and Language*. (Cambridge, MA: MIT Press).

Weiner, N.t (1948). *Cybernetics*. (New York: John Wiley and Sons, Inc.).

Wickens, C.D. (1992). *Engineering Psychology and Human Performance*, 2nd edn. (Harper Collins: New York).

Winograd, T. (1996). *Bringing Design to Software*. (New York, NY: ACM Press).

Chapter 2

Empirical Model of Team Collaboration Focus on Macrocognition

Norman W. Warner and Michael Letsky

Introduction

Over the past twenty years, advances in communication and information technology have changed the face of team collaboration activity. Collaboration technology is developing at an astonishing rate. Use of collaboration tools has become commonplace in today's industry, often changing the very structure of companies and organizations that use them. Such changes include 94,000 companies currently use data/web conferencing systems worldwide; 79 million people use data/web conferencing systems; 31 billion e-mails are sent daily and 93 percent of individuals receive at least 20 e-mails per day; a US$13 billion dollar market for real time collaboration tools, in addition to *integrated* collaboration tools, by 2009, including e-mail servers, shared work spaces, instant messaging, videoconferencing and web conferencing (Drakos *et al.* 2005; Coleman 2006). Despite these improvements, however, collaboration technology is not without its challenges. Today's military, for example, gathers intelligence through a wide variety of diverse sources (Internet search engines, observations from field operatives, databases, inter-agency communications, historical data, videoconferences, face-to-face meetings, the media, e-mails). Because of the tremendous amount of data available, military strategists are often faced with information overload. To complicate the collaborative effort, much of that information comes from open sources such as the Internet. As a result, knowledge uncertainty becomes a concern. Military intelligence is constantly changing and therefore strategists and operational personnel must continually monitor the ever-changing flow of information to ensure accurate and timely mission planning and execution (Wroblewski and Warner 2005). Geographically distributed collaboration teams face additional burdens. Co-located teams have the advantage of real-time collaboration, while distributed teams often receive information asynchronously. In addition, the widening realm of contributors often results in teams that represent a disparity of experience, knowledge and cultural backgrounds. To ensure continued effectiveness, it is imperative that this advancing technology enhance the collaborative effort rather than impede it. 'Somehow we must digest all this data and information and organize it into meaningful and useful knowledge, and then through wisdom make intelligent decisions and judgments about what to do' (Nunamaker *et al.* 2001). Unfortunately, current trends in collaboration tool development base these enhancements on intuition, rather than guide development with empirical research. As a result, the success of a particular tool is often the result of the dynamics of a specific collaborative team, rather than of the tool itself. Without focusing on the principles of cognitive psychology, human factors and team behavior, with validation through supporting experimentation, collaboration tools will not develop to their full potential. Specifically what is needed is a *deeper understanding of how teams collaborate* when solving one-of-kind, time critical collaborative problems along with identifying and understanding the macrocognitive processes used during collaborative problem solving. Given this level of understanding, team collaboration tools can be developed that will support these cognitive processes resulting in more effective team collaboration. To gain this deeper understanding of team collaboration and the macrocognitive processes involved, an empirical model of team collaboration is required. The purpose of this chapter is to describe an initial empirical model of team collaboration emphasizing the macrocognitive processes that humans use. The model includes the collaborative problem domain characteristics (Chapter 1 this volume), the stages of team collaboration, the meta and macrocognitive processes together with the mechanisms for achieving the stages and cognitive processes.

There have been numerous models of team collaboration (Hollan, Hutchins and Kirsh 2000; Orasanu and Salas 1992; Rogers and Ellis 1994; McNeese, Rentsch, Perusich 2000; Hurley 2002; Noble 2002; Klein *et al.* 2003; Hosek 2004; Cooke 2005; Stahl 2006) each focusing on different aspects of human decision-making while describing those aspects at various levels of detail. What is lacking is a unified scientific theory of the cognitive processes used during team collaboration. Chapter 1 provided a detail discussion of the early theories of cognitive psychology together with a discussion of current multidisciplinary theories of human cognition. Each theory provides unique insight and empirical data to explain various aspects of human cognition and its processes. However, even with a multidisciplinary approach to human cognition including philosophy, psychology, linguistics, computer science, anthropology, cognitive neuroscience and artificial intelligence there still is no generally recognized unified theory. This lack of a unified theory is partially due to the recent integration of the various disciplines in studying human cognition along with insufficient *objective metrics* to measure the identified cognitive processes. The challenge in representing the cognitive components in a collaboration model is deciding what theoretical approaches to utilize during initial model development. There is insufficient empirical research describing the actual cognitive processes that teams use especially under one-of-a-kind, time-critical, asynchronous, distributed collaborative problem-solving.

Empirical model of team collaboration

The initial model development began with a *synthesis* of the literature in team collaboration, human information processing and team communication together with the results obtained during the annual Office of Naval Research workshops on Collaboration and Knowledge Management. The model evolution is illustrated in Figure 2.1 The first synthesized model (Warner and Wroblewski 2003) integrated several theories (Rawlings 2000; McGrath 1991; Stahl 2000) together with associated metrics for measuring team collaboration and knowledge-building.

Rawling's TeamWise Collaboration theory discussed the key requirements for effective team collaboration along with the necessary interplay between tasks and teams. McGrath's time, interaction and performance propositions provided information on the nature of groups, temporal factors in group collaboration and the group interaction process. Stahl's knowledge-building process offered the theory of how teams build and share knowledge during collaborative problem-solving. Together these theories and associated metrics provided the first attempt to integrate the major team collaboration components and to understand, from a cognitive perspective, how teams collaborate. Using Warner's integrated team collaboration model as a baseline during the 2003 Collaboration and Knowledge Management workshop (Letsky 2003), 12 initial conceptual models were produced each providing some unique and overlapping information. The models varied in their approach and included information-processing, team recognition-primed decision-making, transactive memory, discovery and innovation, and hybrids such as multi-stage and process models. An integrated framework of descriptive stages and processes was then derived based on their ability to address and support one-of-a-kind, time-critical collaborative problem-solving. As a result, the Preliminary Conceptual Model of Collaboration was developed. This model provided a more descriptive illustration of

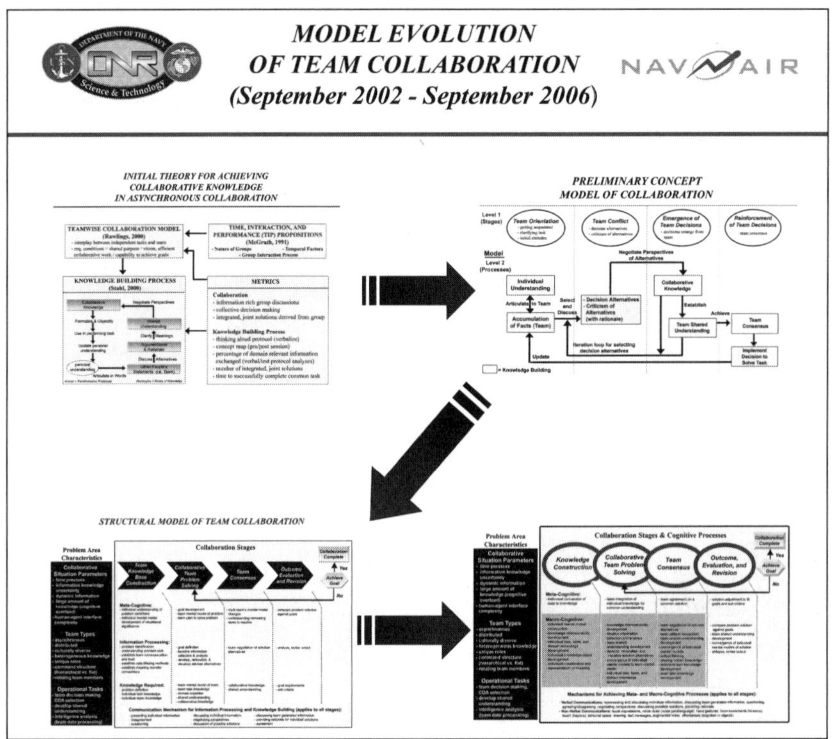

Figure 2.1 Model evolution of team collaboration

team collaboration stages along with the cognitive processes of knowledge-building and sharing. The model continued to be developed (Warner, Letsky and Cowen 2004) by further defining the specific problem area domain that the model addressed along with a better distinction between the collaboration stages. Also, the cognitive processes of team collaboration were described across several levels, metacognitive, information processing, knowledge building and the communication mechanisms necessary for the team to process information and build knowledge. During the 2005 and 2006 Collaboration and Knowledge Management Workshops (Warner, Letsky and Cowen 2005; Letsky 2005, 2006), the specific macrocognitive processes along with associated metrics were refined resulting in the current version of the model. Figure 2.2 presents the current empirical model of team collaboration.

The cognitive mechanisms in this model are described at a *macro level* (e.g., knowledge interoperability development, team shared understanding development, convergence of individual mental models to team mental model) rather than at the *micro level* (e.g., aspects of information processing, attention, perception, memory, neural–cognitive factors) because of the limitations in using microcognitive process to explain higher-order decision-making mechanisms.

'Macrocognition is defined as the internalized and externalized high-level mental processes employed by teams to create new knowledge during complex, one-of-a-kind, collaborative problem-solving' (Chapter 1 this volume). High-level is defined as the process of combining, visualizing, and aggregating information to resolve ambiguity in support of the discovery of new knowledge and relationships. *Internalized processes* (processes occurring inside the head) are those higher-level cognitive processes that occur at the individual or the team level, are not expressed externally (e.g., writing, speaking, gesture), and can only be measured indirectly via

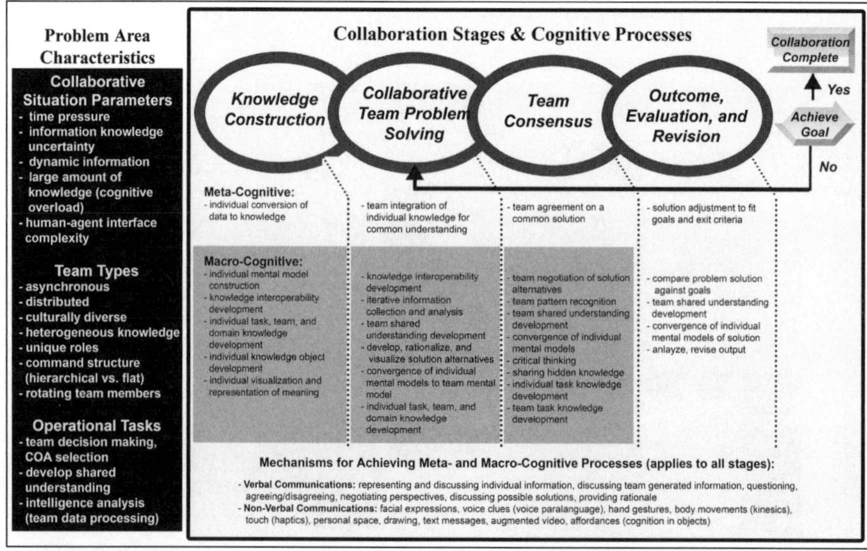

Figure 2.2 Current empirical model of team collaboration

qualitative metrics (e.g., questionnaires, cognitive mapping, think aloud protocols, multidimensional scaling), or surrogate quantitative metrics (e.g., pupil size, galvanic skin response, fMRI). These processes can become either fully or partially externalized when they are expressed in a form that relates to other individual's reference/interpretation systems (e.g., language, icons, gestures, boundary objects). A sample of internalized process metrics could include frequency counts for each behavior type, time duration of each behavior, commonality of utterance meanings and transition probabilities from one process state to the next. *Externalized processes* (processes occurring outside the head) are those higher-level cognitive processes that occur at the individual or the team level, and which are associated only with actions that are observable and measurable in a consistent, reliable, repeatable manner or through the conventions of the subject domain have standardized meanings. Externalized processes can be measured by various techniques each with their associated metrics, which include discourse analysis (Hobbs 1993; Brown and Yule 1983), process tracing (Woods 1993), automated latent semantic analysis (Landauer, Foltz and Laham 1998), automated communication flow analysis (Kiekel, Cooke, Foltz and Shope 2001) and dynamic modeling of communication data (Gorman, Cooke and Kiekel 2004). All current subjective and objective analysis techniques have advantages and disadvantages, therefore requiring the use of multiple techniques to ensure the most valid analysis and conclusions. When discussing internalized and externalized macrocognitive processes throughout the remainder of the chapter, it will be implied that these are the high level processes as described above.

Model components

The model consists of general *inputs* (e.g., task description), *collaboration stages* that the team goes through during the problem-solving task, the *cognitive processes* used by the team and final team output(s) (e.g., selected course of action). The cognitive processes include the metacognitive and the macrocognitive processes. Also described are the communication activities for enabling the meta and macrocognitive processes.

Model inputs

This information includes such items as:

1. a description of the problem
2. team member expertise
3. organizational structure
4. roles and responsibilities of each team member, and
5. projected events and future information.

This representative domain information is provided to the team during team formation.

Collaboration stages and cognitive processes

The model has four unique but interdependent stages of team collaboration. The stages are:

- knowledge construction
- collaborative team problem-solving
- team consensus and outcome
- evaluation and revision.

There is also a feedback loop for revising team solutions. Teams will typically start in the Knowledge construction stage and proceed into collaborative team problem-solving, team consensus and finally outcome, evaluation and revision. The stages are *not sequential* as they appear in Figure 2.2. Because team communication activity is very dynamic, the flow of communication can follow virtually any path. The cognitive processes within each stage are represented at two levels: *metacognitive*, which guides the overall problem-solving process, and *macrocognitive*, which supports team member's activities (e.g., knowledge-building, development of solution options, reaching agreement on a course of action) within the respective collaboration stage. In addition, there are various communication activities (i.e., verbal and non-verbal) for developing the metacognitive and macrocognitive processes.

Knowledge construction, the first stage in team collaboration, begins with team members building individual task knowledge, constructing team knowledge along with creating knowledge interoperability. In this model, data, information, knowledge and understanding are defined and used according to the definitions and principles of Bellinger *et al.* (2004). *Data* represents a fact or statement of event without relation to other things. *Information* embodies the understanding of a relationship of some sort, possibly cause and effect. *Knowledge* represents a pattern that connects and generally provides a high level of predictability as what is described or what will happen next. *Understanding* is a cognitive, analytical and probabilistic process that takes current knowledge and synthesizes new knowledge from previously held knowledge. It is understanding that supports the transition from data, to information, to knowledge. The metacognitive process during knowledge construction is the awareness by each team member that knowledge needs to be developed from data and information in order to solve the collaborative problem. The focus of all the macrocognitive processes in the knowledge construction stage is to support individual and team knowledge development along with knowledge interoperability. This knowledge will be used during collaborative team problem-solving to develop solution alternatives to the problem.

During *collaborative team problem-solving*, team members communicate data, information and knowledge to develop solution options to the problem. The metacognitive process during this stage is the awareness by the team that individual knowledge needs to be integrated for common team representation of the problem. Team mental representation can change during the course of solving the problem. Changes can occur as the team gains more complete understanding of the problem elements, goals or overlooked information (McComb 2007). The focus of the macrocognitive processes in this stage is to support development of solution options for the collaborative problem. Some of the macrocognitive processes under this stage are the same as found in the knowledge construction stage, although the focus of these processes differs. For example, knowledge interoperability under knowledge construction centers on the exchange of knowledge for the creation of new knowledge among the team. In the collaborative problem-solving stage, the emphasis is on exchanging knowledge to develop solution options.

Team consensus is the stage where the team negotiates solution options and reaches final agreement by all team members on a particular option. During team consensus, the metacognitive process is the team's awareness to reach agreement on a common solution. The macrocognitive processes support the team in reaching agreement on the final solution to the problem.

During the *outcome, evaluation and revision* stage, the team evaluates the selected solution option against the problem-solving goal and revises the solution option if that option does not meet the goal. The metacognitive process is the team's awareness to have the final solution option meet the problem-solving goals. The team as a whole compares the complete solution option against the goal. The focus of the macrocognitive processes in outcome, evaluation and revision is to support the team in comparing the final solution option against problem goals and to revise the solution, if necessary.

Macrocognitive processes

Based on our current research and thinking there are 14 macrocognitive processes in the current model of team collaboration each having internalized *and* externalized components. The definition of each macrocognitive process along with an illustration of the respective internalized and externalized processes is provided below.

Individual mental model construction Defined as individual team members using available information and knowledge to develop their mental picture of the problem situation. The internalized processes are the cognitive activities of combining, sorting, filtering new information and knowledge and then internally developing the mental model. The externalized processes are the cognitive activities of exchanging available information and knowledge about the collaborative problem with the other team members through verbal and non-verbal communication.

Knowledge interoperability development The process of exchanging and modifying a mental model or knowledge object such that agreement is reached among team members with respect to a common understanding of the topic's meaning. The internalized processes are the cognitive activities of combining, sorting, filtering new information and then internally modifying the mental model or knowledge object. The externalized processes are the cognitive activities of exchanging information about the modified mental model or knowledge object with the other team members through verbal and non-verbal communication.

Individual task knowledge development Defined as individual team members collecting and processing data or information, or responding to clarification requested by other team members about their assigned task. The internalized processes are the cognitive activities of collecting, sorting, filtering task information and then internally synthesizing the information into new task knowledge. The externalized processes are the cognitive activities of exchanging task knowledge with the other team members through verbal and non-verbal communication.

Team knowledge development Defined as all team members participating to clarify information for building common team knowledge. The internalized processes for each team member are the cognitive activities of collecting and internally synthesizing the new team knowledge into their current knowledge structure. The externalized processes are the cognitive activities of exchanging information with the other team members through verbal and non-verbal communication.

Individual knowledge object development Pictures, icons, representations or standard text developed by an individual team member or the whole team to represent a standard meaning. The internalized processes support individual team member knowledge object development and are the cognitive activities of collecting, sorting, filtering information and then internally synthesizing the information to represent a standard meaning. The standard meaning is internally represented as a knowledge object (i.e., a picture, icon, unique representation or standard text). The externalized processes support team-level knowledge object development and are the cognitive activities of exchanging filtered and synthesized information with the other team members to develop an object that represents a standard meaning. This exchange of filtered and synthesized information between team members occurs through verbal and non-verbal communication.

Individual visualization and representation of meaning Visualization is defined as the use of knowledge objects (e.g., graphs, pictures) by individual team members to transfer meaning to other team members. Representations are artifacts (e.g., note pads) used by individual team members to sort data and information into meaningful chunks. Visualization has externalized cognitive processes of exchanging knowledge objects between team members through verbal or non-verbal communication. Representations are developed by internalized cognitive processes of filtering information, and sorting the information into meaningful chunks then externalized as artifacts.

Iterative information collection and analysis Defined as collecting and analyzing information, in an iterative fashion, to come up with a solution with no specific solution mentioned. The internalized processes are the cognitive activities of individual team members of collecting, sorting, filtering information and then internally synthesizing the information. The externalized processes are the cognitive activities of requesting information from other team members through verbal and non-verbal communication.

Team shared understanding development Defined as the synthesis of essential data, information or knowledge, held collectively by some (complementary understanding) and/or all (congruent understanding) team members working together to achieve a common task. Within each team member, the internalized processes are the high-level cognitive activities of collecting, then internally synthesizing new data, information or knowledge into each team member's current knowledge structure. The externalized processes are the cognitive activities of exchanging the synthesized data, information or knowledge with the other team members through verbal and non-verbal communication.

Develop, rationalize and visualize solution alternatives Defined as a team member or the whole team using knowledge to describe a potential solution. Within each team member, the internalized processes are the cognitive activities of synthesizing existing knowledge to develop possible solutions. The externalized processes are the cognitive activities of exchanging knowledge with the other team members through verbal and non-verbal communication to develop possible solutions.

Convergence of individual mental models to team mental model Defined as convincing other team members to accept specific data, information or knowledge with the emergence of a single common mental model that all team members accept or believe. The externalized processes are the cognitive activities of exchanging data, information or knowledge between team members through verbal and non-verbal communication resulting in communication of a common mental model accepted or believed by all team members. The internalized processes are the cognitive activities of individual team members of collecting, filtering and comparing new data, information and/or knowledge against their own current individual mental model.

Team negotiation Team negotiation is defined as the give and take process whereby team members agree on a common solution or issue. The internalized processes are the cognitive activities of collecting, combining, sorting, filtering new information and/or knowledge provided by other team members. The externalized processes are the cognitive activities of iteratively exchanging information and/or knowledge with other team members through verbal and non-verbal communication resulting in all team members agreeing on a solution.

Team pattern recognition Defined as the team, as a whole, identifying a pattern of data, information or knowledge. Within each team member, the internalized processes are the cognitive activities of collecting, filtering and comparing new data, information or knowledge against current data/information/knowledge patterns resulting in updated pattern structures. The externalized processes are the cognitive activities of the team which involves the team exchanging knowledge about patterns in the data, information or knowledge through verbal and non-verbal communication.

Critical thinking Defined as the team working together toward a common goal, whereby goal accomplishment requires an active exchange of ideas, self-regulatory judgment, and systematic consideration of evidence, counterevidence and context in an environment where judgments are made under uncertainty, limited knowledge and time constraints. The internalized processes are the cognitive activities of individual team members collecting, sorting, filtering and comparing new information and knowledge. The externalized processes are the cognitive activities of the team iteratively exchanging knowledge with the other team members through verbal and non-verbal communication resulting in ideas to achieve a common goal.

Sharing hidden knowledge Defined as individual team members sharing their unique knowledge through prompting by other team members. The internalized processes, for each team member, are the cognitive activities of filtering and comparing current knowledge to determine unique knowledge. The externalized processes are the cognitive activities of requesting and sharing knowledge among team members through verbal and non-verbal communication.

Model outputs The product output will vary depending on the problem domain addressed by the team. This model focuses on three types of products:

1. team decision-making, course of action selection
2. developing shared understanding, and
3. intelligence analysis (team data processing).

Empirical model of team collaboration example

To illustrate how this model represents team collaboration in a collaborative problem-solving environment, a Noncombatant Evacuation Operation (NEO) scenario (Warner, Wroblewski and Shuck 2003) will be described. This example is designed to show how the Operation Planning Team (OPT) (Warner, Wroblewski and Shuck 2003) proceeds through the various collaboration stages along with the cognitive processes used to support each stage. Team collaboration during each collaboration stage and the various cognitive processes are very dynamic. It is beyond the scope of this example to capture *all* the team dynamics throughout the problem-solving scenario but this example will illustrate major team collaborative behaviors and processes. In addition, team collaboration behavior and performance is influenced by the type and extent of collaboration technology employed when solving the collaboration problem (e.g., e-mail, chatrooms, white boards, integrated text/video/audio tools, agent-based collaboration tools). This example will use basic e-mail/chat room/video and web technology as a collaboration environment for solving the NEO scenario problem.

The Noncombatant Evacuation Operation scenario is planned and executed by the Operation Planning Team (OPT) at the United States Pacific Command (PACOM). The OPT is responsible for planning all military operations in the

PACOM theater of operations. This includes 44 countries covering 51 percent of the earth's surface: 75 percent of all natural disasters occur in the PACOM Theater. The NEO decisions are handled by the OPT Key Planners, which has eight members who are geographically separated. The eight members are senior staff specialists from different PACOM directorates. The Key Planners have access to 59 OPT personnel consisting of representatives from special staff (legal, medical) and Joint Task Force components as well as the five OPT Core Group members who collect input and assessment information and post to the OPT web page. In addition to the OPT web site, the OPT members meet in a Virtual Situation Room, to learn more about the unfolding situation, military assets available, distances between key places, cargo, weather, terrain, local news, intelligence reports about threats and capability/availability of our assets. The OPT members meet in the Virtual Planning Room to draft the execution plan and to monitor and evaluate the plan during execution.

Inputs to the team

The OPT Key Planners are given the following problem: increasing tensions on the Southern Pacific island of Drapo have resulted in clashes between Government military forces and rebel insurgents. As the scenario begins, a group of American Red Cross workers traveling a jungle road near the city of Dili in East Drapo becomes caught in crossfire between government and insurgent forces, and takes refuge in an abandoned church. Luckily, their cellular phone still works, and they contact the American Red Cross office. As word of the workers' plight becomes known, the Red Cross contacts the US Department of State to alert the US Government and seek assistance. Department of State notifies the National Command Authority (NCA) and Chairman of the Joint Chiefs of Staff, alerting them to the situation, which results in an Inter-agency Working Group being formed to support the NCAs decision-making responsibilities. The members of the Inter-agency Working Group include the Department of Defense, the National Security Council, the Department of State, the Central Intelligence Agency, and the US Agency for International Development, along with the Commander in Chief of the US Pacific Command (PACOM), whose area of responsibility includes the South Pacific. Additional agencies may join the Inter-agency Working Group as the crisis develops and the NCA considers different options.

Problem area characteristics

This scenario illustrates a distributed, asynchronous team with heterogeneous knowledge working in a flat command structure collecting information and generating alternatives, in performing a COA selection task under time pressure, with information/knowledge uncertainty, large amounts of knowledge and dynamically changing information. Team interactions of interest are defined by four unique but interdependent stages of team collaboration:

1. knowledge construction,
2. collaborative team problem-solving,
3. team consensus, and
4. outcome evaluation and revision.

Collaboration stages and cognitive processes

Knowledge construction

During this first stage of collaboration, the OPT members have two objectives to achieve: (1) to convene as a team (using the virtual situation room) where they will be notified that a warning order has been received and their assistance is needed to develop and evaluate Course of Action options, and (2) to understand the situation and mission as a team (using the virtual situation room).

Cognitive processes

In order to achieve the two objectives the OPT members need to develop a team understanding of the problem conditions and their own individual mental model of the situation. Both team understanding and individual mental models are developed as the team builds knowledge about the problem, and as their individual and team knowledge develops. Problem and task knowledge develops through *communication* among OPT members using the information available in the Virtual Situation Room. This communication occurs in an asynchronous, distributed fashion. Such information includes major events that have occurred, components of the situation that might require military resources, location of the situation, groups and organizations involved in the situation, topography, climate and weather, transportation, telecommunications, politics, coalition capabilities and enemy capabilities. The communication mechanisms involved for building problem and task knowledge include a subset of those items in Figure 2.2. Such items include presenting and discussing individual information, questioning, agreeing/disagreeing, and discussing team generated information. After the OPT members develop the initial problem and individual and team knowledge, the *team* can then use that knowledge to identify the specific problem to be addressed (i.e., three American Red Cross workers trapped in Dili church surrounded by fighting government and rebel forces) along with having a good understanding of the problem task (i.e., develop several COA's to evacuate three Red Cross workers near the city of Dili within the next 24 hours). The team applies their knowledge of the problem, their individual and team knowledge to support the next collaboration stage, collaborative team problem-solving.

Collaborative team problem-solving

The OPT has one objective at this stage, which is to develop COA's that fulfill requirements in the warning order. The OPT will use the Virtual Situation Room for team member communication for collecting and analyzing the necessary information to develop the various COA's. The Virtual Planning Room will be used by the OPT to develop the various COA options, provide COA rationalization, COA visualization, and to monitor execution of the selected COA.

Cognitive processes

For effective team collaboration at this stage the OPT needs to develop a team goal, and develop a team mental model of the problem. The goal in this scenario is given to the OPT, which is to develop and evaluate various COA's that successfully meet the warning order (i.e., evacuate the three American Red Cross Workers within 24 hours) and to recommend the best COA for execution. The team mental model will evolve as the OPT collects and analyzes information along with developing, rationalizing and visualizing COA options. In order for the OPT to develop COA options there are several types of knowledge required, individual task knowledge, team knowledge, domain expertise, and team shared understanding. The OPT members have developed *some* individual and team knowledge from the earlier stage along with the domain expertise they bring. However, there is little team shared understanding at this point because the OTP has been gathering information on the problem with little collaboration. As the OPT collects, analyzes, and discusses information relevant to COA options team shared understanding will increase. Individual and team knowledge will also increase. The OPT's mental model of the problem will be further developed as the team goes through the iterative information collection, analysis, discussion process. The type of information collected, analyzed and discussed includes combat forces required, force movement, staging locations, destination, required delivery date of forces and effect of enemy COA on the success of our COA option. The OPT will use their shared understanding, and task knowledge to develop, rationalize and visualize the various COA options. The mechanism for building the knowledge types and the COA options are the same communication mechanisms listed in Figure 2.2 (e.g., discussions of possible solutions, negotiating perspectives, questioning, agreeing). The final outcome at this collaboration stage is a listing of viable COA's with advantages and disadvantages for each option. For this scenario a list of viable COA's for evacuation of the Red Cross Workers could include:

1. *Marine Force Pacific* consisting of one AH-1 attack helicopter for air support and one CH-53 helicopter for transporting the evacuees. CH-53 personnel include one-armed squad of ten marines for support, two medical corpsman, five support aircrew, and pilot/co-pilot. All assets are stationed on CV-65 (Enterprise), which is located near Drapo island. Advantages include: air and personnel assets located 100 miles from Drapo island, CH-53 holds 40 people, sufficient air support (i.e., AH-1) against enemy threat, sufficient armed ground support against potential enemy attack, no re-fueling of air assets required, travel time to designation is one hour (one hour return). Disadvantages include: air and personnel assets will not be ready for deployment for 10 hours.

2. *US Army Pacific* consisting of one Apache attack helicopter for air support and one UH-60 Blackhawk helicopter for transporting evacuees. The UH-60 includes one armed squad of five infantry personnel, two medical corpsman, four support crew and pilot/co-pilot. All assets are stationed on an Army base located 600 miles from Drapo. Advantages include: air and personnel assets are immediately available for deployment, UH-60 has sufficient room for all aircrew personnel and evacuees, sufficient air support against enemy threat. Disadvantages include: UH-60 and Apache needs to refuel twice, once inbound, once outbound, minimum armed ground support against enemy attack, travel time to designation is 6 hours (also 6 hour return).

Team consensus

The objective at this stage is for the OPT to agree on which COA promises to be the most successful in accomplishing the mission. The OPT will negotiate COA alternatives through their team discussion using the virtual planning room. The virtual situation room may be used by the OPT if additional information is needed by the team to support negotiation.

Cognitive processes

For the OPT to achieve effective collaborative negotiation and reach consensus on the best COA, the team needs to keep track of the changes in the team's mental model as they conduct COA negotiations. The OPT also needs to understand the remaining items to resolve when trying to reach consensus on the best COA. The OPT will use their team shared understanding, which has been developing throughout the collaboration process, to update their team mental model and monitor remaining items to resolve. The OPT uses the communication mechanisms in Figure 2.2 (e.g., negotiating perspectives, providing rationale for individual solutions, questioning) during team negotiation to reach team consensus of the best COA. During team negotiation each COA (i.e., marine option versus army option) is discussed with respect to (a) the enemy's capability to adversely affect execution of our COA, and (b) the strengths and weaknesses of each COA and its probability of success within constraints of operational factors. This negotiation is a very dynamic cycle of team discussion which leads to increase team shared understanding and a richer team mental model that in turn results in deeper team discussions. This dynamic team negotiation cycle needs to occur to reach team consensus and produce the most effective COA. The final outcome of the team consensus stage is a selection of one COA, with rationale, that has the highest probability of success within the constraints of the mission. For this scenario example, the OPT agreed that the Marine Force Pacific course of action was the best option to meet mission requirements.

Outcome evaluation and revision

The objective at this stage is to evaluate the selected COA (i.e., Marine Force Pacific) and revise as necessary. Before executing the selected Marine Force Pacific option, the OPT conducted several what-if simulations to determine any problem areas and how robust the COA is in meeting the mission. No major problems were found and the Marine Force Pacific option was executed. The OPT used the Virtual Planning Room to conduct the what-if simulations and to monitor the COA during execution.

Cognitive processes

As part of the process of comparing the selected Marine Force Pacific option against the mission goals, the OPT needs to have a clear understanding of the mission goals. At this point in the collaboration process the mission goal information has become part of the team's mental model and the OPT will use its mental model during the COA evaluations. In addition, in monitoring the execution of the COA the team will use its shared understanding to determine problem areas and possible revisions to the COA. The communication mechanisms for the team to evaluate and revise the COA are presented in Figure 2.2.

Empirical validation of model

Two major empirical experiments (Warner and Wroblewski 2004; Warner and Wroblewski 2003) were conducted to examine the validity of the collaboration stages and cognitive processes within the model. Table 2.1 presents the collaboration stages and the macrocognitive processes that were found to be significant based on the two empirical experiments.

Cognitive Process States	Knowledge Construction	Team Problem Solving	Team Consensus	Outcome, Evaluation, and Revision
Individual mental model construction	✓			
Individual task knowledge development	✓	✓	✓	
Visualization & representation of meaning	✓	✓		
Knowledge interoperability	✓	✓		
Iterative information collection	✓		✓	
Team shared understanding	✓			
Develop solution alternatives	✓	✓		
Convergence of mental models	✓	✓	✓	
Team negotiation of solution alternatives			✓	
Team pattern recognition				✓

Table 2.1 Significant macro cognitive processes within each collaboration stage (experiments I and II)

These research findings in Table 2.1 provide a starting point for understanding the important macrocognitive processes that support team collaboration during one-of-a kind, time-critical problem-solving. These initial studies also provide some preliminary insight into the characteristics of macrocognition in collaborative teams including:

- The unit of analysis includes both the individual team member and the whole team because of the unique macrocognitive processes operating at the individual and team level
- Cognitive activities are analyzed at a high level because of the limitations in using microcognitive processes to explain higher order decision-making mechanisms
- Focus on both internalized and externalized mental processes employed by team members during complex, one-of-a-kind, collaborative problem-solving
- Macrocognitive processes can be empirically studied in the lab and in operational field settings given domain-rich collaborative problem-solving scenarios
- Macrocognitive processes (i.e., internalized and externalized) occur during team member interaction (i.e., socially and collaboratively mediated) and are influenced by the artifacts in the environment
- Macrocognitive processes develop and change over time
- Macrocognitive processes are domain-dependent and collaboration environment-dependent (e.g., face-to-face versus asynchronous, distributed collaboration tools).

Conclusions

The model of team collaboration presented in this chapter together with the empirical validation results from the two experiments is a starting point for obtaining a deeper understanding of how teams collaborate when solving one-of-a-kind, time-critical collaborative problems along with identifying and understanding the respective macrocognitive processes.

In order to expand our understanding of how teams collaborate and the supporting macrocognitive processes future research is required into the *consistency* and *contribution* of these macrocognitive processes across different problem-solving domains (i.e., NEO, intelligence analysis, and emergency management). By continuing to use a model-based approach in future experiments, the current model can be refined by empirically examining the consistency and contribution of the 14 macrocognitive processes across different problem domains. Also, in future experiments more *sensitive metrics* need to be developed and validated not only for measuring the overall collaboration stages and macrocognitive processes but the internalized and externalized components as well. For example, to measure the existence of macrocognitive processes current metrics (e.g., discourse analysis, content analysis) are marginal. What is needed is an integration of selected current metrics with new metrics such as the development of composite measures integrating verbal protocol analysis, content analysis, communication flow analysis and dynamic modeling of communication all of which are recorded and analyzed automatically. In addition, new metrics need to be developed and validated for the internalized and externalized components of each macrocognitive process. For instance, expanding current techniques like multidimensional scaling to capture the internal processes of combining, sorting, filtering new information and knowledge to build an individual mental model. Finally, future experiments need to explore macrocognition in team collaboration from a multi-method experimentation approach. Such investigations into macrocognition should include multiple approaches such as classic laboratory studies, computer modeling and simulations, ethnographic studies and operational field studies. Results from these various types of studies need to be compared to develop a more accurate understanding of how macrocognitive processes support team collaboration. By improving our methods and metrics a deeper understanding of macrocognition will be achieved. Given this better understanding, team collaboration tools can be developed that will support these cognitive processes resulting in more effective team collaboration.

References

Bellinger, G., Castro, D. and Mills, A. (2004). *Data, Information, Knowledge and Wisdom*, http://www.systems-thinking.org/dikw/dikw.htm, accessed 31 August 2004.

Brown, G. and Yule, G. (1983). *Discourse Analysis.* (Cambridge: Cambridge University Press).

Coleman, D. (2006). *Real Time Collaboration (RTC) Report 2006: A Market in Transition.* http://reports.collaborate.com. Collaborative Strategies.

Cooke, N.J. (2005). 'Communication in Team Cognition', *Collaboration and Knowledge Management Workshop Proceedings*, pp. 378–443. (San Diego, CA: Office of Naval Research).

Drakos, N., Smith, D., Mann, J., Cain, M., Burton, G. and Gartner, B. (2005). *RAS Core Research Note G00131098*. 13 October 2005. http://mediaproducts.gartner.com/gc/webletter/alcatel/issue3/articles/article2.html.

Gorman, J., Cooke, N. and Kiekel, P. (2004). 'Dynamical Perspectives on Team Cognition', *Proceedings of the Human Factors and Ergonomics Society 48th Annual Meeting*, pp. 673–677. (Santa Monica, CA: Human Factors and Ergonomics Society).

Hobbs, Jerry R. (1993). 'Summaries from Structure', in B. Endres-Niggemeyer, J. Hobbs and K. Sparck Jones (eds), *Proceedings Workshop on Summarizing Text for Intelligent Communication*, pp. 121–135. (Germany: Schloss Dagstuhl).

Hollan, J., Hutchins, E. and Kirsh, D. (2000). 'Distributed Cognition: Toward a New Foundation for Human-Computer Interaction Research', *ACM Transactions on Computer-Human Interaction* 7(2), 174–196.

Hosek, J. (2004). 'Group Cognition as a Basis for Supporting Group Knowledge Creation and Sharing', *Journal of Knowledge Management* 8(4), 55–64.

Hurley, John (2002). 'Cognitive Elements of Effective Collaboration', *Proceedings from Towards a Cognitive Organizational Framework for Knowledge Management*, pp. 78–89. (San Diego, CA: Office of Naval Research/SPAWAR Systems).

Kiekel, P.A., Cooke, N.J., Foltz, P.W. and Shope, S.M. (2001). 'Automating Measurement of Team Cognition Through Analysis of Communication Data', in M.J. Smith, G. Salvendy, D. Harris and R.J. Koubek (eds), *Usability Evaluation and Interface Design*, pp. 1382–1386. (Mahwah, NJ: Lawrence Erlbaum Associates).

Klein, G., Ross, K., Moon, B., Klein, D., Hoffman, R. and Hollnagel, E. (2003) .'Macrocognition', *IEEE Computer Society Intelligent Systems* 18(3), 81–85.

Landauer, T.K., Foltz, P.W. and Laham, D. (1998). 'Introduction to Latent Semantic Analysis', *Discourse Processes* 25, 259–284.

Letsky, M. (2003). 'Overview of Collaboration and Knowledge Management Program', *Collaboration and Knowledge Management Workshop Proceedings*, pp. 1–15. (Arlington, VA: Office of Naval Research).

Letsky, M. (2005). 'Overview of Collaboration and Knowledge Management Program', *Collaboration and Knowledge Management Workshop Proceedings*, pp. 1–12. (Arlington, VA: Office of Naval Research).

Letsky, M. (2006). 'Overview of Collaboration and Knowledge Management Program', *Collaboration and Knowledge Management Workshop Proceedings*, pp. 1–12. (Arlington, VA: Office of Naval Research).

McComb, S.A. (2007). 'Mental Model Convergence: The Shift from Being an Individual to Being a Team Member', in F. Dansereau and F. J. Yammarino (eds), *Research in Multi-level Issues: Multi-level Issues in Organizations and Time*, pp. 22–35. (Amsterdam: JAI Press/Elsevier).

McGrath, J. (1991). 'Time, Interaction, and Performance (TIP) a Theory of Groups', *Small Group Research* 22(2), 147–174.

McNeese, M.D., Rentsch, J.R. and Perusich, K. (2000). 'Modeling, Measuring and Mediating Teamwork: The Use of Fuzzy Cognitive Maps and Team Member Schema Similarity to Enhance BMC3I Decision Making', *IEEE International Conference on Systems, Man and Cybernetics*, pp. 1081–1086. (New York: Institute of Electrical and Electronic Engineers).

Noble, D. (2002). 'Cognitive-Based Guidelines for Effective use of Collaboration Tools', *Cognitive Elements of Effective Collaboration. Collaboration and Knowledge Management Workshop Proceedings*, pp. 42–55. (Arlington, VA: Office of Naval Research).

Nunamaker, J.F., Romano, N.C. and Briggs, R.O. (2001). 'A Framework for Collaboration and Knowledge Management', *Proceedings of the 34th Hawaii International Conference on Systems Science*, pp. 233–245. (Honolulu, Hawaii).

Orasanu, J. and Salas, E. (1992). 'Team Decision Making in Complex Environments', in G. Klein, J. Orasanu and R. Calderwood, (eds), *Decision Making in Action: Models and Methods*, pp. 331–352. (Norwood, NJ: Ablex Publishing Corp.).

Rawlings, D. (2000). 'Collaborative Leadership Teams: Oxymoron or New Paradigm?', *Consulting Psychology Journal* 52, 36–48.

Rogers, Y. and Ellis, J. (1994). 'Distributed Cognition: An Alternative Framework for Analyzing and Explaining Collaborative Working', *Journal of Information Technology* 9(2), 119–128.

Stahl, G. (2000). 'A Model of Collaborative Knowledge-Building', in B. Fishman and S. O'Connor-Divelbiss, (eds), *Fourth International Conference of the Learning Sciences*, pp. 70–77. (Mahwah, NJ: Erlbaum).

Stahl, G. (2006). *Group Cognition: Computer Support for Building Collaborative Knowledge*. (Cambridge, MA: MIT Press).

Warner, N. (2003). 'Collaborative Knowledge in Asynchronous Collaboration', *Research Collaboration and Knowledge Management Workshop Proceedings*, pp. 25–36. (Arlington, VA: Office of Naval Research).

Warner, N. and Wroblewski, E. (2003). 'Achieving Collaborative Knowledge in Asynchronous Collaboration: Murder Mystery Collaboration Experiment', *Collaboration and Knowledge Management Workshop Proceedings*, pp. 327–340. (Arlington, VA: Office of Naval Research).

Warner, N. and Wroblewski, E. (2004). 'Achieving Collaborative Knowledge in Asynchronous Collaboration: NEO Collaboration Experiment', *Collaboration and Knowledge Management Workshop Proceedings*, pp. 63–72. (Arlington, VA: Office of Naval Research).

Warner, N., Letsky, M. and Cowen, M. (2004). 'Overview of Collaboration and Knowledge Management Program', *Collaboration and Knowledge Management Workshop Proceedings*, pp. 1–14. (Arlington: Office of Naval Research).

Warner, N., Letsky, M. and Cowen, M. (2005). 'Cognitive Model of Team Collaboration: Macro-Cognitive Focus', *Proceedings of the Human Factors and Ergonomics Society 49th Annual Meeting*, pp. 269–273. (Orlando, FL: Human Factors and Ergonomics Society).

Warner, N., Wroblewski, E. and Shuck, K. (2003). *Noncombatant Evacuation Operation Scenario*. (Patuxent River, MD: Naval Air Systems Command).

Woods, D.D. (1993). 'Process-tracing Methods for the Study of Cognition Outside of the Experimental Psychology Laboratory', in G.A. Klein, J. Orasanu, R. Calderwood and C.E. Zsambok, (eds), *Decision Making in Action: Models and Methods*, pp. 21–31. (Norwood, NJ: Ablex Publishing Corporation).

Wroblewski, E. and Warner, N. (2005). 'Achieving Collaborative Knowledge in Asynchronous Collaboration', *Collaboration and Knowledge Management Workshop Proceedings*, pp. 45–52. (Arlington, VA: Office of Naval Research).

Chapter 3

Shared Mental Models and Their Convergence[1]

Sara A. McComb

Mental models are simplified characterizations humans create of their worlds (Johnson-Laird 1983). Individuals use them to describe, explain, and predict their surroundings (Rouse and Morris 1986). They are comprised of content and any relationships or structure among the content (Mohammed *et al.* 2000). When team members interact, their mental models converge, resulting in a scenario where the individual team members' mental models become similar to, or shared with, that of their teammates' mental models (McComb 2007; McComb and Vozdolska 2007). Thus shared mental models can be defined as:

> Knowledge structures held by members of a team that enable them to form accurate explanations and expectations for the task, and in turn, to coordinate their actions and adapt their behavior to demands of the task and other team members.
>
> (Cannon-Bowers *et al.* 1993)

Shared mental models are a critical construct for consideration when examining teams facing operational tasks such as team decision making, shared understanding development, or intelligence analysis. As Warner, Letsky and Wroblewski point out in Chapter 2, such operational tasks are effectively accomplished through macrocognitive processes. Warner and colleagues identify individual mental model construction and the convergence of individual mental models as important macrocognitive processes. Moreover, macrocognitive processes, such as individual knowledge object development, may aid in the mental model convergence process. Shared mental models, in turn, may aid other macrocognitive processes like team negotiation. Thus, understanding the role of shared mental models in completing these types of operational tasks is paramount.

The purpose of this chapter is to examine shared mental models research and the role of shared mental models in the macrocognitive processes of teams. To this end, the chapter begins with a brief overview of shared mental models terminology and empirical research results demonstrating their essential role in effective team functioning. The next section describes my mental model convergence

1 Research described in this chapter was supported in part by Grant N000140210535 and N000140610031 from the Office of Naval Research, and Grant 0092805 from the National Science Foundation.

conceptualization and provides preliminary evidence of mental model convergence. I then turn my attention to mental model measurement and offer some suggestions for advancing this critical aspect of mental model research. Finally, I examine the relationship between macrocognitive processes and shared mental models. The chapter ends with implications for research and practice relating to the study of shared mental models.

Shared mental models

Terminology

The terminology regarding shared mental models has not coalesced. The reasons for this lack of consistency may be attributed to the relative immaturity of the field and/or the complexity of the phenomenon under investigation. For instance, shared mental models, team mental models, and compatible mental models have been introduced in the literature. While shared mental models is the most commonly used term, the others have been attributed to very specific types of mental models. Langan-Fox and colleagues (2001) propose that shared mental models represent dyadic cognitive relationships, whereas team mental models represent collective cognition in groups of three or more members. To date, this distinction has not been adapted by the field. Similarly, Kozlowski and Klein (2000) introduced compatible mental models to describe those mental models that are comprised of information distributed among team members (versus information held commonly by members). As the field continues to expand, our understanding of the complex nature of shared mental models will also expand. We may need to begin adapting these more specific definitions to better articulate exactly what we are examining.

In addition to the terminology used to describe mental models, several closely related fields also inform our research in this area. These fields include cognitive consensus (e.g., Mohammed and Ringseis 2001), group mind (e.g., Weick and Roberts 1993), interpretive schema (e.g., Bartunek 1984; Dougherty 1992), intersubjectivity (e.g., Eden *et al.* 1981), shared beliefs (e.g., Cannon and Edmondson 2001), shared cognition (e.g., Cannon-Bowers and Salas 2001; Tan and Gallupe 2006), shared meaning (e.g., Smircich 1983), shared strategic cognition (e.g., Ensley and Pearce 2001), sociocognition (e.g., Gruenfeld and Hollingshead 1993), team cognition (e.g., Fiore and Salas 2004), teamwork schema agreement (e.g., Rentsch and Klimoski 2001), and transactive memory (e.g., Brandon and Hollingshead 2004).

Antecedents and consequences of shared mental models

Over the past decade, the number of empirical studies examining shared mental models has grown. As I highlight the results of these studies in this section, the collective evidence motivates our continued examination of shared mental models. Studies have identified antecedents to and consequences of shared mental models. Antecedents may be team demographics such as education similarity, organization level similarity, percentage of experienced team members, percentage of recruited team members, and team size (negative correlation) (Rentsch and Klimoski 2001). Behaviors have been identified that positively influence shared mental models including cognitive conflict (Ensley and Pearce 2001), cognitive process behaviors (Mohammed and Ringseis 2001), unanimity (versus majority rule) (Mohammed and Ringseis 2001), high-quality planning (Stout *et al.* 1999), team learning behaviours (Van den Bossche *et al.* 2006), and task cohesion (Van den Bossche *et al.* 2006). Finally, organizational interventions may prove beneficial, such as coaching (Cannon and Edmondson 2001), clear direction (Cannon and Edmondson 2001), and training (Marks *et al.* 2002; Smith-Jentsch *et al.* 2001).

Many studies have linked shared mental models to team performance (e.g., Cannon and Edmondson 2001; Edwards *et al.* 2006; Lim and Klein 2006; McComb 2007; McComb and Vozdolska 2007; Rentsch and Klimoski 2001; Webber *et al.* 2000). Others have found links between shared mental model accuracy and team performance (e.g., Edwards *et al.* 2006; Lim and Klein 2006) and shared mental model consistency and team performance (Smith-Jentsch *et al.* 2005). In addition to team performance, shared mental models have been linked to team processes (Mathieu *et al.* 2000; Mathieu *et al.* 2005), team member satisfaction (McComb and Vozdolska 2007; Mohammed and Ringseis 2001), and safety (Smith-Jentsch *et al.* 2005).

Not all relationships studied have been simple ones between independent and dependent variables. Mathieu and colleagues (2005) found that mental model accuracy moderates the relationship between teamwork shared mental models and team processes. Additionally, several mediated relationships have been reported. Specifically, team processes have been shown to mediate the relationship between shared mental models and team performance (Marks *et al.* 2002; Mathieu *et al.* 2000); team mental model accuracy partially mediates the relationship between member ability and team performance (Edwards *et al.* 2006); teamwork schema mediates the relationships between percentage of experienced team members and member growth, percentage of experienced team members and team viability, and percentage of team members recruited and team viability (Rentsch and Klimoski 2001); and mutually shared cognition mediates the relationship between team learning behavior and team performance (Van den Bossche *et al.* 2006).

The empirical results have also identified some very interesting patterns. For instance, multiple mental models exist simultaneously (Mathieu *et al.* 2000; Mathieu *et al.* 2005; McComb 2007; McComb and Vozdolska 2007; Smith-Jentsch *et al.* 2005). The content of these mental models shifts over time from generic to specific (McComb 2007). Fowlkes and colleagues (2000) identified mental model content shifts from procedural knowledge to strategic knowledge as experience increases. Experience also leads to more abstract, consolidated cognitive maps (Rentsch *et al.* 1994). Interestingly, while shared beliefs may exist within groups, these shared beliefs vary across groups within the same organization (Cannon and Edmondson 2001). Mental models also vary between managers/experts and team members when they are confronted with unstructured tasks (Langan-Fox *et al.* 2001). Finally, when working on unstructured tasks, more shared mental model development activity can be observed (Waller *et al.* 2004).

Mental model convergence

A conceptualization of mental model convergence is shown in Figure 3.1. Team members develop shared mental models by proceeding through the three phases of mental model convegence: orientation, differentiation, and integration. These three phases stem from group development and information processing research. Whether the phases of group development are called forming, storming, and norming (Tuckman 1965); team finding, designing, and transforming (Uhl-Bien and Graen 1992); or some other variation found in the literature; the generalized process is basically the same (Tuckman and Jensen 1977). Members:

1. orient themselves to their unique domain,
2. create their own view of the situation, which may or may not be similar to their fellow team members' views; and
3. allow their individual perspective to evolve into a team view.

Likewise, information processing occurs when individuals differentiate among available alternatives and subsequently reconcile, or integrate, similarities and differences among the alternatives to determine a course of action (Driver and Streufert 1969; Schroder *et al.* 1967).

My conceptualization of mental model convergence exploits the synergies between group development and information processing and represents a bottom-up, or emergent, process (Kozlowski and Klein 2000). The process begins at the onset of team activity when each individual team member has a unique, independent view of the team, its assignment and its context. The team members then begin to orient themselves to the team situation. Orientation is retained from the research on group development to represent the phase where the team members collect information about their unique domain that will be used to create shared mental models. Differentiation and integration stem from information processing (Driver and Streufert 1969; Schroder *et al.* 1967). Differentiation occurs as team members interpret their situation in the second phase of group development. From an information-processing perspective, interpretation occurs as team members sort through information they have collected. This sorting process allows them to differentiate among their fellow team members' knowledge and beliefs. As a result, the mental models held by individual team members depict the differences among team members' knowledge and beliefs about the team and its task.

In the integration phase of mental model convergence, the individual team members' mental models shift from having an individual-level focus to having a team-level focus as the differences sorted out during differentiation are transformed into a converged representation of the collective views of the team members. Thus, integration can be viewed as a transformational process (Dansereau *et al.* 1999). Specifically, this process takes place as the team members integrate their perspectives of the team by identifying and strengthening the interrelationships among themselves in order to achieve unity of effort. I do not intend to imply that all team members will hold an identical set of mental models at the conclusion of the integration phase. Rather, the degree of integration (i.e., the strength of the interrelationships that are

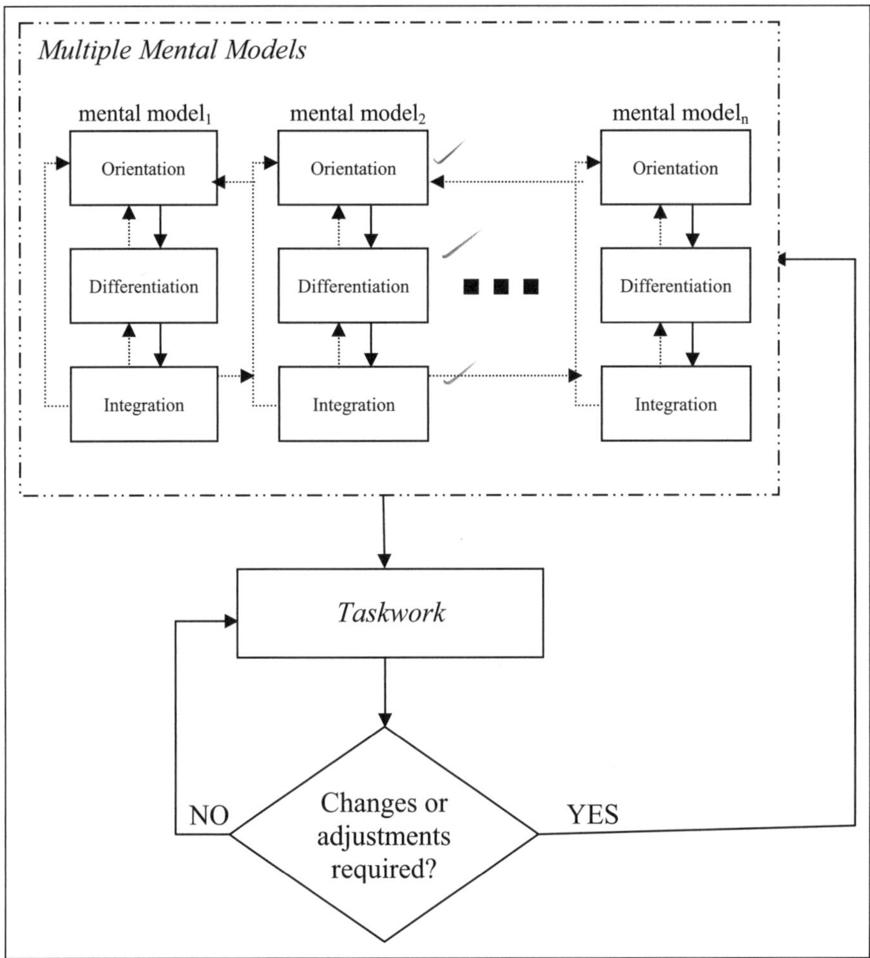

Figure 3.1 The mental model convergence process. Reprinted from Research in Multi-Level Issues: Multi-Level Issues in Organizations and Time, Volume 6, McComb, S.A., Mental model convergence: The shift from being an individual to being a team member, page 131, 2007, with permission from Elsevier

developed among team members), which will be discussed later in this chapter, must be carefully considered.

As shown in Figure 3.1, multiple mental models may develop simultaneously, although the progression may occur at different speeds for each one. The speed may be dependent, in part, upon the amount of previously held knowledge team members possess that is applicable to the current situation. For example, if a majority of team members have worked together on previous projects, logic suggests that the time required for attaining mental model convergence will be much shorter than the time required for a set of individuals with no previous experience working together.

Additionally, the information processed for one mental model may influence the development cycle of others. As such, while the development phases are depicted as occurring linearly, new information attained regarding one mental model may have ramifications for other mental models. The team may need to regress to an earlier phase and revise the affected mental model(s), accordingly. The dotted arrows in Figure 3.1 allow for this iterative process to occur. Upon creation of shared mental models, team members then shift their focus to their assigned taskwork until a time when one or more shared mental models need revision. The decision point in Figure 3.1 represents this mental model maintenance process. A more detailed description of the mental model convergence process can be found in McComb (2007).

Empirical evidence of mental model convergence

The literature contains limited empirical evidence examining mental model convergence over time. Indeed, I am aware of only six such studies that directly examine convergence over time and the results of those studies are conflicting. Edwards and colleagues (2006) and Mathieu and colleagues (2000, 2005) reported that mental models did not converge over time. Alternatively, Levesque *et al.* (2001), McComb (2007), and McComb and Vozdolska (2007) found evidence that convergence changed over time. At one level, the studies appear to be similar since all six were conducted using student teams. Levesque and colleagues (2001) and McComb (2007) report using student teams completing a semester-long course, the other studies were conducted in a laboratory environment. Additionally, all who examined the relationship between shared mental models and team performance reported that it was positive. Levesque *et al.* (2001) did not examine this relationship.

The primary difference between these two sets of studies is the focus on mental model content versus mental model structure. Those studies that did not find convergence tested for convergence of mental model structure using paired comparisons and measures of matrix commonality (e.g., Pathfinder C, UCINET's QAP correlation). The three studies identifying convergence changes, examined converging mental model content. Specifically, McComb (2007) used a sentence stem completion test where respondents finished sentences such as 'The goals of this project are ...'. and 'Our team has discussed using the following modes of communication to exchange information ...'. In Levesque *et al.* (2001) and McComb and Vozdolska (2007), respondents completed five-point Likert scale questionnaire scales. The commonality across responses was captured via r_{WG} and $r_{WG(j)}$, respectively. McComb (2007) and McComb and Vozdolska (2007) report increasing convergence over time. Levesque and colleagues (2001) reported a decline in mental model similarity over time, which they attributed to a reduction in interaction among team members over the course of the semester.

This difference in the way in which shared mental models were operationalized may be the reason for these differential findings. While mental model content is embedded in some structure within an individual's head, commonality across mental model structure may take longer to develop than commonality across mental model content. Moreover, commonality across team members' mental model structures may not be important for certain types of tasks. For instance, decision-makers reviewing options for a search and rescue operation may not need to possess identical linkages within their mental models, as these identical linkages imply that they would filter new information through their mental models in an identical manner. Indeed, if they did have identical structures, they may not consider all possible options before making their decisions. Instead, these decision-makers may only need to possess shared mental model content about issues such as the current situation, the mission objective, and the processes the team will use to distribute information and make decisions. Alternatively, the search and rescue team, such as a Navy Seal team, that executes the decision-makers' plan may need to have shared structure so that they react predictably in a dynamic situation where their lives may be in danger. To achieve this shared structure, these types of teams complete extensive amounts of training.

The studies by Edwards *et al.* (2006) and Mathieu *et al.* (2000, 2005) conducted in the laboratory over a relatively short period of time may not have given the team members adequate time to develop a shared structure. The results of other studies support this notion. First, Marks *et al.* (2002) and Smith-Jentsch *et al.* (2001) found that training increased levels of shared mental model structure across team members. Second, Rentsch and colleagues (1994) found that individuals with more team experience had different knowledge structures than those with less experience and they were able to represent their knowledge structures consistently using different methods for capturing mental model structure. Taking this body of research together suggests that team members may be able to establish shared mental model content rather quickly. Through training and experience, their mental model structures will begin to converge. This conclusion suggests that many cycles through the mental model convergence process may be necessary to attain shared mental model structure. Cannon-Bowers (2007) suggests training interventions that may help expedite mental model convergence during the different phases.

Measuring shared mental models

Measuring mental models is one of the largest issues facing the field. Several methods have been used successfully. Langan-Fox *et al.* (2000), Mohammed *et al.* (2000) and Cooke *et al.* (2004) provide overviews of various techniques available for capturing mental model content and structure. Thus, I will not be redundant here. I will point out, however, that in addition to the techniques they discuss, three additional techniques have been reported recently: (1) observation (e.g., Fowlkes *et al.* 2000; Waller *et al.* 2004), (2) repertory grid (e.g., Tan and Gallupe 2006; Wright 2004), and (3) sentence stem completion (e.g., McComb 2007). Observation and sentence stem completion capture mental model content; repertory grid captures mental model structure. While several techniques are available, at least two efforts are needed to provide clarity with respect to measurement: (1) specifying the multidimensionality of shared mental models and (2) triangulating multiple measurement methods within the same studies.

Shared mental model's multidimentionality Shared mental models are complex constructs that are multidimensional. Each mental model can be represented as containing content that is organized in some structure. The content and structure of each mental model can be characterized with respect to its accuracy, its degree of integration and its revision history. Thus, a mental model has at least five dimensions, namely content, structure, accuracy, degree of integration and revision, that are briefly described in the following paragraphs. As the field continues to mature other dimensions may be uncovered. Researchers have examined content, structure and accuracy to date. Additionally, these dimensions, with the exception of accuracy, have been studied in isolation. Further research that focuses on the unique characteristics of each dimension is needed. Moreover, research is needed to identify various configurations of these dimensions (Rentsch and Small 2007). Identifying various optimal configurations for specific domains may be the result.

Mental model content Mental model content represents the information that comprises a given mental model. Evidence suggests that multiple mental models exist simultaneously. Studies have identified teamwork and taskwork mental models (Mathieu et al. 2000, 2005; Smith-Jentsch et al. 2005) as well as multiple teamwork mental models (McComb 2007; McComb and Vozdolska 2007). Further research is needed to identify a comprehensive list of the relevant mental model content that might be functioning simultaneously regardless of team domain and mental model content only relevant to specific domains.

Mental model structure Mental model structure depicts the way in which the content is organized. Structure may be similar or complementary, following the 'to have in common' or 'to divide' definition of sharing, respectively. The research on mental model structure to date has focused on assessing similarity across dyads and teams (e.g., Carley 1997; Langon-Fox et al. 2001; Tan and Gallupe 2006). While this research is important, additional research is needed for capturing the structure within mental models that are complementary across team members.

Accuracy Accuracy captures how well an individual's mental model matches that of an expert's. The accuracy of mental model content and/or structure can be measured. Webber and colleagues (2000) conducted the only study I found that assessed content accuracy. The majority of research studies that assess accuracy have focused on structural accuracy (e.g., Edwards et al. 2006; Lim and Klein 2006; Smith-Jentsch et al. 2001). Smith-Jentsch and colleagues (2005) examined two different types of accuracy: agreement and consistency. Agreement represents the degree to which respondents are interchangeable (i.e., they have identical mental model content and structure). Consistency represents the degree to which respondents identify the same relative importance when making paired comparisons. Their findings will be presented in the next section of this chapter when I discuss various efforts to triangulate measurement techniques.

Degree of integration During the integration phase of mental model convergence, individuals' mental models converge and shared mental models emerge. When the shared mental models are similar, the degree of integration represents the amount of detail commonly held by the team members. Alternatively, when shared mental models are complementary, the degree of integration represents the amount of uniqueness in the mental models. A more comprehensive discussion about the degree of integration can be found in McComb (2007). To my knowledge, no empirical studies have examined degree of integration. Research is needed to identify how much integration is optimal for various mental model content and structure in various team domains. For example, a team determining search and rescue options may find that too much or too little integration of mental model content inhibits their ability to perform effectively. On the one hand, if too much detail is shared, the team members may all view the task from a very similar perspective and, therefore, may not arrive at an innovative solution. On the other hand, if too little integration exists, the team may not be able to achieve the process and performance benefits attributed to shared mental models. Alternatively, the military team conducting the mission may find themselves unable to coordinate their actions if their shared mental model content and structure are not identical.

Revision Revision captures any shifts in content or structure over time. My conceptualization of mental model convergence (see Figure 3.1) includes an iterative process for mental model maintenance. Given the dynamic environment in which teamwork transpires, revisions to mental models will be necessary as new information becomes available. Indeed, as previously mentioned, mental model content has been observed shifting over time from generic to specific (McComb 2007) and from procedural knowledge to strategic knowledge as experience increases (Fowlkes et al. 2000). To my knowledge, the impact of mental model revision on team performance has not been studied. Research is needed, for example, to ascertain if team members are revising their mental models as necessary to maintain accurate mental models, to identify how disruptive revision is to team functioning, and to determine how best to trigger mental model maintenance cycles.

Triangulating multiple measurement techniques

The second effort required to add clarity to shared mental model measurement is triangulation. Triangulation research has been conducted comparing the following measurement dimensions: (1) shared mental model accuracy versus shared mental model similarity (Edwards *et al.* 2006; Lim and Klein 2006; Smith-Jentsch *et al.* 2001) and (2) shared mental model agreement versus shared mental model consistency, where agreement reflects identical structure (measured as averaged squared Euclidian space) and consistency (measured as average correlations) reflects the same relative structure (Smith-Jentsch *et al.* 2005). The findings suggest that, with respect to performance, shared mental model similarity and accuracy are positively related to team performance (Edwards *et al.* 2006; Lim and Klein 2006), but the relationship is stronger for accurate shared mental models (Edwards *et al.* 2006). Both shared mental model similarity and accuracy can be improved by training (Smith-Jentsch *et al.* 2001). Finally, consistent structures are more important than identical structures (Smith-Jentsch *et al.* 2005).

With respect to measurement methods, the following have been compared within the same study: (1) multidimensional scaling of paired comparisons and concept maps as two possible measures of mental model structure (Rentsch *et al.* 1994) and (2) Cronbach alpha and r_{WG} as measures of mental model content (Webber *et al.* 2000). As previously mentioned, individuals with more experience working on teams may be better able to represent their mental model structure consistently regardless of measurement method (Rentsch *et al.* 1994). The two methods used to assess content were not found to be equivalent (Webber *et al.* 2000). Specifically, alpha correlated with accuracy, but not with performance; r_{WG} correlated with performance, but not with accuracy.

While these comparisons are good first steps, more triangulation research is imperative. Indeed, studies incorporating multiple dimensions and/or methods will allow for extensive comparisons across dimensions and methods. For example, an examination of multiple measures of shared mental model content and structure within the same study will help to advance our understanding of how these two dimensions interact.

Shared mental models and macrocognition

Shared mental models are at the heart of the collaboration stages and macrocognitive processes introduced in Chapters 1 and 2. Macrocognitive processes either inform the mental model convergence process (e.g., individual visualization and representation of meaning, team knowledge development) or are influenced by shared mental models (e.g., develop, rationalize and visualize solution alternatives). These processes, both macrocognitive processes and the mental model convergence process, drive the cognitive processes undertaken by team members throughout the collaboration stages.

During the knowledge construction stage, the macrocognitive processes invoked by the individual team members include developing a team understanding of the problem conditions, creating their own mental models about the situation, and sharing hidden knowledge. In order to create their own mental models about the situation, team members will traverse through the orientation and differentiation phases of mental model convergence (see Figure 3.1). The outcome will be a set of mental models that may or may not be shared across team members, but serve to describe the situation in which the team is embedded from each individual team member's perspective. Some of these mental models will be integrated as the team develops a common understanding of the problem conditions. One of the most critical macrocognitive processes during this collaboration stage may be the sharing of hidden information. Indeed, pooling unshared information is one of the primary incentives for creating teams (Wittenbaum and Stasser 1996). Tapping the hidden information, therefore, becomes critical for the creation of accurate mental models and shared understanding (McComb 2007).

Much of the integration phase of mental model convergence will happen during the collaborative team problem-solving stage. This stage represents the onset of team collective activities. Thus, the macrocognitive processes shift from the individual to the team, and thus, the focal unit of the individual team members' mental models shift from the individual to the team through the integration phase (McComb 2007). As this shift occurs, the knowledge contained in the mental models becomes interoperable across team members. Moreover, each team member begins to have a better understanding of the unique role they will play on the team. One result of this phase should be a set of shared mental models that can be used to reduce uncertainty (Klimoski and Mohammed 1994) and govern team behavior (Cannon-Bowers *et al.* 1993).

In addition to mental model convergence, the process of applying mental models to macrocognitive processes will begin during the team problem-solving stage. Specifically, as the team begins to develop, rationalize and visualize solution alternatives, the information and knowledge contained in their mental models will influence how they work through this collaboration stage. Similarly, this application of mental models to macrocognitive processes will continue during the team consensus stage and the outcome, evaluation and revision stage. For instance, the team's mental model about their goal, and the timing associated with completing the mission, will motivate the entire process. Their mental model about team member expertise will help them with the rationalization of various alternatives because the team can look to particular members with unique information to help them ascertain the viability of the various alternatives from varying perspectives.

Throughout the collaboration stages, new information will become available to the team. Thus, the mental model convergence process will need to be revisited to incorporation this new information. In other words, they will need to conduct mental model maintenance. The process may occur subconsciously. It will, however, be more effective if the team switches back to actively thinking about their mental models, if only for a short time (McComb 2007). This switch to active thinking may not occur effortlessly (Beyer *et al.* 1997), particularly under stressful situations (Dutton 1993; Gersick and Hackman 1990). Training may help teams to identify points in their life cycle when they need to make this shift from subconsciously adjusting and applying mental models to actively maintaining them and collectively incorporating new information. In particular, training teams to conduct intra-team performance monitoring and feedback, such as guided team self-correction, may be most beneficial (Cannon-Bowers 2007).

Implications

The purpose of this chapter was to briefly examine the extant research on shared mental models and connect this research to macrocognitive processes. Several implications for researchers were highlighted throughout the chapter. Most significant are the preliminary results that link shared mental models to effective team processes and team performance. These results provide the motivation to extend this field of inquiry. In particular, empirical research is needed to:

1. validate my mental model convergence conceptualization
2. examine the multidimensionality of shared mental models

3. triangulate various dimensions and measurement techniques within the same study, and
4. explore how shared mental models influence macrocognition in teams.

This field of inquiry also has implications for practice. For instance, evidence suggests that successful teams put significant effort into developing protocols for working together as a team, such as an agreed upon approach for conducting the requisite work (Katzenbach and Smith 1999). In other words, successful teams focus explicitly on creating teamwork shared mental models. This explicit focus on teamwork at the onset of team activity is critical for at least two reasons. First, many organizations provide the most support for team development early in the team's life cycle (Druskat and Pescosolido 2002). During this early developmental phase, training focusing on expediting mental model convergence should be considered. Second, team members will develop protocols (implicitly or explicitly) for collective work (i.e., the teamwork phase of the team's life cycle) (Smircich 1983); moreover, they will develop them very quickly and sustain them for extended periods of time (Gersick 1988). Even when mental model convergence occurs implicitly as individual team members actively consider teamwork issues relevant to their own personal agendas, the phases proposed will still be relevant. The results, however, will be inconsistent across team members. Further, the accuracy and comprehensiveness of the resultant shared mental models may be questionable. These issues of making the necessary adjustments in each individual team members' mental models occur throughout the life cycle of teams.

To expedite the mental model convergence process and ensure accuracy, tools and training may be necessary. For instance, software tools that encourage team members to discuss explicitly the mental model contents necessary at a particular stage of team development may be beneficial. Specific training exercises, such as training aimed at uncovering the relevant expertise each team member brings to the task, may be useful as well. Prior to the developing these types of tools and training programs, more research is necessary to expand the body of empirical data. While the extant empirical evidence links shared mental models to team performance in general terms, more specific data is needed across all the aforementioned dimensions of mental models, namely mental model content, mental model structure, accuracy, degree of integration and revision. Moreover, this data will be useful in creating computational models of the mental model convergence process that will be useful for examining what-if scenarios. In sum, more research is needed to ensure that any tools and training programs developed will indeed facilitate team members' efforts in expediently creating and maintaining shared mental models that are appropriate, accurate, and comprehensive.

References

Bartunek, J.M. (1984). 'Changing Interpretive Schemes and Organizational Restructuring: The Example of a Religious Order', *Administrative Science Quarterly* 29, 355–372.

Beyer, J. *et al.* (1997). 'The Selective Perception of Managers Revisited', *Academy of Management Journal* 40, 716–737.

Brandon, D.P. and Hollingshead, A.B. (2004). 'Transactive Memory Systems in Organizations: Matching Tasks, Expertise, and People', *Organization Science* 15, 633–644.

Cannon, M.D. and Edmondson, A. (2001). 'Confronting Failure: Antecedents and Consequences of Shared Beliefs about Failure in Organizational Work Groups', *Journal of Organizational Behavior* 22, 161–177.

Cannon-Bowers, J.A. (2007). 'Fostering Mental Model Convergence Through Training', in F. Dansereau and F.J. Yammarino (eds), *Multi-Level Issues in Organizations and Time*, pp. 149–158. (Oxford: Elsevier Science Ltd).

Cannon-Bowers, J.A. and Salas, E. (2001). 'Reflections on Shared Cognition', *Journal of Organizational Behavior* 22, 195–202.

Cannon-Bowers, J.A. *et al.* (1993). 'Shared Mental Models in Expert Team Decision Making', in N.J. Castellan, Jr. (ed.), *Individual and Group Decision Making: Current Issues*, pp. 221–246. (Hillsdale, NJ: Lawrence Erlbaum).

Carley, K.M. (1997). 'Extracting Team Mental Models Through Textual Analysis', *Journal of Organizational Behavior* 18, 533.

Cooke, N.J. *et al.* (2004). 'Advances in Measuring Team Cognition', in E. Salas and S.M. Fiore, (eds), *Team Cognition: Understanding the Factors that Drive Process and Performance*, pp. 83–106. (Washington, DC: American Psychological Association).

Dansereau, F. *et al.* (1999). 'Multiple Levels of Analysis from a Longitudinal Perspective: Some Implications for Theory Building', *Academy of Management Review* 24, 346–357.

Dougherty, D. (1992). 'Interpretive Barriers to Successful Product Innovation in Large Firms', *Organization Science* 3, 179–202.

Driver, M.J. and Streufert, S. (1969). 'Integrative Complexity: An Approach to Individuals and Groups as Information-Processing Systems', *Administrative Science Quarterly* 14, 272–285.

Druskat, V.U. and Pescosolido, A.T. (2002). 'The Content of Effective Teamwork Mental Models in Self-Managing Teams: Ownership, Learning and Heedful Interrelating', *Human Relations* 55, 283–314.

Dutton, J.E. (1993). 'Interpretations on Automatic: A Different View of Strategic Issue Diagnosis', *Journal of Management Studies* 30, 339–337.

Eden, C. *et al.* (1981). 'The Intersubjectivity of Issues and Issues of Intersubjectivity', *Journal of Management Studies* 18, 35–47.

Edwards, B.D. *et al.* (2006). 'Relationships Among Team Ability Composition, Team Mental Models, and Team Performance', *Journal of Applied Psychology* 91, 727.

Ensley, M.D. and Pearce, C.L. (2001). 'Shared Cognition in Top Management Teams: Implications for New Venture Performance', *Journal of Organizational Behavior* 22, 145–160.

Fiore, S.M. and Salas, E. (2004). 'Why we Need Team Cognition', in E. Salas and S.M. Fiore, (eds), *Team Cognition: Understanding the Factors that Drive Process and Performance*, pp. 235–248. (Washington, DC: American Psychological Association).

Fowlkes, J.E. *et al.* (2000). 'The Utility of Event-Based Knowledge Elicitation', *Human Factors* 42, 24–35.

Gersick, C.J.G. (1988). 'Time and Transition in Work Teams: Toward a New Model of Group Development', *Academy of Management Journal* 31, 9–41.

Gersick, C.J.G. and Hackman, J.R. (1990). 'Habitual Routines in Task-Performing Groups', *Organizational Behavior and Human Decision Processes* 47, 65–97.

Gruenfeld, D.H. and Hollingshead, A.B. (1993). 'Sociocognition in Work Groups: The Evolution of Group Integrative Complexity and its Relation to Task Performance', *Small Group Research* 24, 383–405.

Johnson-Laird, P.N. (1983). *Mental Models: Towards a Cognitive Science of Language, Inference, and Consciousness.* (Cambridge, MA: Harvard University Press).

Katzenbach, J.R. and Smith, D.K. (1999). 'The Discipline of Teams', *Harvard Business Review* 71, 111–120.

Klimoski, R. and Mohammed, S. (1994). 'Team Mental Model: Construct or Metaphor?', *Journal of Management* 20, 403–437.

Kozlowski, S.W.J. and Klein, K.J. (2000). 'A Multilevel Approach to Theory and Research in Organizations', in K.J. Klein and S.W.J. Kozlowski, (eds), *Multilevel Theory, Research, and Methods in Organizations: Foundations, Extensions, and New Directions*, pp. 3–90. (San Francisco, CA: Jossey-Bass).

Langan-Fox, J. *et al.* (2000). 'Team Mental Models: Techniques, Methods, and Analytic Approaches', *Human Factors* 42, 242–271.

Langan-Fox, J. *et al.* (2001). 'Analyzing Shared and Team Mental Models', *International Journal of Industrial Ergonomics* 28, 99–112.

Levesque, L.L. *et al.* (2001). 'Cognitive Divergence and Shared Mental Models in Software Development Project Teams', *Journal of Organizational Behavior* 22, 135.

Lim, B. and Klein, K.J. (2006). 'Team Mental Models and Team Performance: A Field Study of the Effects of Team Mental Model Similarity and Accuracy', *Journal of Organizational Behavior* 27, 403.

Marks, M. *et al.* (2002). 'The Impact of Cross-Training on Team Effectiveness', *Journal of Applied Psychology* 87, 3–13.

Mathieu, J.E. *et al.* (2000). 'The Influence of Shared Mental Models on Team Process and Performance', *Journal of Applied Psychology* 85, 273–283.

Mathieu, J.E. *et al.* (2005). 'Scaling the quality of Teammates Mental Models: Equifinality and Normative Comparisons', *Journal of Organizational Behavior* 26, 37–56.

McComb, S.A. (2007). 'Mental Model Convergence: The Shift from Being an Individual to Being a Team Member', in F. Dansereau and F.J. Yammarino (eds), *Multi-Level Issues in Organizations and Time*, pp. 95–148. (Oxford: Elsevier Science Ltd).

McComb, S.A. and Vozdolska, R.P. (2007). 'Capturing the Convergence of Multiple Mental Models and their Impact on Team Performance', Paper presented at the Southwestern Academy of Management, March 13–17, 2007.

Mohammed, S. *et al.* (2000). 'The Measurement of Team Mental Models: We have no Shared Schema', *Organizational Research Methods* 3, 123–165.

Mohammed, S. and Ringseis, E. (2001). 'Cognitive Diversity and Consensus in Group Decision Making: The Role of Inputs, Processes, and Outcomes', *Organizational Behavior and Human Decision Processes* 85, 310–335.

Rentsch, J.R. *et al.* (1994). 'What you Know is What you Get from Experience: Team Experience Related to Teamwork Schemas', *Group and Organization Management* 19, 450.

Rentsch, J.R. and Klimoski, R. (2001). 'Why do Great Minds Think Alike?: Antecedents of Team Member Schema Agreement', *Journal of Organizational Behavior* 22, 107–120.

Rentsch, J.R. and Small, E.E. (2007). 'Understanding Team Cognition: The shift to Cognitive Similarity Configurations', in F. Dansereau and F.J. Yammarino, (eds), *Multi-level Issues in Organizations and Time*, pp. 159–174. (Oxford: Elsevier Science Ltd).

Rouse, W.B. and Morris, N.M. (1986). 'On Looking into the Black Box: Prospects and Limits in the Search for Mental Models', *Psychological Bulletin* 100, 349–363.

Schroder, H.M. *et al.* (1967). *Human Information Processing.* (New York: Holt, Rinehart and Winston).

Smircich, L. (1983). 'Organizations as Shared Meanings', in L.R. Pondy *et al.* (eds), *Organizational symbolism*, pp. 55–65. (Greenwich, CT: JAI Press).

Smith-Jentsch, K.A. *et al.* (2001). 'Measuring Teamwork Mental Models to Support Training Needs Assessment, Development, and Evaluation: Two Empirical Studies', *Journal of Organizational Behavior* 22, 179–194.

Smith-Jentsch, K.A. *et al.* (2005). 'Investigating Linear and Interactive Effects of Shared Mental Models on Safety and Efficiency in a Field Setting', *Journal of Applied Psychology* 90, 523–535.

Stout, R.J. *et al.* (1999). 'Planning, Shared Mental Models, and Coordinated Performance: an Empirical Link is Established', *Human Factors* 41, 61.

Tan, F.B. and Gallupe, R.B. (2006). 'Aligning Business and Information Systems Thinking: A Cognitive Approach', *IEEE Transactions on Engineering Management* 53, 223–237.

Tuckman, B.W. (1965). 'Developmental Sequence in Small Groups', *Psychological Bulletin* 63, 384–399.

Tuckman, B.W. and Jensen, M. (1977). 'Stages of Small-Group Development Revisited', *Group and Organization Studies* 2, 419–427.

Uhl-Bien, M. and Graen, G.B. (1992). 'Self-Management and Team-Making in Cross-Functional Work Teams: Discovering the Keys to Becoming an Integrated Team', *The Journal of High Technology Management Research* 3, 225–241.

Van den Bossche, P. *et al.* (2006). 'Social and Cognitive Factors Driving Teamwork in Collaborative Learning Environments', *Small Group Research* 37, 490–521.

Waller, M.J. *et al.* (2004). 'Effects of Adaptive Behaviors and Shared Mental Models on Control Crew Performance', *Management Science* 50, 1534.

Webber, S.S. *et al.* (2000). 'Enhancing Team Mental Model Measurement with Performance Appraisal Practices', *Organizational Research Methods* 3, 307–322.

Weick, K.E. and Roberts, K.H. (1993). 'Collective Mind in Organizations: Heedful Interrelating on Flight Decks', *Administrative Science Quarterly* 38, 357–381.

Wittenbaum, G.M. and Stasser, G. (1996). 'Management of Information in Small Groups', in J.L. Nye and A.M. Brower, (eds), *What's Social About Social Cognition?*, pp. 3–28. (Thousand Oaks, CA: Sage).

Wright, R.P. (2004). 'Mapping Cognitions to Better Understand Attitudinal and Behavioral Responses in Appraisal Research', *Journal of Organizational Behavior* 25, 339–374.

Chapter 4

Communication as Team-level Cognitive Processing

Nancy J. Cooke, Jamie C. Gorman and Preston A. Kiekel

Teams can be defined as two or more people who interact interdependently with respect to a common goal and who have each been assigned specific roles to perform for a limited lifespan of membership (Salas, Dickinson, Converse and Tannenbaum 1992). To the extent that team performance takes place in cognitively demanding environments, team cognition is an important construct underlying team performance.

In this chapter we focus on team cognition, defined as cognitive processing on the part of the team (e.g., team-level thinking, problem-solving and attending) and distinguish it from individual cognition. The relationship between individual and team cognition is controversial and is also addressed in this chapter. Both team and individual cognition can be defined using macrocognitive constructs and thus, macrocognition (Chapter 2, this volume), encompasses both individual and team cognition.

The current state of team cognition research can be roughly divided into two approaches. 'In the head' (ITH) approaches localize team cognition in the sum of individual (internalized) team member inputs (e.g., individual knowledge contributions). Alternatively, 'between the heads' (BTH) approaches localize team cognition as the interaction processes involved in team activity (e.g., team communication and coordination). The BTH approach involves externalized macrocognitive processes (Chapter 2, this volume). For any given team-level cognitive processing such as team planning there is cognition operating at both the level of individuals and teams, but the question here is how best to characterize the team level (ITH or BTH) for predicting team performance. Thus, the question is not the relative value of individual or team cognition for effective collaboration, rather it is a question how team cognition should be measured (ITH or BTH or some hybrid).

The ITH approach to measurement of team cognition focuses on extending individual cognitive constructs and measures to teams, whereas the BTH approach focuses on team interaction processes, such as communications. Figure 4.1 shows a popular framework for group information processing (Input–Process–Output; IPO; Hackman 1987). With respect to the I–P–O framework, ITH approaches attempt to understand and predict team outcomes by focusing on individual-level inputs (e.g., knowledge). BTH approaches attempt to understand and predict team outcomes by measuring team processes (e.g., team coordination, communication).

Causal models, therefore, that are strictly based on ITH approaches tend to include performance predictors that are aggregates of individual-level inputs (e.g., shared mental models).

These causal models have had some success at predicting team performance and process in various domains (Converse, Cannon-Bowers and Salas 1991; Mathieu, Goodwin, Heffner, Salas and Cannon-Bowers 2000). Causal models that are based on BTH approaches tend to include performance predictors that are related to team processes (e.g., team coordination). In our research on team cognition, primarily within a command-and-control domain, the BTH constructs and associated measures have tended to account for the more reliable patterns of results than ITH measures (Cooke *et al.* 2004). In this chapter we focus on team communication processes and emphasize, therefore, the contribution of the BTH approach to measuring and understanding team cognition.

An important distinction that can be made between ITH versus BTH measures is related to classical information processing theory versus ecological psychology. Classical information processing theory is centered on the notion that cognition involves transformations of internal states represented in the head. Ecological psychology, on the other hand, is centered on the dynamics of person–environment relations, focusing on 'not what's inside your head, but what your head is inside of' (Mace 1977). The latter theory suggests that cognitive phenomena are directly observable in the form of changing person–environment relations while the former theory assumes that cognitive phenomena are only indirectly observable and are independent of the dynamics of person–environment relations. Internalized versus externalized macrocognition also captures this distinction. Focusing on the dynamics of team-member relations rather than internal representational states leads to different assumptions about how cognitive processes can most directly be observed and measured. For instance, team communication allows for direct observation of team cognition from a BTH approach.

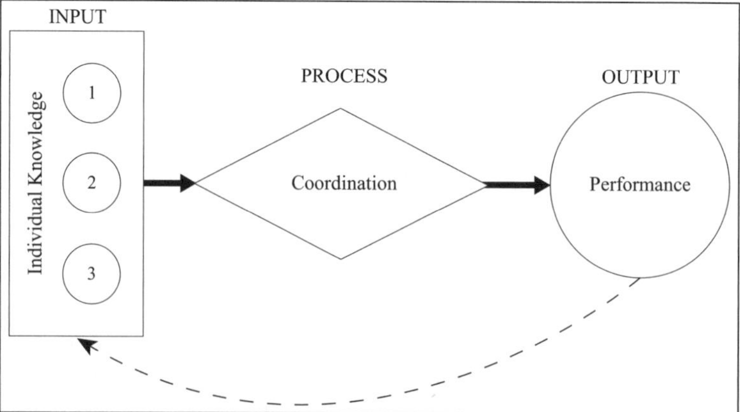

Figure 4.1 The Input-Process-Output framework of team information processing; examples of input, process, and output variables are shown

In addition to dynamic team-member relations, context is an important factor related to the fundamental BTH ecological assumption that actor and environment are complimentary and dependent notions; that is, one cannot be described without the otherand that metric units of actor–environment relations are nested. The implications for team cognition are that (1) an actor is a team-member only in the context of team activityand that (2) 'atomic' units of team cognition are nested within a larger context of team cognition, although atomisms can be different based on the particular domain. Holistic measures (versus aggregate measures) of team cognition (Cooke, Salas, Cannon-Bowers and Stout 2000) that incorporate team interaction processes have addressed assumption (1). However, understanding how isolated team-member relations are embedded in a larger context of team-member interactions is necessary in order to address assumption (2). In particular, what a team-member is talking about currently is often correlated with what the team was talking about in the past and will be talking about in the future (Gorman 2005). These sorts of 'long memory' phenomena are common in individual-level cognitive processing (Gilden, Thornton and Mallon 1995; Van Orden, Holden and Turvey 2003), but present a relatively recent reconsideration of how BTH cognitive processing can represent important new insights into how current activity is contextualized over long sequences of actor–environment interactions. Later in the chapter we present some dynamical systems analysis of coded communication data that speak to the temporally extended nature of team cognition. A particularly powerful method for understanding and measuring the dynamics of actor–environment relations, where teammates account for the structure of the environment, is via communication analysis.

Team communication and its relation to team cognition

When taking the BTH approach to measuring team cognition, focus turns to the interactions among team members and how those interactions pertain to team cognition. Team researchers have developed methods and tools that make use of structured observations and self report of team process behaviors (e.g., Fowlkes, Lane, Salas, Franz and Oser 1994). However, these measures can be cumbersome because they rely on subjective opinions or are administered outside of the task context. Communication is one of the predominant forms of team interaction and one that affords unobtrusive measurement in the context of the task.

Communication itself can take many forms, including non-verbal communication such as gestures and eye contact and verbal communication that ranges from written or digital text (letters, notes, chat, text messages and email) and voice communications. Communication can also occur among individuals who are distributed across time and space and thus can be asynchronous, synchronous, face-to-face, or at a distance. Communication also carries with it much richness in meaning from the sheer identity of the individuals engaged in communication to the content of what they are saying and the tone used to say it. In this chapter we use the term 'communication' broadly to include all types and nuances of discourse. Communication is a largely observable rich team-level cognition (Stahl 2006).

Just as individual cognition is reflected in the behavior of the individual, team cognition is reflected in the behavior of the group. In fact, there are parallels between individual think-aloud protocols or verbal reports and communication. It is fortuitous though that for teams, these verbal reports occur 'in the wild' and do not require prompting or reminding on the part of an experimenter. Thus, communication might provide the same information about team cognition that verbal reports provide about individual cognition. However, we take this line of reasoning one step further and propose that communication is not just a window into team cognition, but that communication *is* cognitive processing at the team or group level. That is, communication is an externalized team-level macrocognitive process – one that is very rich, and in most cases its richness is untapped.

Communication analysis has been conducted that investigates the sequential patterns of communication and the flow of communication among team members (e.g., Bowers, Jentsch, Salas and Braun 1998), however, until recently such analysis was prohibitive in terms of the resources required to transcribe and code communication data manually. Estimates of up to 28 hours of analysis per hour of communication data (Emmert 1989) were enough to seriously thwart the full exploitation of these data. In the best case, communications were coded by type and the frequency of each type noted with some sequential dependencies. Recent efforts to automate communication analysis through the use of speech recognition and algorithms to detect patterns have opened the doors to these data.

Automating communication analysis

Before discussing methods for automating the analysis of communication data, it is useful to address the various features of the data that may be the focus of analysis. Both flow and content analysis can be performed on either static or sequential communication data (Kiekel, Cooke, Foltz *et al.* 2002). Static communication data are represented by summary measures of communication content or flow, over a series of communication events. Number of communication events over a specified interval of time is an example of a *static flow measure*, whereas frequency of content codes is an example of a *static content measure*. Dominance, a lagged cross-correlation between team-member communications (Budescu 1984; Dillon, Madden and Kumar 1983; Wampold 1984) is an example of a measure that takes into account *sequential communication flow*, whereas lagged LSA (Latent Semantic Analysis) cosine between communication events (Gorman, Cooke and Kiekel 2004) is an example of a measure that takes into account *sequential communication content*. In terms of automation, and ultimately, real time analysis, sequential measures present a more challenging problem. For example, the need for handling complexities such as fluctuations in communication volume necessitates something like the analysis of 'moving windows' of different sizes. In addition, content measures are more challenging to automate than flow methods in the case of voice communication.

In the remainder of this section we discuss the long-term objective of automating communication analysis, first in terms of current research and future directions for automating analysis of communication content or 'what was said' and then in terms of automating analysis of communication flow, or 'who talks to whom, when and for how long.' We follow this discussion with some data that compare the various methods and with a discussion of interpreting communication patterns.

Automating content analysis

At least two content methods for which automation can be envisioned include LSA and keyword indexing. LSA (Landauer, Foltz and Laham 1998) is a statistical/mathematical technique for representing words and vectors of words (e.g., utterances) in a high-dimensional semantic space, where the semantic space can be generated using either domain-specific or general background data (corpora). For example, using a UAV (unmanned aerial vehicle)-specific corpus, communication events (utterances) generated by teams performing the UAV task can be measured relative to the semantic space itself (vector length, static) or relative to other utterances (cosine, sequential). LSA has been applied to communication analysis to reveal content density over task segments (static), lag coherence across utterances (sequential), performance prediction based on mission-level cosines (static)and designing algorithms for automatically generating content codes for novel utterances (sequential). For example, there is some evidence that automatically tagging utterances for content codes (e.g., Question, Response, etc.) can be facilitated using the LSA cosine measure (Foltz, Martin, Abdelali *et al.*, 2006).

Keyword indexing is a relatively simple technique for representing words and vectors of words in a k-dimensional space, where k is the number of keywords. Similar to LSA, keyword indexing can be performed using a UAV-specific corpus, where the keywords are the k most frequent words used in the task. Also like LSA, communication events (utterances) can be measured against the keyword space (vector length, static) or relative to other utterances (cosine, sequential). Although there are some similarities between LSA and keyword indexing, they are fundamentally different. LSA incorporates an inverse weighting procedure to help reduce the influence of raw word frequency on the semantic space and ultimately communication measures. Keyword indexing measures, on the other hand, are taken directly over the most frequent words, excluding articles and conjunctions.

Because both LSA and keyword indexing incorporate 'automatic' measures of communication content (vector length, cosine), both can be envisioned as likely candidates for automation. However, both are also based on transcribed or at least textual input in order to measure content. This presents a limitation for automating the analysis of spoken communications. Research on the use of automatic speech recognition (ASR) for generating LSA input in real time has been promising (Foltz, Laham and Derr 2003). For keyword indexing, the problem of transforming raw speech into textual input may be a simpler one: any algorithm for transforming raw speech into textual input need only identify certain keywords, which may often represent a relatively small and unique-sounding subset of the total lexicon of communication behavior.

Automating flow analysis

It is generally easier to automate communication flow methods, simply because that is the nature of lower-level data. For communication flow analysis, we have developed a suite of methods that capture both sequential and static aspects of communication data. For this chapter, we focus on two of our sequential methods, which show the most promise. These are ProNet and Dominance.

ProNet (Procedural Networks; Cooke, Neville and Rowe 1996) is a form of lag sequential analysis, based on the Pathfinder algorithm. Pathfinder is used to identify linkages between nodes, using the set of pairwise proximities among nodes. Nodes can be defined as events in a potential sequence and proximity can be defined as a transition probability among these events. ProNet is the application of Pathfinder to such a sequence of nodes.

In this case, we use ProNet to identify representative sequences of discrete communication flow events in our time-stamped sequential log of speech utterance events, or ComLog (communication log). We define six nodes, one for each team member (in the AVO, PLO, or DEMPC role) beginning and ending a speech act (i.e., Abeg, Pbeg, Dbeg, Aend, Pend, Dend). This is the most basic unit of communication. The output of a ProNet analysis is a set of statistically detectable chains of these six nodes within the discourse.

With these six nodes, many possible patterns can be identified. We use these nodes to define three types of sequences. First, an *Xloop* means that person X begins and ends an uninterrupted utterance, then begins and ends again. This may indicate repetition. Second, an *XYcycle* means that person X speaks a complete utterance, then person Y does. Third, *XiY* means that person X interrupts person Y. We chose these sequence patterns because other chains were either too specific to be generally applicable (e.g., specific types of interruption), or were too short to be very meaningful (e.g., X begins then ends a speech event). We measure both the ProNet detection of these chainsand their frequency of occurrence within the ComLog.

Another measure we use from the ProNet output is the descriptive statistics computed on the set of chain lengths for each team-at-mission discourse. For example, if a team's regular communication sequences tend to be short, then the mean length of their set of statistically detectable chains will be small. In that case, the team tends to be pithy in their speech patterns, or that they generally make use of more communication patterns of a shorter duration. This implies adaptability, rather than regularity, in the team's communication style. On the other hand, if the mean chain length is long, it implies that the team's discourse tends to come in long, predictable sequences. This implies a high degree of communication stability in the team's discourse.

ProNet, like Pathfinder, is limited in the sense that the multiple-link paths represented in the network structure are only certain to exist on a pairwise basis. ChainMaster is a software tool that implements the ProNet algorithm, but that extends it by doing tests for the existence of chains at multiple lags. With these tests, the likely multiple-link paths can be highlighted.

In addition to ProNet and ChainMaster for analyzing flow data, we defined a measure representing the influence a team member's communication flow exhibits over the communication flow of other team members, in terms of amount of speech. Building on work by Budescu (1984), we defined a dominance measure based on the cross-correlations of speech quantity in the ComLog, among all pairs of speakers. Correlations among sequences of speech quantity were computed for each lag, among each pair of speakers. This constitutes the set of cross-correlation functions. Then, because we were not concerned with directionality of influence, we squared each correlation. Because we were working with R^2, Fisher's R to Z transformation was not required, before performing arithmetic computations. Next, we took the weighted average of each squared cross-correlation function, with weighting being the inverse of the lag. This gives greater emphasis to influence revealed at early lags (i.e., speech events that are closer in time).

This process leaves us with a squared correlation for each pair of speakers, representing the capacity of each speaker's speech quantity behavior to predict each other speaker's. To obtain a relative dominance value for each speaker pair, we computed ratios of squared cross-correlations. So, for speakers X and Y, we took the correlation of X predicting Y (R^2_{xy}), divided by Y predicting X (R^2_{yx}), to obtain a dominance ratio. These ratios indicate dominance of each team member over each other.

In the above ratio, the dominance value ranges from 0 to 1 if Y is dominant (i.e., if R^2_{yx} is larger than R^2_{xy}), but from 1 to positive infinity if X is dominant. To convert this to a symmetrical scale, we took the natural log of the dominance ratio. This yields a pairwise dominance measure ranging from negative infinity to 0 when Y is dominant and from 0 to positive infinity when X is dominant. However, due to the variance stabilizing aspect of log transformations, the distribution of the pairwise dominance measure is approximately normal, with a mean of 0. Under these conditions, it is appropriate to perform arithmetic operations. To get a total dominance score for each team member, we averaged the log of the ratios. The three dominance scores obtained in the context of our three person UAV task are Adom, Pdomand Ddom, associated with the three team roles of AVO, PLOand DEMPC, respectively.

Comparisons of methods

LSA measures and keyword indexing measures have been compared on very basic levels. Correlations based on over 20,545 utterances in four UAV experiments between LSA vector length, keyword vector length and word counts across four UAV experiments are presented in Table 4.1. These results indicate that for this data set it is highly likely that LSA vector length derives most of its variance from the length, in number of words, of utterances. This could result from one of two things. First, LSA vector *is* a measure of word counts. Or second, and most probably, task-related talk comprises the lion's share of the utterances in our transcripts. Interestingly, however, keyword vector length was found to share less variance with straight word counts.

In addition, various clusters of communication measures (including both flow and content measures) developed over series of experiments using the UAV task have been evaluated in large regression models to predict team performance. In this manner, partial correlations with team performance (ignoring experimental effects)

Table 4.1 Correlations between LSA density component and other content metrics

	LSA vector length	Keyword vector length
Word count	0.944*	0.699*
LSA vector length	–	0.725*

Note: $N = 20{,}545$; * $p < 0.001$.

of these clusters were estimated for each of our communication measures. This analysis revealed that the most predictive cluster of communication measures included an LSA measure, lag coherenceand a keyword measure, communication density (zero-order correlations were $r\,(197) = 0.28$, $p < 0.001$ and $r\,(197) = 0.74$, $p < 0.001$, respectively). A flow measure, mean ProNet chain length, was also included in this cluster (zero-order correlation $r\,(197) = -0.19$, $p < 0.01$). This last result is promising because flow methods present perhaps the best opportunity for automating communication analysis in the near-term.

Interpreting communication patterns

The power of the flow methods is that, being very low level data, they can be collected and analyzed in real time. The general purpose of this approach would be to identify predictive communication patterns, without communication content. Certain communication event patterns may be indicative of specific communication content issues. Several examples come to mind. Arguments tend to be associated with increased interruptions (detected with ProNet). Discourse tends to decrease during periods of high workload (detected with simple frequency counts of speech utterances). Leadership shifts may be detected by changes in the dominance pattern (detected with the dominance measure). Chapter 16 of this volume describes an example of how flow patterns in electronic mail communication were mapped to meaningful Enron events.

We have developed the capability to analyze a running record of the ComLog during actual team discourse. It may be possible to use this method to identify communication events in team discourse, before they actually occur and without actually having a transcript of the discourse. This can allow for appropriate design interventions, such as rerouting tasks to alternative teams when workload communication patterns suggest that workload may be high. In addition, the flow patterns may be used to identify communication windows that are good candidates for a deeper content analysis. This approach to team communication analysis raises many questions, which require extensive empirical research in order to answer effectively.

Foremost, interpretation is difficult. It requires extensive data mining methods, such as cluster analysis, regression models and so on. As with all data mining problems, it would be most appropriate to start with as many a priori hypotheses as possible. Overfitting is a ubiquitous problem in model building. It needs to be addressed by replication, split-half reliability tests and so on.

Most importantly, interpretation hinges on having access to other indices of team performance and cognition. Communication patterns can be mapped to these indices and interpreted in terms of them. Initial forays into the domain may allow communication patterns to be mapped to team performance such that good and bad teams can be discriminated by communication alone. The ProNet regression results presented previously address this objective. Later analyses mapping patterns to more diagnostic indices can begin to explain the behavior or cognition underlying the good or bad performance. Importantly, interpretation requires not only criterion data, but also the rules for mapping patterns to these data which are likely specific to the domain. A domain expert may be able to state these rules or observe connections over time. Automation of the interpretation process requires machine learning approaches by which the computer learns over many instances to associate certain communication patterns with specific performance or cognitive parameters.

Another question raised with all momentary measurement is that of sample size. In the limit, momentary measurement takes place at a single point in time. This fails to serve the basic function of statistics, which is to reduce data to meaningful patterns. However, employing data across the entire duration of the discourse is too course of a granularity for any real time prediction measure. What is the balance? This is an empirical question. We are testing several different approaches to the sample size issue. The question largely hinges on that of how one meaningfully segments a time series. Complex pattern detection algorithms exist for doing this (e.g., in DNA research, in cluster analysis techniques, etc.). A more parsimonious approach is to choose a window of fixed size and update the model as each time segment of that size elapses. Should this be done with a moving window, or a segmented one? Should the model build continuously on all information from the beginning of the discourse, or only on a smaller segment at a time? Though our flow methods are descriptive in nature, they still require adequate sample sizes to be meaningful. For example, a common rule of thumb for time series methods, on which our dominance measure is based, is to have at least 200 cases in the data set.

These issues and others need to be resolved by hypothesis-driven empirical research. The final conclusion will largely be a pragmatic one, supported only by extensive replication.

Applications of communication analysis

The regression analysis described in the previous section suggested that communication flow techniques like ProNet can be used to discriminate high- from low-performing teams. In this section we describe some results that indicate that flow patterns can also be used to differentiate co-located from geographically distributed teams. Then we describe ongoing efforts to temporally extend our communication analysis through dynamical modeling approaches which are then applied to automated tagging of macrocognitive processes.

Differentiating co-located versus distributed teams by team process

We collected data in two UAV experiments in which teams were either in the same room, though communicating over head sets (i.e., co-located) or in different rooms, out of visual range, but still communicating over headsets (i.e., distributed). Results indicated minimal performance effects of the environmental manipulation. However, there were team processing differences observed (and rated) by experimenters. We conclude that teams adapted to their different environments by adjusting their cognitive processing strategies. Communication analysis methods were applied to understand the exact nature of the differences.

If distributed teams have a weaker hierarchical structure, then the discourse dominance should generally be weaker in distributed teams. Dominance measures were defined for each of the three team members per team, labeled by role (i.e., AVO [Adom], PLO [Pdom] and DEMPC [Ddom]). Also, there were subscales for each person's speech quantity predicting each other person's (e.g., CrossR^{2}AP for AVO predicting PLO).

Distributed teams had positive Adom values (M_d = 0.391, SE_d = 0.079), while co-located teams had negative Adom values (M_c = –0.220, SE_c = 0.086), $F(1, 16.78) = 16.41$, $p = 0.001$, $\eta^2 = 0.49$. Distributed teams had negative Ddom values (M_d = –0.535, SE_d = 0.075), while co-located teams have positive Ddom (M_c = 0.092, SE_c = 0.081), $F(1, 16.51) = 12.84$, $p = 0.002$, $\eta^2 = 0.44$. There was no co-location main effect for Pdom.

Examining the average squared cross-correlation function variables, we see that Distributed teams had greater AVO-to-DEMPC predictability (M_d = 0.012, SD_d = 0.009) than co-located teams (M_c = 0.005, SD_c = 0.009), CrossR^{2}AD, $F(1, 16.18) = 5.39$, $p = 0.034$, $\eta^2 = 0.25$. Distributed teams also had greater PLO-to-DEMPC predictability (M_d = 0.005, SE_c = 0.0005) than did co-located teams (M_c = 0.004, SE_c = 0.0005), CrossR^{2}PD, $F(1, 16.93) = 3.62$, $p = 0.074$, $\eta^2 = 0.18$.

Taken together, these dominance results mean that the speech behavior of distributed teams was dominated by the AVO's speech behavior, with the DEMPC's speech being largely reactive to other speakers. For co-located teams, the opposite pattern was observed. AVO was reactive and DEMPC tended to be (moderately) dominant. Thus, although the results did not support the hypothesis of weaker discourse dominance among distributed teams, they did identify different dominance patterns for the two conditions.

ProNet chain analysis also identified several effects of co-location. Perhaps the most interesting aspect of these findings is not the specific chains themselves, but rather the fact that co-location impact decreases over time for several measures. This suggests that, with task experience, the impact of communication mediation becomes less important.

Co-located (M = 16.82, SE = 0.86) teams have more uninterrupted AVO utterances (Aloops) than distributed teams (M = 8.70, SE = 0.79; $\eta^2 = 0.31$, $F(1, 16.19) = 7.36$, $p = 0.015$). However, this is one of the measures that shows a decrease in co-location effect over missions (co-location by mission interaction: $R^2 = 0.21$, $F(6, 89) = 3.87$, $p = 0.002$). Hence, co-located AVOs speak more adjacent utterances than distributed, but this difference declines with task experience.

Other measures showing co-location effects, but with mission interactions suggesting a general decline in their impact over time, include AVO interrupting DEMPC (AiD; $\eta^2 = 0.15$, $F(6, 89) = 2.66$, $p = 0.020$), PLO interrupting AVO (PiA; $\eta^2 = 0.14$, $F(6, 89) = 2.33$, $p = 0.039$)and DEMPC interrupting AVO (DiA; $\eta^2 = 0.18$, $F(1, 16.23) = 3.54$, $p = 0.078$).

For every detected pattern, co-located teams exhibited that pattern more than distributed teams. This is at least partly due to the fact that the measures are counts of speech events and co-located teams spoke more than distributed teams. However, the finding that reappears for several of the co-location effects is that the co-location effect decreased over time. This is not because of a simple drop in communication quantity over time on the part of co-located teams, because communication quantity did not change significantly over time.

These findings paint a fairly straightforward picture that discriminates co-located from distributed teams in early missions. In co-located teams (but not in distributed teams) DEMPCs interrupt AVOsand AVOs repeat themselves. In terms of interruption, for AVO interrupting DEMPC and PLO interrupting AVO, the impact of co-location decreases over time, but not until after teams have reached performance asymptote.

Tagging macrocognitive processes

It has been hypothesized that macrocognitive processes can be categorized in terms of constructs such as knowledge interoperability development, team consensus and revision/analysis of collaborative solutions. Some of our most recent communications analysis work has been directed toward a better understanding of these constructs. Communications data from a simulated non-combatant evacuation operation (NEO; Warner, Wroblewski and Shuck 2003) was collected and each unit of communication (e.g., utterance) was coded (i.e., tagged) in terms of a particular construct. The coded data revealed that some of the tagged processes occurred more frequently than others. The temporal structure of the coded data was investigated using dynamical systems techniques. This analysis revealed that not only do some top-down tags occur more frequently than others, but that deterministic sequences of tagged processes were present in the data. The purpose of using a dynamical systems approach with these (or other interaction process) data is to quantify phenomena that are temporally extended. Ultimately the goal is to automate the tagging process through communication analysis.

The coded NEO data were found to be relatively patterned (approximately 30 percent determinism) compared to a random noise process (e.g., white noise was 0.01 percent deterministic). Although this analysis gave us a rough idea of how temporally patterned the coded NEO data are, relative to other known processes, a more sensitive modeling approach was also applied. The latent dynamics of the coded collaboration data were unfolded using an approach called phase-space reconstruction (Abarbanel 1995). The latent dimensionality gives an idea of the dynamical degrees of freedom (ddf; dimensionality $\geq$ ddf) of the modeled process, in this case the coded collaboration data. The dimensionality of the 'collaboration space' therefore contains any temporally extended macrocognitive processes, where any particular dimension can be loosely interpreted as one macrocognitive degree of freedom, or more specifically as one dimension of macrocognitive behavior.

The results of the analysis suggested at least six dimensions of macrocognitive behavior. Further, these behaviors ranged up to thirty collaboration events in length. The conclusion that we draw from the results of these analyses is that macrocognitive processes are temporally extended.

This example demonstrates a top-down application of dynamical analysis to manually coded sequential data. Efforts are also underway to apply communications analysis methods (including dynamical approaches) to uncoded data in a bottom-up fashion in order to identify patterns in the data that can then be tied to macrocognitive processes.

Conclusions

Communication in its many different forms is a ubiquitous source of data on externalized macrocognition. The very richness of these data affords extensive information about macrocognition and, specifically, team cognition, but has also been until recently the showstopper due to the resource intense manual analysis required. Advances in the automation of flow and content analysis allow tapping this rich resource while minimizing the expense of manual analysis. Flow techniques in particular are amenable to instrumentation and automation and do not require the laborious transcription and coding processes associated with content data. Thus, our research plan has been to determine to what extent flow patterns alone can be exploited to assess and diagnose team performance and the macrocognitive processes associated with that performance. Then to the extent necessary, the flow methods could be supplemented with the more extensive content methods. For instance, the flow methods may provide an automated and quick way to detect a change in team cognition and content analysis can explore deeper to identify the nature of the change.

Although there may be general patterns of flow that are indicative of conflict, problem-solving, or knowledge sharing, we anticipate that flow patterns will be to a great extent specific to a domain and a task within that domain. What may be a pattern indicative of good team performance in an aviation task may be indicative of poor situation awareness in a team planning setting. This is why it is critical that communication patterns be interpreted in light of the domain, task and other performance measures. The process of acquiring meaningful communication–macrocognition mappings also has the potential for automation through machine learning approaches.

Apart from automation, the development of algorithms that detect patterns in flow and content data also contributes to the theoretical understanding of macrocognition and practical applications. Specifically, meaningful communication patterns can shed light on macrocognitive constructs and eventually help to better discriminate one construct from another and even tag utterances. In addition, the patterns serve as quantitative indices for assessing and diagnosing team performance which can be useful in team experiments as well as team training and operational environments. In fact the operational vision is to be able to monitor team communication continuously and unobtrusively, extracting communication patterns and flagging potential problems and opportunities for intervention. Another application is in the evaluation of products designed to enhance team macrocognition. The methods described in this chapter can be applied to such an evaluation to determine not only if the product is successful at improving team performance, but to understand its precise impact on macrocognitive processes.

Acknowledgments

Work described in this chapter was supported by Grant N000140310580 and N000140510625 from the Office of Naval Research. The authors are grateful to the contributions of Nia Amazeen, Ben Schaub, Steven Shope and the chapter's reviewers.

The views, opinions and findings contained in this article are those of the authors and should not be construed as official or as reflecting the views of the Department of Defense.

References

Abarbanel, H.D.I. (1995). *Analysis of Observed Chaotic Data*. (New York: Springer-Verlag Inc.).

Bowers, C.C., Jrensch, F., Salas, E. and Braun, C.C. (1998). 'Analyzing Communication Sequences for Team Training Needs Assessment', *Human Factors* 40, 672–679.

Budescu, D.V. (1984). 'Tests of Lagged Dominance in Sequential Dyadic Interaction', *Psychological Bulletin* 96, 402–414.

Converse, S., Cannon-Bowers, J.A. and Salas, E. (1991). 'Team Member Shared Mental Models', *Proceedings of the 35th Human Factors Society Annual Meeting*, pp. 1417–1421. (Santa Monica, CA: Human Factors and Ergonomics Society).

Cooke, N.J., DeJoode, J.A., Pedersen, H.K., Gorman, J.C., Connor, O.O. and Kiekel, P.A. (2004). *The Role of Individual and Team Cognition in Uninhabited Air Vehicle Command-and-Control*. Technical Report for AFOSR Grant Nos. F49620-01-1-0261 and F49620-03-1-0024. (Mesa, AZ: ASU Polytechnic).

Cooke, N.J., Neville, K.J. and Rowe, A.L. (1996). 'Procedural Network Representations of Sequential Data', *Human–Computer Interaction* 11, 39–68.

Cooke, N.J., Salas, E., Cannon-Bowers, J.A. and Stout, R. (2000). 'Measuring Team Knowledge', *Human Factors* 42, 151–173.

Dillon, W.R., Madden, T.J. and Kumar, A. (1983). 'Analyzing Sequential Categorical Data on Dyadic Interaction: A Latent Structure Approach', *Psychological Bulletin* 94, 564–583.

Emmert, V.J. (1989). 'Interaction Analysis', in P. Emmert and L.L. Barker, (eds), *Measurement of Communication Behavior*, pp. 218–248. (White Plains, NY: Longman, Inc).

Foltz, P.W., Laham, D.and Derr, M. (2003). 'Automatic Speech Recognition for Modeling Team Performance', *Proceedings of the Human Factors and Ergonomics Society*, pp. 673–675. (Santa Monica, CA: Human Factors and Ergonomics Society).

Foltz, P.W., Martin, M.J., Abdelali, A., Rosenstein, M. and Oberbreckling, R. (2006). 'Automated Team Discourse Modeling: Test of Performance and Generalization'. *Proceedings of the 28th Annual Conference of the Cognitive Science Society*, pp. 1317–1322.

Fowlkes, J.E., Lane, N., Salas, E., Franz, T. and Oser, R. (1994). 'Improving the Measurement of Team Performance: The TARGETs Methodology', *Military Psychology* 6, 47–61.

Gilden, D.L., Thornton, T. and Mallon, M.W. (1995). 'Low Frequency Noise in Human Cognition', *Science* 267, 1837–1839.

Gorman, J.C. (2005). 'The Concept of Long Memory for Assessing the Global Effects of Augmented Team Cognition'. Paper presented at the Human–Computer Interaction International Conference, 22–27 July, Las Vegas, NV.

Gorman, J. C., Cooke, N. J.and Kiekel, P. A. (2004). Dynamical perspectives on team cognition. *Proceedings of the Human Factors and Ergonomics Society 48th Annual Meeting*, pp. 673–677. (Santa Monica, CA: Human Factors and Ergonomics Society).

Hackman, J.R. (1987). 'The Design of Work Teams', in J.W. Lorsch (ed.), *Handbook of Organizational Behavior*, pp. 315–342. (Englewood Cliffs, NJ: Prentice Hall).

Kiekel, P.A., Cooke, N.J., Foltz, P W., Gorman, J.C. and Martin, M.J. (2002). 'Some Promising Results of Communication-Based Automatic Measures of Team Cognition', *Proceedings of the Human Factors and Ergonomics Society 46th Annual Meeting*, pp. 298–302. (Santa Monica, CA: Human Factors and Ergonomics Society).

Landauer, T., Foltz, P. and Laham, D. (1998). 'Introduction to Latent Semantic Analysis', *Discourse Processes* 25, 259–284.

Mace, W.M. (1977). 'James J. Gibson's Strategy for Perceiving: Ask Not What's Inside Your Head but What Your Head is Inside of', in R. Shaw and J. Bransford, (eds), *Perceiving, Acting and Knowing*, pp. 43–65. (Hillsdale, NJ: Erlbaum).

Mathieu, J.E., Goodwin, G.F., Heffner, T.S., Salas, E. and Cannon-Bowers, J.A. (2000). 'The Influence of Shared Mental Models on Team Process and Performance', *Journal of Applied Psychology* 85, 273–283.

Salas, E. Dickinson, T.L., Converse, S.A. and Tannenbaum, S.I. (1992). 'Toward an Understanding of Team Performance and Training', in R.W. Swezey and E. Salas, (eds), *Teams: Their Training and Performance*, pp. 3–29. (Norwood, NJ: Ablex).

Stahl, G. (2006). *Group Cognition*. (Cambridge, MA: MIT Press).

Van Orden, G.C., Holden, J.G. and Turvey, M.T. (2003). 'Self-organization of Cognitive Performance', *Journal of Experimental Psychology: General* 132, 331–350.

Wampold, B.E. (1984). 'Tests of Dominance in Sequential Categorical Data', *Psychological Bulletin* 96, 424–429.

Warner, N., Wroblewski, E. and Shuck, K. (2003). 'Noncombatant Evacuation Operation Scenario: Red Cross Rescue Scenario'. Paper presented at the Collaboration and Knowledge Management Workshop, San Diego, CA.

Chapter 5

Collaboration, Training, and Pattern Recognition

Steven Haynes and C.A.P. Smith

Introduction

The human cognitive system has acute information processing limitations; these limitations can directly impact task performance (Wickens 1984). For example, Miller (1956) suggested that most people can remember only seven things at any given moment. Another limited cognitive resource is attention. Humans have difficulty dividing attention among several tasks, or attending to all the information provided by our senses (Broadbent 1958; Treisman 1969). As Nobel Laureate Herb Simon has said, ‘a wealth of information creates a poverty of attention’ (Varian 1995). Furthermore, our cognitive resources for perception are also limited. Often, we do not perceive most of the information that is available to us (Lavie 1995). To address some of these limitations, decision support tools have been developed to support individual cognition and these tools are successful because they strengthen specific cognitive strategies (Kaempf, Klein and Wolf 1996). Kaempf *et al.* found that individual experts make almost 90 percent of their decisions by ‘feature matching’ between the current situation and one from prior experience.

Keltner (1989) found that teams, rather than individuals, make many decisions. He observed that most important economic, political, legal, scientific, cultural and military decisions are made by groups. Thus, considerable effort has been devoted to the development of groupware to support these domains (DeSanctis and Gallupe 1987; Jessup and Valacich 1993; McGrath and Hollingshead 1994; Nunamaker 1997 to name just a few). Many studies have shown that groupware-supported distributed (or virtual) teams can outperform face-to-face teams (Schmidt, Montoya-Weiss and Massey 2001 as merely one example). However, within the large body of literature on groupware, we are not aware of any systems specifically designed to optimize the utilization of human cognitive resources in group decision situations; most systems have addressed behavioral issues associated with human interaction. In this chapter we address this shortcoming by proposing and testing a model of group cognition for collaboration drawn from the literature on individual cognition. Our research objective is to examine whether a collaboration tool that is closely aligned with our cognitive structures leads to improved team performance. Specifically, we will explore team pattern recognition in a visuospatial environment. However, in order to study pattern-recognition in teams, we must first train the participants to recognize visuospatial patterns. The issue of cognitive alignment may apply equally to the training regimen: to control for effects of training, we develop two pattern-training methods. One training method divides the patterns into discrete elements, and the other method presents patterns as complete chunks.

Prior research

Multiple resource theory proposes that there are separate and finite reservoirs of cognitive resource, each available for different purposes (Card, Moran and Newell 1986; Wickens 2002). With respect to cognitive resources for working memory, there appear to be two such finite reservoirs: visuospatial and articulatory loop memory (Baddeley 1992, 1998; Baddeley, Chincotta and Adlam 2001). To overcome this limitation, it has been theorized and shown that experts create a cognitive structure called a 'chunk', where many related pieces of data are aggregated (Chase and Simon 1973a, b). The use of these cognitive structures is vital for encoding domain knowledge, pattern-recognition of situations and selective search techniques such as those used by chess experts.

The advantages of chunking have been demonstrated for the performance of skilled chess players who are able to encode and accurately recall unfamiliar chess positions shown to them for only a few seconds (Chase and Simon 1973b; Gobet and Simon 1996a). Interestingly, these same chess players were unable to accurately recall pieces randomly distributed on the board (in no known pattern) and to account for this Chase and Simon proposed that chess players stored chunks in long-term memory (LTM) corresponding to patterns of the pieces.

Acquiring, sharing and processing information are critical activities for decision-making in a group setting. Ideally, groups should be less affected by the cognitive limitations of their members. Consider a team attempting to recall the names of all the US state capitals. It is likely that there will be several capitals that every member knows. This is information that the group holds in common. One would also expect that the gaps in the members' knowledge would complement each other to some extent, so that the collective recollection of the group would be superior to the average individual recollection.

Hutchins' (1991, 1995) work on distributed cognition defines the unit of cognitive analysis as a distributed sociotechnical system. The cognitive properties of this system are produced by an interaction between the structures internal to individuals and structures external to individuals. In particular, that portion of cognition that governs the coordination of the elements of a task might be represented in the external environment, and be available for inspection. By making this representation 'public', the group can share it. Hutchins (1995) describes how a group's use of a technical artifact can transform a complex computational task into a simple perceptual task. For example, a team of aircraft pilots sets the airspeed 'bug' on the desired airspeed so they can share perceptions of whether the aircraft is being flown at, above or below the target airspeed (they don't have to remember the actual target speed). This publicly displayed structure is a pointer to a memory item representing a critical piece of procedural domain knowledge. In this case, a pointer to the script of how to configure the aircraft during this particular phase of flight: the task is changed from a difficult cognitive one to an easier perceptual task. Effective crew resource management requires that well-trained teams such as a flight crew develop chunks that are common to all.

Klein's (1993) Recognition Primed Decision Making model (RPD) also represented the naturalistic decision process of each individual that Hutchins describes in the distributed cognition environment. Klein emphasizes situation assessment through pattern recognition; human experts perform 'feature-matching' within the task, which triggers recall of similar situations or patterns from memory. Cognitive science researchers would call this a 'macrocognitive' process, because it occurs at the cortical network level, not at the neural level (which would be 'microcognitive'). Bara (1995), in the context of connectionism, referred to macrocognition as:

> those processes, such as reasoning and communication, where analysis does not take place at the level of the single processing unit. At these high level cognitive domains, analysis and simulation do not concern the individual behavior of each neuron, but the functioning of the mind as a whole.

Cacciabue and Hollnagel (1995) expanded this definition to include the study of the role of cognition in realistic tasks while interacting with the environment. Klein *et al.* (2003) have continued to insist that cognitive processes be studied in these natural settings, as opposed to laboratory settings. To them, microcognition occurs in laboratory experiments on puzzle-solving, etc., and only informs us about the building blocks of cognition, or the processes that are invariant across individuals. Macrocognition focuses on the emergent phenomena that appear when the decision-maker is faced with uncertainty, complexity, time pressure, high stakes or risk and ill-defined or conflicting goal sets. Klein and others (Cooke and Gorman 2006; Warner, Letsky and Cowen 2005) also contend that macrocognitive functions are generally performed in collaboration with others. There is both internal and external communication between processing units, within and without the individual, and these processes can occur in asynchronous, distributed environments – the exact naturalistic settings that many difficult situations face.

Hayne, Smith and Turk (2003) incorporated this into their Team Recognition Primed Decision-Making macrocognitive model (see Figure 5.1), adapting Klein's model of individual decision-makers to capture how chunk stimulating structures can be utilized to compensate for cognitive limitations in a group decision-making situation. In the team version of RPD, we propose that teams perform essentially the same steps as individuals in Klein's model. Then they share these situation assessments (patterns) among members, and the individual team members select a response by adapting a strategy from their previous experience. Finally, they execute their plan, and observe the results.

Note that situation assessment and response selection occur *within* the individual, whereas pattern communication and execution occur *between* the individual team members. Individuals use an internal cognitive process to perform situation assessment, but must use a different process when collaborating with the team. The stimulating structures used in communicating patterns may also trigger

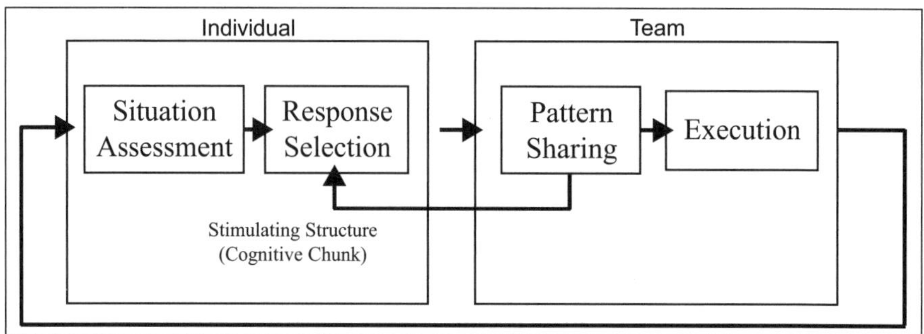

Figure 5.1 Model of individual and team collaboration

internal cognitive processes. Finally, when acting as a team executing the response, individuals make incremental actions that feedback to the team.

Kaempf, Klein and Wolf (1996) found that individual experts spent most of their time scanning the environment and developing their situation assessments. Relatively little time was spent selecting and implementing responses. If the situation assessment task has the same relative importance for teams as for individuals, then the initial focus for team decision support should be directed towards the development of tools to support collective situation assessment. For individual members, these tools should be designed to reduce the cognitive effort required to perceive patterns, attend to the highest priority tasks, and remember the most important features of the task environment. For the team, these tools should facilitate sharing of assessments through placement of public representations.

Public representations have been studied before in other contexts, including distributed cognition (as above). Grassé (1959) coined the term stigmergy, referring to a class of mechanisms, or stimulating structures, that mediate animal–animal interactions. The concept has been used to explain the emergence, regulation, and control of collective activities of social insects (Susi and Ziemke 2001). Social insects exhibit a coordination paradox: they seem to be cooperating in an organized way. However, when looking at any individual insect, they appear to be working independently as though they were not involved in any collective task. The explanation of the paradox provided by stigmergy is that the insects interact indirectly by placing stimulating structures in their environments. These stimulating structures trigger specific actions in other individuals (Theraulaz and Bonabeau 1999). Stigmergy appears to be the ultimate example of reduction of cognitive effort because social insects, having essentially no cognitive capability, are able to perform complex collaborative tasks.

We advocate that this concept can be applied to human teams, i.e., when a stimulating structure is placed in the external environment by an individual, other team members can interpret it and take appropriate action, without the need for specific communication or coordination (see Figure 5.1). Stigmergy in its current form is complementary to distributed cognition, because the stimulating structure may or may not have any cognitive properties, e.g., the heading 'bug' by itself has no direct cognitive properties. Its relative placement on the airspeed indicator triggers the cognition and actions in the pilot.

So what are the appropriate structures to be shared for human group cognition? We believe a chunk representation is the most appropriate stimulating structure for collective pattern recognition and communication due to its superior cognitive alignment. However, it is possible that any portion of the recognized pattern might be a suitable stimulating structure. For example, if one were trying to communicate about a room in a house, using the word 'kitchen' would facilitate the memory retrieval of a 'kitchen' chunk, complete with sink and all the appliances. However, another way to represent this room might be to communicate sink, refrigerator, stove, dishwasher, cupboard, etc; the discrete elements. Both imply similar amounts of information, yet a difference in cognitive alignment and retrieval from memory. Research suggests that the chunk representation can be retrieved faster and with less effort, because it uses fewer cognitive resources.

At this point we must acknowledge that recent intensive research in skilled memory has shown that parts of the original chunking model must be incorrect and a new theory is emerging. For example, in contrast to the usual assumptions about short-term memory, chess masters are relatively insensitive to interference tasks (Charness 1976; Frey and Adesman 1976) and can recall several boards that have been presented successively (Cooke, Atlas, Lane and Berger 1993; Gobet and Simon 1996a). Experts can also increase the size of their chunks based on new information; effectively increasing short-term memory (Gobet and Simon 1998). Because an explanation of this performance based on chunking requires holding far too many chunks simultaneously in working memory, Chase and Ericsson (1982) and Staszewski (1990) proposed that these subjects have developed structures ('retrieval structures') that allow them to encode information rapidly into long-term memory. Such structures have been used at least since classical times, when rhetoricians would link parts of a speech to a well-known architectural feature of the hall in which the speech was to take place, to facilitate recall (Yates 1966). Retrieval structures are an essential aid to expert memory performance. Gobet and Simon (1996b) went on to demonstrate that the time required to encode and retrieve chunks is much shorter than previously believed. These results lead to the somewhat controversial refinement of chunk theory into Template Theory.

Template Theory assumes that many chunks develop into more complex structures (templates), having a 'core' of data to represent a known pattern of information, and 'slots' for variable data to enhance the core (Gobet and Simon 1996a, 2000). Templates have been referred to by various other names in other non-cognitive domains, i.e., schemas (Bartlett 1932), frames (Minsky 1977), prototypes (Goldin 1978; Hartston and Wason 1983), etc. In the domain of chess, templates allow rapid encoding and retrieval from long-term memory of more data than chunks (10–40 items as opposed to 4–5 items). For example, when a chess position pattern is recognized (say, as a King's Indian defense), the corresponding stored representation of the chess board provides specific information about the location of a number of core pieces (perhaps a dozen) together with slot data which may possess default values ('usual' placements in that opening) that may be quickly revised. Templates are cued by salient characteristics of the position, and are retrieved from long-term memory in a fraction of the time than other memory structures (Gobet and Simon 1996b).

Vessey (1991) coined the term 'cognitive fit' to describe enhanced performance when there is a good match between the information emphasized in the representation type and that required by the task type. Similarly, we suggest that effectiveness and efficiency of the decision process will increase when there is a good 'fit' between cognitive memory structures and the software artifact. Thus, our first hypotheses are:

> H1a: In a pattern-recognition task, the outcome quality of teams supported with a chunk-sharing tool will be greater than the outcome quality of teams supported with a discrete-item tool.

> H1b: In a pattern-recognition task, the time to decision of teams supported with a chunk-sharing tool will be faster than that of teams supported with a discrete-item tool.

If a team member is trying to communicate to their team members that they are seeing pattern resembling a 'kitchen' and use the label 'kitchen' to evoke the shared memory, then only one message need be sent to signal recognition of the pattern. However, if that team member is communicating all the discrete elements of a 'kitchen', then several more messages will need to be sent. Sharing the words 'sink' and 'cupboard' might not be sufficient to confirm the kitchen pattern. Thus, we expect that the amount of sharing will be less in the chunk treatment.

The sharing of knowledge by teams members maps nicely into the knowledge interoperability development construct proposed by Warner, Letsky and Cowen (2005). The internal templates initially created by each team member should have the same 'core' features, but probably vary as to the slot data. The knowledge is externalized by sharing a chunk (label) and each individual's mental model (template) may be internally modified or updated depending on the 'consensus' of the team. For example, if each team member shares that they see a 'kitchen' pattern, the template for each team member will be re-inforced. However, if a majority of the team shares a 'bathroom' pattern, the remaining team members may have to update their template. This process is essentially team knowledge development (Warner, Letsky and Cowen 2005).

> H1c: In a pattern-recognition task, the amount of pattern sharing in teams supported with a chunk-sharing tool will be less than for teams supported with a discrete-item tool.

The literature on chunking encompasses many different areas of research, and the concept of a chunk has consequently diversified in its meaning. The literature itself can be divided into two broad areas (Gobet 2005), based on how and when chunking is assumed to occur: the first assumes a deliberate, conscious control of the chunking process (explicit), and the second a more automatic and continuous process of chunking during perception (implicit). A key assumption of chunk-based models is that perceptual skills, anchored in concrete examples, play a central role in the development of expertise, and that conceptual knowledge is later built on such perceptual skills. One of the theoretical reasons behind this claim is that perceptual cues offer an efficient way for LTM knowledge to be retrieved, and that, without such cues, much LTM knowledge would remain 'inert' (Whitehead 1929). Some of the clearest evidence for perceptual chunking is found in how primitive stimuli are grouped into larger conceptual groups, such as the manner by which letters are grouped into words, sentences or even paragraphs (Simon 1974).

Categorization into chunks is a critical skill that every organism must possess in at least a rudimentary form, because it allows them to respond differently, for example, to nutrients and poisons and to predators and prey. There is much recent evidence that human category learning relies on multiple systems (e.g., Ashby, Alfonso-Reese, Turken and Waldron, 1998; Ashby and Ell 2001; Erickson and Kruschke 1998; Smith, Patalano and Jonides 1998). In all cases in which multiple systems have been proposed, it has been hypothesized that one system uses explicit (i.e., rule-based) reasoning and at least one other system involves some form of implicit learning. Nevertheless, there is still much disagreement.

One distinctive feature of procedural learning, which sets it apart from perceptual learning, is its association with motor performance (e.g., Squire, Knowlton and Musen 1993; Willingham, Nissen and Bullemer 1989). While visuospatial computer-based collaboration tasks do not direct specific motor responses, the objects on the screen may have to be manipulated corresponding to a physical pattern. Consequently, we predict that explicit (procedural) pattern training will lead to higher performance than implicit training. In explicit training, the chunk structures are provided. In contrast, implicit training allows the participants to form their own chunks:

> H2a: In a pattern-recognition task, the outcome quality of the teams with explicit training will be greater than the outcome quality of the teams with implicit training.

> H2b: In a pattern-recognition task, the time to decision of the teams with explicit training will be less than that of the teams with implicit training.

Implicit training should result in the participants developing chunks that differ somewhat in their composition. As each participant in the implicit treatment may develop their own unique chunk representation of the pattern, we expect that they will need to share more to achieve the goal of confirming the recognition of the pattern to their team members:

> H2c: In a pattern-recognition task, the pattern sharing in the teams with explicit training will be less than that in the teams with implicit training.

Method

For this study, we used 49 3-person teams of undergraduate students enrolled in a junior level course at a state university in the western United States. The data collection sessions took place in a large computer lab equipped with 40 workstations. Participant seating locations were randomly assigned so as to physically separate team members as much as possible. A maximum of 2 hours were available for each data collection session. Except for one, all teams finished well within the allotted time.

Decision task and experimental procedure

In order to validate our model and test our hypotheses, we created a collaborative game consisting of a Java applet client incorporating a pattern-communicating tool and real-time play. The game is an extension of McGunnigle, Hughes and Lucas' (2000) two-player game to teams of three players playing independent decision scenarios or trials. The game is played against a computer opponent (not another team) on a shared computerized board consisting of three large circles with the intersecting areas representing 'regions' (see Figure 5.2). The game presents a partially revealed pattern, and requires the participants to place their resources on the game board to match the pattern. This software platform has been used in many experiments published elsewhere (Hayne, Smith and Turk 2003; Hayne and Smith 2005; Hayne, Smith and Vijayasarathy 2005a). Please see these citations for a full description of the study.

All subjects received scripted training in the patterns and in the use of the system immediately prior to the game. During six practice sessions, the subjects were shown the results of their decisions, and informed of the pay-off that they would have received if the practice scenarios had been real. After the practice sessions,

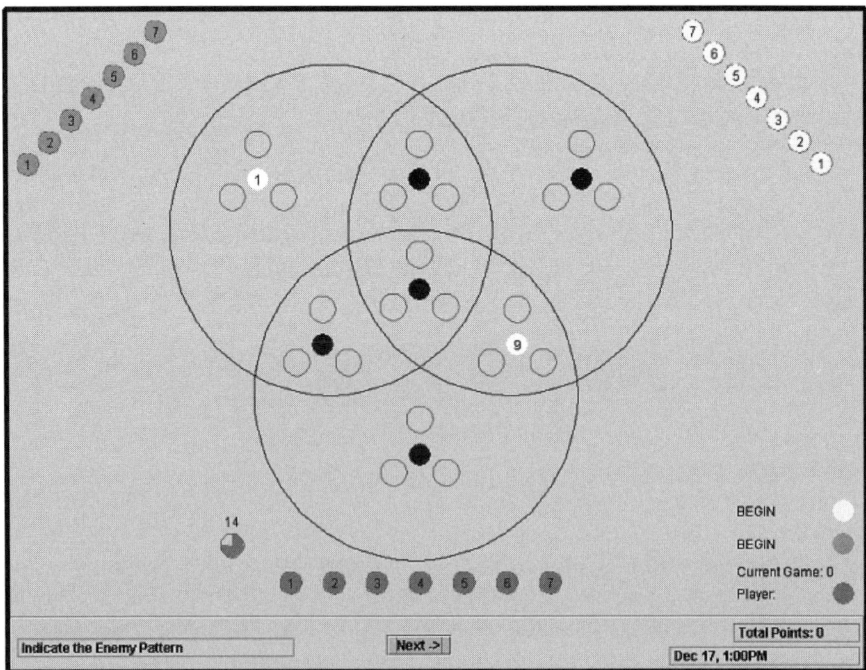

Figure 5.2 Multiplayer collaborative game

subjects were given one more opportunity to ask questions about experimental procedures. Following that, they participated in at least 24 paid experiment trials. At the completion of the last trial, the subjects filled out a survey, and were debriefed, paid and dismissed.

Manipulation

We used a 2*2 factorial design for our study. The two factors are (a) type of training – implicit versus explicit, and (b) type of sharing tool – item-sharing (IS) versus chunk-sharing (CS). Subjects in the implicit training (IT) treatment were trained on the three patterns in terms of each pattern's discrete components. After the training was completed, an informal testing procedure was conducted in which the trainer would point at the missing components of a partially revealed pattern, and ask each participant 'what value goes HERE?' The training in the explicit (ET) treatment was similar in that the discrete components were described, but in the ET treatment the patterns were given descriptive labels (i.e., pattern 9, pattern 10 or pattern 14). The informal testing was conducted in the same fashion, with the exception that the participant probes were phrased, 'which pattern is THIS?' In each treatment, the informal testing procedure was repeated until every participant was able to correctly identify each of the three patterns.

Participants in our experiment played the collaborative game in two phases; in the first phase they used a tool to share pattern information, and in the next phase they moved their tokens. Both the IS and CS team members could see their colleagues' pattern indications in real time. All token movements, pattern sharing, and scoring data was captured with millisecond timestamps as appropriate.

Incentives

Subjects were externally motivated to take these experiments seriously and to behave 'as if' they were making real allocation decisions (Cox, Roberson and Smith 1982). This was accomplished by instituting a salient monetary pay-off function directly related to the teams' outcome quality, as measured by the number of regions won in every trial:

Individual payoff/trial (US$) = (correct regions – 5) * $0.50

Subjects were informed of this function and told that money would be paid to each team member in cash at the end of the experiment. The incentive money was displayed to encourage them to believe they would indeed be paid. Subjects were also paid $5 to show up on time for the session. Individual participants typically earned $15–26 for the two-hour session.

Data analysis and results

There were 20, 10, 6 and 13 teams in the IT/IS, ET/IS, IT/CS and ET/CS treatments respectively. The majority of the participants were male (73.7 percent). The average age was 22.98 years (ranging from 19 to 38 with the mode = 21) with no significant differences across treatments. Subjects were paid over $3,600 for their participation and performance. Since our training manipulation explicitly focused on training subjects to identify specific patterns, our analysis includes only those trials that had definitive patterns (i.e., trials in which the information revealed was sufficient to ascertain the underlying pattern).

Outcome quality

Analysis of variance (ANOVA) was conducted to test for differences in outcome quality by treatments. Outcome quality was measured as the points (number of regions won) earned by each team per trial. The results (Table 5.1) show that outcome quality differs significantly between the two types of sharing tools ($F = 4.81, p < 0.029$) thus providing support for Hypothesis 1a, that the chunk-sharing tool would afford greater performance. Hypothesis 2a was also supported; there was a significant difference in outcome quality based on the type of training ($F = 6.51, p < 0.011$): on average, the explicit chunk training was associated with higher performance. Interestingly, there is an interaction effect between type of training and type of tool ($F = 25.52, p < 0.000$). In particular, the complementary combinations of training and tool-type were associated with higher performance than the congruent combinations. That is, the groups that received implicit item-level training combined with the chunk-level tool for sharing performed very well. Similarly, groups that received explicit chunk-level training combined with the item-level sharing tool also performed very well.

Time to decision

From Table 5.2, we can see that decision time was affected by both treatments ($F > 9.26, p < 0.003$). H1b was supported; teams using the item-level sharing tool used more time to make their decisions. H2b was also supported; teams having the implicit training used more time than teams having the explicit chunk training. Once again, there was an interaction between the training and tool such that the teams in the IT/IS condition took significantly longer than the other conditions (150.7 sec/trial). The teams who finished fastest were in the ET/IS condition (80.6 sec/trial).

Sharing

We counted the total number of items (or chunks depending on treatment) shared by using the tool, in each trial (see Tables 5.3a and 5.3b). We found much more sharing in the IS condition than in the CS condition ($F = 221.4, p < 0.000$). H1c

Table 5.1 ANOVA results for outcome quality descriptive statistics

	IT (Implicit Training)	ET (Explicit Training)
IS (Item Sharing)	6.261 (.786, 119)	6.891 (.315, 55)
CS (Chunk Sharing)	6.861 (.543, 36)	6.654 (.599, 78)

Standard deviation and trial counts are in parentheses.

ANOVA results

Source	df	Mean Square	F	Sig.
Training (IT vs. ET)	1	2.67	6.51	.011
Tool (IS vs. CS)	1	1.97	4.81	.029
Training * Tool	1	10.44	25.52	.000
Error	284	0.008		

Table 5.2 Time to decision in seconds per trial

	IT (Implicit Training)	ET (Explicit Training)
IS (Item Sharing)	150.5 (62.9)	80.6 (34.3)
CS (Chunk Sharing)	90.3 (33.0)	102.0 (40.7)

Standard deviation is in parentheses.

ANOVA results

Source	df	Mean Square	F	Sig.
Training (IT vs. ET)	1	50982	20.8	.000
Tool (IS vs. CS)	1	22685	9.26	.003
Training * Tool	1	99478	40.6	.000
Error	284	2451		

Table 5.3a Sharing tool usage counts per trial

	IT (Implicit Training)	ET (Explicit Training)
IS (Item Sharing)	8.91 (4.1)	8.47 (3.6)
CS (Chunk Sharing)	1.85 (0.73)	2.46 (0.75)

Standard deviation is in parentheses.

ANOVA results – all sharing

Source	df	Mean Square	F	Sig.
Training (IT vs. ET)	1	0.4	0.04	.835
Tool (IS vs. CS)	1	2162	221.4	.000
Training * Tool	1	13.9	1.42	.234
Error	260	9.767		

Table 5.3b Correct sharing tool usage counts per trial

	IT (Implicit Training)	ET (Explicit Training)
IS (Item Sharing)	6.64 (3.7)	8.15 (3.5)
CS (Chunk Sharing)	1.27 (0.83)	1.82 (1.1)

Standard deviation is in parentheses.

ANOVA results – correct sharing

Source	df	Mean Square	F	Sig.
Training (IT vs. ET)	1	53.65	6.36	.012
Tool (IS vs. CS)	1	1731	205.2	.000
Training * Tool	1	11.57	1.37	.243
	260	8.437		

was supported. We found no difference in total sharing frequency between training conditions (F = 0.04, $p < 0.8$). H2c was not supported.

However, restricting this data to only sharing what 'ex post' was determined to be 'correct' given the revealed information, we see similar results for differences in sharing frequency by tool type (F = 205.2, $p < 0.000$). Yet, when looking at the sharing frequency of only correct information, we found that the training was also associated with significant differences (F = 6.36, $p < 0.012$), in the *opposite* direction predicted in H2c. In contrast with our prediction about the effects of training on sharing, we found that the participants who received the explicit chunk training used the sharing tool *more* than the other participants.

Bumping

As mentioned earlier, the game permitted participants to displace another teammate's token with their own, thus 'bumping' the original token. Bumped tokens could then be replayed as desired. Examining our process data, we counted the total number of bumps in each trial. The bumping data are presented in Table 5.4. We found that the type of tool made a difference in the amount of bumping, with users of the chunk sharing tool generally bumping more (0.370 vs 0.235) (F = 10.73, $p < 0.001$). There was also a highly significant interaction in which the participants having the

Table 5.4 Average number of bumps per trial

	IT (Implicit Training)	ET (Explicit Training)
IS (Item Sharing)	.202 (.507)	.022 (.221)
CS (Chunk Sharing)	.269 (.635)	.312 (.855)

Standard deviation is in parentheses

ANOVA results – bumping

Source	df	Mean Square	F	Sig.
Training (IT vs. ET)	1	0.025	0.04	0.84
Tool (IS vs. CS)	1	6.406	10.733	.001
Training * Tool	1	12.90	21.606	.000
Error	1223	0.597		

Table 5.5a Average number of bumps per trial in core locations

	IT (Implicit Training)	ET (Explicit Training)
IS (Item Sharing)	.202 (.507)	.022 (.221)
CS (Chunk Sharing)	.269 (.635)	.312 (.855)

Standard deviation is in parentheses.

Table 5.5b Average number of bumps per trial in slot locations

	IT (Implicit Training)	ET (Explicit Training)
IS (Item Sharing)	.434 (.808)	.060 (.240)
CS (Chunk Sharing)	.117 (.415)	.677 (1.42)

Standard deviation is in parentheses.

ANOVA results – bumping by core/slot

Source	df	Mean Square	F	Sig.
Training (IT vs. ET)	1	0.012	0.02	0.89
Tool (IS vs. CS)	1	4.457	7.448	.006
Core/Slot	1	10.93	18.262	.000
Error	1223	0.598		

complementary conditions of training and tool engaged in much less bumping than the other participants (F = 21.61, $p < 0.000$).

In order to explore the implications of Template Theory on our task, we further categorized the bumps as being applied to either 'core' or 'slot' data. The core regions are defined as those containing the {9, 10, 19, 20} and the slot regions are those containing the {1, 14} (Gobet 2003, private communication). The results are shown in Tables 5.5a and 5.5b. We found that the participants were more likely to bump on slot data than core data (0.420 vs 0.223) (F = 18.3, $p < 0.000$).

Discussion

Klein's original model of individual Recognition Primed Decision-Making contends that response selection is contingent on situation assessment and pattern recognition. In a team environment where individuals have to collaborate to achieve high performance, the communication of individual situation assessment is crucial. Our results imply that by providing a stimulating structure, individual members were able to effectively communicate their assessment with each other, and therefore choose responses that collectively increased their team performance. Essentially, this stimulating structure created a public representation for distributed cognition and transformed situation awareness from an effortful memory task into a simpler perceptual task. The team member's limited cognitive resources were conserved (Wickens 2002) so they could apply them to the resource allocation phase of the task.

With regard to way structures were shared, providing the participants with a tools that labeled the chunks lead to better outcomes. Chunks are a more efficient cognitive memory representation and a 'chunk' label (Gobet and Simon 1996a) creates a quick pointer to LTM. The chunk tool used in this study facilitated cognitive alignment of memory retrieval structures; thus creating a good cognitive fit between the problem representation and memory. This reduction in effort resulting from cognitive alignment left the team members with a greater proportion of cognitive resources available for the allocation task. Communicating this chunk label appeared to align the entire team (Doumont 2002).

Explicit training in chunk representations of the patterns was also associated with higher performance. At least for a short-term memorization task, the explicit training method appears to have an advantage over implicit pattern training. However, the interactions between training-type and tool-type were unexpected. Although the main effects for the advantages of chunk representations were strong and significant, we observed that complementary combinations of training- and tool-type were associated with the highest performance among our participants. We initially suspected that the interactions might have been related to the differences in the ways that the sharing tools were used between treatments. However, when we compared the frequency and type of objects shared, we found no differences between the complementary and congruent treatments with respect to the use of the sharing tools. We conclude that the interactions were related to differences in the cognitive representations rather than differences in the way the tools were used. The issue of interactions between training and tool type is interesting and warrants further study.

Template Theory gains support from our analysis of the bumping data, because it is more likely that the core features of the pattern will be learned by each team member, regardless of the exact template. However, the extent of learning of the slot features may differ.

We observed an interaction between the training method and the type of pattern-sharing tool. Participants in the complementary conditions outperformed the participants having both explicit chunk training and a chunk-sharing tool, which we hypothesized would be the most favorable condition. These results demonstrate the complexity of pattern learning mechanisms. Ashby, Ell and Waldron (2003) found similar interactions in a procedural learning task, and concluded that explicit and implicit category-learning systems compete throughout training. As a consequence it becomes difficult to predict a priori which form of training will be best for which type of task. However, these results are important because they lend further support for the notion that pattern learning is mediated by two or more cognitive systems, as suggested by several recent studies (Ashby *et al.* 1998; Ashby and Ell 2001, 2002; Erickson and Kruschke 1998; Pickering 1997; Waldron and Ashby 2001). In the context of macrocognition, these results suggest that it is important for future research to distinguish between implicit and explicit learning, and the potential interactions with macro- and microcognitive processes.

The interaction between implicit and explicit learning for individuals is especially relevant to team macrocognitive processes. Macrocognitive processes such as knowledge interoperability are founded on a shared interpretation of the situation, roles and resources available to the team. The degree of common ground achieved by the team affects their performance on tasks that require collaboration and coordination. Typically, development of common ground and shared mental models are achieved implicitly through long experience working together. Moreover, it is desirable to reduce the time required for the development of common ground, for example when new members are added to the team. In this case, new members can be brought up to speed quickly through the use of explicit template learning. Such explicit template training has the benefit of quickly introducing new members to the stimulating structures and coordinating representations in use by the team. Stimulating structures and coordinating representations typically reduce the time and effort required for coordination, and lead to higher performance in time-critical tasks.

However, explicit template learning does not completely replace the common ground learned through long experience. Conversely, implicit learning strategies seek to provide new members with ways to build the common ground and shared mental models in use by the existing team members. Such implicit template learning strategies seek to strengthen the long-term congruence of the shared cognitive structures, without addressing the immediate collaboration and coordination mechanisms. In our study we found that when we provided teams with tools that supported complimentary strategies for learning and execution, for example tools that supported explicit template learning and implicit template communications, the performance was better than when pure strategies were employed.

Thus we speculate that, for teams engaged in high-stakes, time-critical tasks, team macrocognitive processes might be improved through the use of tools that promote both implicit learning of common ground and explicit learning of coordinating representations.

Implications for practice

We believe that organizations can benefit by analyzing their group tasks; when groups face scenarios for which there are patterns, the patterns should be labeled and tools created to allow groups to communicate these labels with minimal task interruption to the sender and receiver. We expect that group members who share these labels will create shared cognition, have knowledge interoperability and increase their performance (Cannon-Bowers and Salas 2001; Nosek and McNeese 1997; Warner, Letsky and Cowen 2005).

There are many team task domains that can benefit from tools that support macrocognition by promoting improved communication of pattern information. In military domains, tasks such as intelligence analysis and operational planning are typically performed by heterogeneous teams. In these scenarios, the use of stimulating structures to convey template labels may facilitate rapid self-synchronization among team members, and relatively effortless assimilation of new team members. Similar opportunities to increase performance by communicating chunk information exist for project management teams in both the public and private sector, e.g., construction, software development, product design, bid review, or engineering teams that face detectable patterns.

Conclusion

In our study we have demonstrated that a tool designed to promote sharing a template or chunk label, associated with a pattern, led to superior performance due to superior cognitive alignment. When the training on the patterns is also cognitively aligned, performance improves even further. We believe that our model can be applied to other settings characterized by high stakes and time pressure based on other results published (Hayne, Smith and Vijayasarathy 2005b). These results demonstrate the importance of macrocognitive processes for team performance.

Future research

In naturalistic environments, decision-makers are faced with dynamic situations in which the patterns are constantly changing. In our study, the patterns were static for the duration of a trial. We are in the process of developing a version of our strategy game in which the patterns change during the course of a game trial. Once again, we believe that a pattern-sharing tool should reduce the cognitive loads on the team members, and promote improved performance.

Humans have limited cognitive resources for perception (e.g., of patterns), attention, and memory. The pattern-sharing tool seemed to improve collective perception and collective recall of patterns by allowing the team to bring its collective cognitive resources to bear on the task. In naturalistic environments, team members are often required to perform multiple simultaneous tasks. In multi-tasking environments, tools that support attention management will be especially important. Thus, another possible extension of this research involves the development of tools that provide cognitive support for attention management.

Template Theory as applied to collaboration deserves more investigation, i.e., when might sharing core data be more important than sharing slot data? As decision-makers scan their environment they are directing attention to their perceptual processes. The information they perceive is sorted through their template discrimination net, and when core items are noticed, the appropriate templates are swiftly retrieved. In other words, core data items activate recognition of familiar patterns, thus creating situation awareness. As templates are retrieved, the slot data are made available in working memory. These slot data provide additional information to the decision-maker regarding variants of the patterns, and potential strategies for action. In other words, the slot data provide the key to successful responses. If each decision-maker has knowledge about different patterns, what stimulating structure is most appropriate to share?

Acknowledgments

The research described in this chapter was supported by Dr. Mike Letsky at the Office of Naval Research, Grant #66001-00-1-8967.

References

Ashby, F. and Ell, S.W. (2001) 'The Neurobiological Basis of Category Learning', *Trends in Cognitive Science* 5, 204–210.

Ashby, F.G. and Ell, S.W. (2002). 'Single Versus Multiple Systems of Category Learning: Reply to Nosofsky and Kruschke (2002)', *Psychonomic Bulletin and Review* 9, 175–180.

Ashby, F.G., Ell, S.W. and Waldron, E.M. (2003). 'Procedural Learning in Perceptual Categorization', *Memory and Cognition* 31(7), 1114–1125.

Ashby, F.G., Alfonso-Reese, L.A., Turken, A.U. and Waldron, E.M. (1998). 'A Neuropsychological Theory of Multiple Systems in Category Learning', *Psychological Review* 105, 442–481.

Baddeley, A. (1992). 'Working Memory', *Science* 255(5044), 556–559.

Baddeley, A. (1998). 'Recent Developments in Working Memory', *Current Opinion in Neurobiology* 8(2), 234–238.

Baddeley, A., Chincotta, D. and Adlam, A. (2001). 'Working Memory and the Control of Action: Evidence from Task Switching', *Journal of Experimental Psychology: General* 130, 641–657.

Bara, B.G. (1995). *Cognitive Science: A Developmental Approach to the Simulation of the Mind.* (Hove, East Sussex: LEA).

Bartlett, F.C. (1932). *Remembering.* (Cambridge: Cambridge University Press).

Broadbent, D.E. (1958). *Perception and Communication.* (London: Pergamon Press).

Cacciabue, P.C. and Hollnagel, E. (1995). 'Simulation of Cognition: Applications', in J.M. Hoc, P.C. Cacciabue and E. Hollnagel, (eds), *Expertise and Technology: Issues in Cognition and Human-Computer Cooperation*, pp. 55–74. (Hillsdale, NJ: LEA).

Cannon-Bowers, J.A. and Salas, E. (2001). 'Reflections on Shared Cognition', *Journal of Organizational Behavior* 22, 195–202.

Card, S.K., Moran, T.P. and Newell, A. (1986). 'The Model Human Processor: An Engineering Model of Human Performance', in K.R. Boff, L. Kaufman and J.P. Thomas, (eds), *Handbook of Perception and Human Performance*, 2nd edn. (New York: Wiley and Sons).

Charness, N. (1976). 'Memory for Chess Positions: Resistance to Interference', *Journal of Experimental Psychology: Human Learning and Memory* 2, 641–653.

Chase, W.G. and Ericsson, K.A. (1982). 'Skill and Working Memory', in G.H. Bower, (ed.), *The Psychology of Learning and Motivation*, vol. 16, pp. 1–58). (New York: Academic Press).

Chase, W.G. and Simon, H.A. (1973a). 'Perception in Chess', *Cognitive Psychology* 4, 55–81.

Chase, W.G. and Simon, H.A. (1973b). 'The Mind's Eye in Chess', in W.G. Chase, (ed.), *Visual information processing*, pp. 215–281. (New York: Academic Press).

Cooke, N.J., Atlas, R.S., Lane, D.M. and Berger, R.C. (1993). 'Role of High-Level Knowledge in Memory for Chess Positions', *American Journal of Psychology* 106, 321–351.

Cox, J., Roberson, B. and Smith, V. (1982). 'Theory and Behavior of Single Object Auctions', in V.L. Smith, (ed.), *Research In Experimental Economics*, vol. 2, pp. 1–43. (JAI Press, Greenwich, New York).

DeSanctis, G. and Gallupe, B. (1987). 'A Foundation for the Study of Group Decision Support Systems', *Management Science* 33(5), 589–609.

Doumont, J. (2002) 'Magical Numbers: The Seven-Plus-or-Minus-Two Myth', *IEEE Transactions On Professional Communication* 45(2), 123–127.

Erickson, M.A. and Kruschke, J.K. (1998). 'Rules and Exemplars in Category Learning', *Journal of Experimental Psychology: General* 127, 107–140.

Frey, P.W. and Adesman, P. (1976). 'Recall Memory for Visually Presented Chess Positions', *Memory and Cognition* 4, 541–547.

Gobet, F. (2005). 'Chunking Models of Expertise: Implications for Education', *Applied Cognitive Psychology* 19, 183–204.

Gobet, F., Lane, P.C.R., Croker, S., Cheng, P.C-H., Jones, G., Oliver, I. and Pine, J.M. (2001). 'Chunking Mechanisms in Human Learning', *TRENDS in Cognitive Sciences* 5, 236–243.

Gobet, F. and Simon, H.A. (1996a). 'Templates in Chess Memory: A Mechanism for Recalling Several Boards', *Cognitive Psychology* 31, 1–40.

Gobet, F. and Simon, H.A. (1996b). 'Recall of Rapidly Presented Random Chess Positions is a Function of Skill', *Psychonomic Bulletin and Review* 3, 159–163.

Gobet, F. and Simon, H.A. (1998). 'Expert Chess Memory: Revisiting the Chunking Hypothesis', *Memory* 6, 225–255.

Gobet, F. and Simon, H.A. (2000). 'Five Seconds or Sixty? Presentation Time in Expert Memory', *Cognitive Science* 24(4), 651–682.

Goldin, S.E. (1978). 'Memory for the Ordinary: Typicality Effects in Chess Memory', *Journal of Experimental Psychology: Human Learning and Memory* 4, 605–616.

Grassé, P. (1959). 'La Reconstruction du Nid et les Coordinations Inter-Individuelles Chez Bellicositermes Natalensis et Cubitermes sp. La théorie de la Stigmergie: Essai d'interprétation du Comportement des Termites Constructeurs', *Insectes Sociaux* 6, 41–81.

Hartston, W.R. and Wason, P.C. (1983) *The Psychology of Chess*. (London: Batsford).

Hayne, S. and Smith, C.A.P (2005). 'The Relationship Between e-Collaboration and Cognition', *International Journal of e-Collaboration* 1(3), 17–34.

Hayne, S., Smith, C.A.P. and Turk, D. (2003). 'The Effectiveness of Groups Recognizing Patterns', *International Journal of Human Computer Studies* 59, 523–543.

Hayne, S., Smith, C.A.P. and Vijayasarathy, L. (2005a). 'The Use of Pattern–Communication Tools and Team Pattern Recognition', *IEEE Transactions on Professional Communication* 48(4), 377–390.

Hayne, S., Smith, C. and Vijayasarathy, L. (2005b). 'Training for Collaboration and Cognitive Alignment', Americas Conference Information Systems, August. Conference Best Paper.

Hutchins, E. (1991). 'The Social Organization of Distributed Cognition', in L. Resnick, J. Levine and S. Teasdale, (eds), *Perspectives on Socially Shared Cognition*, pp. 283–307. (Washington, DC: American Psychological Association).

Hutchins, E. (1995). 'How a Cockpit Remembers its Speeds', *Cognitive Science* 19, 265–288.

Jessup, L. and Valacich, J. (1993). *Group Support Systems: A New Frontier.* MacMillan, New York.

Kaempf, G.L., Klein, G.A. and Wolf, S. (1996). 'Decision Making in Complex Naval Command-and-Control Environments', *Human Factors* 38(2), 220–231.

Keltner, J. (1989). 'Facilitation: Catalyst for group problem-solving', *Management Communication Quarterly* 3(1), 8–31.

Klein, G.A. (1993). 'A Recognition-Primed Decision (RPD) model of rapid decision making', in G.A. Klein, J. Orasanu, R. Calderwood and C.E. Zsambok, (eds), *Decision Making In Action: Models And Methods*, pp. 138–147. (Norwood, NJ: Ablex).

Klein, G., Ross, K.G., Moon, B.M., Klein, D.E., Hoffman, R.R. and Hollnagel, E. (2003). 'Macrocognition', *IEEE Intelligent Systems* 81–85.

Lavie, N. (1995). 'Perceptual Load as a Necessary Condition for Selective Attention', *Experimental Psychology: Perception and Performance* 21(3), 451–468.

McGrath, J.E. and Hollingshead (1994) *Groups: Interacting with Technology.* (Thousand Oaks, CA: Sage).

McGunnigle, J., Hughes, W. and Lucas T. (2000). 'Human Experiments on the Values of Information and Force Advantage', *Phalanx* 33(4), 35–46.

Miller, G.A. (1956). 'The Magical Number Seven, Plus or Minus Two: Some Limits on Our Capacity for Processing Information', *The Psychological Review* 63, 81–97.

Minsky, M. (1977). 'Frame-system Theory', in P.N. Johnson-Laird and P.C. Wason, (eds), *Thinking. Readings in Cognitive Science*, pp. 355–376. (Cambridge: Cambridge University Press).

Nosek, J. and McNeese, M.D. (1997). 'Augmenting Group Sense-making in Ill-defined, Emerging Situations', *Information Technology and People* 10(3), 241–252.

Nosofsky, R.M. and Kruschke, J.K. (2002). 'Single-System Models and Interference in Category Learning: Commentary on Waldron and Ashby (2001)', *Psychonomic Bulletin and Review* 9, 169–174.

Nunamaker, J.F. (1997). 'Future Research in Group Support Systems: Needs, Some Question, and Possible Directions', *International Journal of Human-Computer Studies* 47(3), 357–385.

Pickering, A.D. (1997). 'New Approaches to the Study of Amnesic Patients: What can a Neurofunctional Philosophy and Neural Network Methods Offer?', in A.R. Mayes and J.J. Downes, (eds), *Theories of Organic Amnesia*, pp. 255–300. (Hove, UK: Psychology Press).

Schmidt, J., Montoya-Weiss, M. and Massey, A. (2001). 'New Product Development Decision-Making Effectiveness: Comparing Individuals, Face-To-Face Teams, and Virtual Teams', *Decision Sciences* 32(4), 575–600.

Simon, H.A. (1974). 'How Big is a Chunk?', *Science* 183, 482–488.

Smith, E.E., Patalano, A.L. and Jonides, J. (1998). 'Alternative Strategies of Categorization', *Cognition* 65, 167–196.

Squire, L.R., Knowlton, B.J. and Musen, G. (1993). 'The Structure and Organization of memory', *Annual Review of Psychology* 44, 453–495.

Staszewski, J. (1990). 'Exceptional Memory: The Influence of Practice and Knowledge on the Development of Elaborative Encoding Strategies', in F.E. Weinert and W. Schneider, (eds), *Interactions Among Aptitudes, Strategies, and Knowledge in Cognitive Performance*, pp. 252–285. (New York: Springer).

Susi, T. and Ziemke, T. (2001). 'Social Cognition, Artefacts, and Stigmergy: A Cooperative Analysis of Theoretical Frameworks for the Understanding of Artefact-mediated Collaborative Activity', *Journal of Cognitive Systems Research* 2, 273–290.

Theraulaz, G. and Bonabeau, E. (1999). 'A Brief History of Stigmergy', *Artificial Life* 5, 97–116.

Treisman, A.M. (1969). 'Strategies and Models of Selective Attention', *Psychological Review* 76, 282–299.

Varian, H. (1995). 'The Information Economy: How Much Will Two Bits be Worth in the Digital Marketplace?', *Scientific American* 200–201.

Vessey, I. (1991) 'Cognitive Fit: A Theory-Based Analysis of the Graphs Versus Tables Literature', *Decision Sciences* 22(2), 219–240.

Waldron, E.M. and Ashby, F.G. (2001). 'The Effects of Concurrent Task Interference on Category Learning: Evidence for Multiple Category Learning Systems', *Psychonomic Bulletin and Review* 8, 168–176.

Warner, N., Letsky, M. and Cowen, M. (2005). 'Cognitive Model of Team Collaboration: Macro-Cognitive Focus', *Proceedings of the 49th Annual Meeting of the Human Factors and Ergonomic Society*, pp. 2–19. (Santa Monica, CA: Human Factors and Ergonomics Society).

Whitehead, Alfred North. (1929). *The Aims of Education*. (New York: The Free Press).

Wickens, C.D. (1984). 'Processing Resources in Attention', in R. Parasuraman and R. Davies, (eds), *Varieties of Attention*, pp. 63–101. (Orlando, FL: Academic Press).

Wickens, C.D. (2002). 'Multiple Resources and Performance Prediction', *Theoretical Issues in Ergonomic Science* 3, 159–177.

Willingham, D.B., Wells, L.A., Farrell, J.M. and Stemwedel, M.E. (2000) 'Implicit Motor Sequence Learning is Represented in Response Locations', *Memory and Cognition* 28, 366–375.

Yates, F. (1966). *The Art of Memory*. (Chicago, IL: University of Chicago Press).

Chapter 6

Toward a Conceptual Model of Common Ground in Teamwork

John M. Carroll, Gregorio Convertino, Mary Beth Rosson and Craig H. Ganoe

Introduction

Testing, refining, and incrementing common ground is a basic function of communication and collaboration. Initially, language researchers defined *common ground* as the sum of mutual, common, or joint knowledge, beliefs, suppositions and protocols shared between people engaged in communications (Clark 1996). They investigated the development of common ground – the grounding process. Communication is a joint activity, in which partners use verbal and non-verbal signs (e.g., gestures, gaze) to coordinate dialogue and actions toward common, continuously tested goals (Clark 1996). Later, researchers of computer-mediated communication found that different communication media present different affordances and cost structures for different parts of the grounding process (Clark and Brennan 1991). We study the grounding process *in the context of computer-supported cooperative work*. Greater common ground enables better communication, coordination, and collaboration. We study how collaboration tools, member and team variables and work conditions affect how collaborators share and coordinate knowledge and how they operate and develop as a team.

In this chapter, we describe a conceptual model for investigating the process and consequences of common ground in teamwork. The model integrates concepts from group process theory and activity theory. We analyze group work as a process; we view work as a social, purposeful and tool-mediated activity. We use the conceptual model to motivate and organize a study of common ground in a team performance task involving emergency planning. We show how the model guides our selection of experimental manipulations and control variables, and how we use the model to operationalize and interpret process and outcome measures for group work.

A conceptual model of common ground in teamwork

In this section, we first describe the theoretical context for activity theory and group theory in Computer-Supported Cooperative Work (CSCW) research. Following that we integrate the ideas and methods from the two sources to guide our study of common ground.

Theoretical context: From cognitive tasks to situated action and process

A major shift occurred in HCI theory over the last three decades: from analysis of cognitive tasks to study of situated action. The initial vision of HCI researchers was to bring cognitive science methods and theories to bear on software development (Carroll 2003). The foundations were given by the paradigm of cognitive psychology of the 1960s and 1970s, which focused on short-term tasks, information-processing models of behavior, methods for detailed task analysis and performance measures (e.g., errors, completion time).

After significant progress in the 1980s guided by the cognitive paradigm, new directions in HCI theory began to appear. Influential research contributions during the 1980s and 1990s pointed to limitations of information-processing models and shifted the analytic focus from cognitive task to situated action. The old view was that humans can be seen like computers: as information-processing units with analogous functioning mechanisms. The new view is that a theory describing and explaining the *context* of human interaction with computers is needed (Kaptelinin 1996).

In the new view, the analysis of tasks and performance measures are supplemented with models of work context and process measures. Recent theoretical contributions (activity theory, distributed cognition, phenomenology, actor–network theory) share foundational ideas: the critical role of tools in human experience; people are not defined by their skin but rather extend into broader functional systems; the mind–body dualism is rejected. In this new theoretical context, we leverage concepts from activity theory, which conceives the individual as a social, purposeful, and technologically empowered agent. Human-specific attributes like intentionality, consciousness, and adaptability are at the core of activity theory (Kaptelinin and Nardi 2006). It also provides concepts for explaining human development in relation to tool-mediated activities.

An additional source of methods and concepts for HCI research is social psychology. During the late 1980s and 1990s research on Computer-Mediated Communication (CMC) and CSCW flourished as software and systems that support communication and shared work became available. Theoretical and methodological contributions from social psychology were integrated into CSCW research (see Kraut 2003 for a review). Two contributions include group process measures and models of group process that account for levels of analysis and group development (e.g., Arrow *et al.* 2000).

Experimental research in social psychology and human factors has focused on communication in groups, contrasting face-to-face and technology-mediated communication. Two clear findings have been that differences due to the medium depend on the type of experimental task used, and that measures of task performance (e.g., errors and time) are not sensitive to subtle changes in technology (Monk *et al.* 1996).

Empirical results reported in CMC research tend to be based on measures designed to tap the process of communication, rather than its outcomes (e.g., Sellen 1995). This is consistent with the theoretical argument that communication via different media can be described in terms of the set of grounding constraints that each medium offers. If the constraints change, the communication structure also changes, fitting the new constraints (Clark and Brennan 1991). We study communication in the context of computer-supported teamwork. We focus on the development of group structures that regulate cooperative work, we use measures of process, and we relate them to measures of outcomes.

Motivation and strategy for our model: From conceptual to empirical

Our conceptual model development was guided by two distinct motivations. First, from the standpoint of research planning and implementation, a conceptual model helps researchers represent and analyze the research problem in a more systematic way (e.g., identifying critical concepts and their relationships). After analyzing and relating elements of a model, the resulting concepts can be mapped to empirical measures to create a corresponding empirical model. In this way a conceptual model acts as a descriptive and explanatory 'reference system' that helps researchers to orient, collect, aggregate and interpret empirical data.

Second, a conceptual model helps in understanding and specifying users' requirements. The practice of design consists of envisioning and building design products, including creation of intermediate representations (e.g., function lists, scenarios, prototypes). Conceptual models help designers to explore, represent, test, record and communicate design ideas and decisions within a development team and with end users (Benyon and Imaz 1999). Rigorous theory development can help the designers of collaboration systems to understand and repeat successful design ideas elsewhere, learn from failures (e.g., Grudin 1989) and avoid them elsewhere. At the same time designers can feed knowledge gained from practical experience back into the theoretical model guiding the design.

Two theoretical sources

Our conceptual model integrates concepts from a model of group process developed to study complex group behavior (McGrath 1984), and from an activity system model developed to study tool-mediated cooperative activities (Engestrom 1990). Our goal is to identify and relate factors that affect sharing and coordination of knowledge in computer-supported teamwork.

The group process model

> 'Statics, the physicists knows, is only an abstraction from dynamics … dynamics, on the other hand, deals with the general case' Karl Popper (1957), cited in Arrow and McGrath (1995).

McGrath (1984) proposed a conceptual model of group process to guide the study of complex group behaviors: he breaks the problem in manageable chunks, examines the evidence about each chunk, and then fits the parts back together again. His model is a general map rather than a substantive theory of groups that reflects a viewpoint or a set of hypotheses. It distinguishes different classes of variables and the logical relationship between those classes, without any details about the specific variables or relationships (Figure 6.1).

The model starts with two classes of variables: properties of the Members and properties of the Environment (physical, sociocultural, technological). When people become interrelated as part of a group they acquire and maintain social relationships that constitute group structures of various sorts. These include group composition, communication structure, division of labor (roles), interpersonal relationship structure and power structure. A third class of variables includes these relationships – the Standing Group. The fourth variable class is Task/Situation, which includes the requirements, demands, possibilities and constraints that relate to the task(s) and situation(s) of a group. The juxtaposition of Standing Group and Task/Situation forms the Behavior Setting, a derived class of variables that we omit in our summary view in Figure 6.1.

All variables are inputs and outputs of the Acting Group, which represents the process variables in play when the members actually interact. Overall, the model has the Acting Group as its centerpiece and two submodels: one submodel of influence

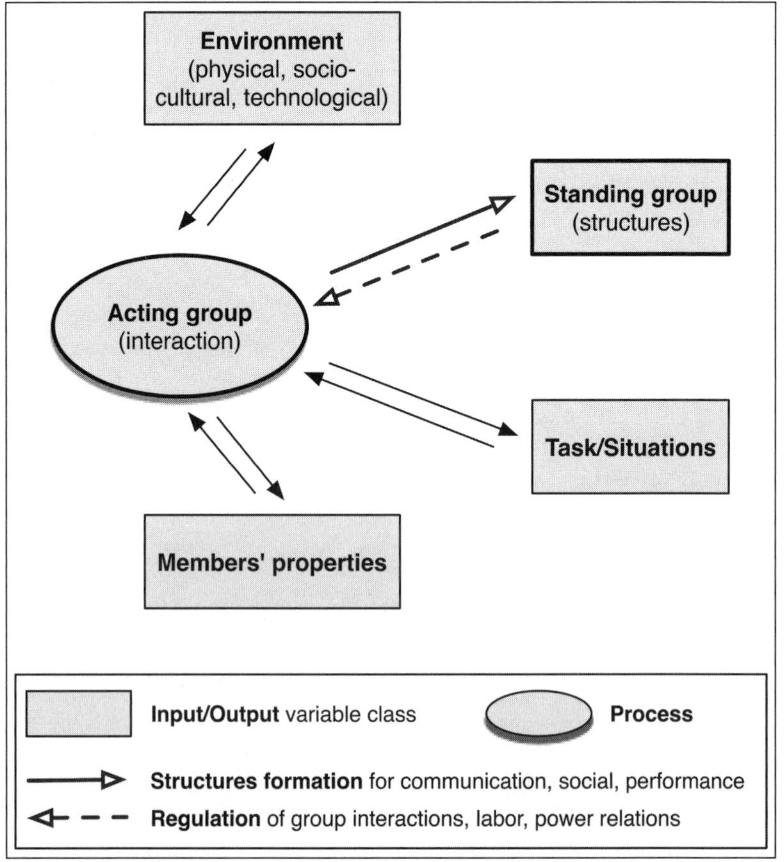

Figure 6.1 Group process model. Adapted from McGrath (1984)

on, and a second of impact from, the Acting Group. In other words, the Acting Group is influenced by but also influences all other variables. Additionally, the group process is affected by variables that are internal to the interaction process. The group process is decomposed into three stages related to the Standing Group (McGrath 1984):

1. The communication process that results in a *communication structure* (1st stage: influence on Standing Group).
2. The task performance process and the social interaction or acquaintance process, which result, respectively, in the *performance structure* (e.g., division of labor) and the *social structure* (e.g., trust relationships) (2nd stage: influence on Standing Group).
3. In turn, the three structures that are part of the *standing group* influence one another and regulate the *acting group* (process) over the subsequent steps of interaction (3rd stage: impact on Acting Group).

Extensions have been made to the original model to better reflect structures and processes of naturally occurring groups (e.g., McGrath 1991; Arrow and McGrath 1995; Arrow *et al.* 2000). These include the introduction of levels of analysis, group functions, modes or task types, and the dimension of time and group development. In the Time, Interaction and Performance (TIP) theory McGrath (1991) explains group behavior in terms of three functions of groups at three different levels of analysis: production, member-support and group well-being. Groups engage in purposeful process at three partially nested levels: project, tasks and steps. Modes or task types (inception, problem-solving, conflict resolution, execution) are classes of tasks that groups perform as part of a longer project.

If we use this conceptual model to explain group behavior in the context of CSCW research two limitations become visible, suggesting useful changes. First, the relationship between group process and technology (tool mediation) is of minor importance – technology is part of the background with other environmental factors. Second, motivational factors (individual and collective) do not play a central role in the model. In more recent versions of this model, the group process is seen as purposeful and directed toward a goal. However, the concept of a goal continues to be merely a descriptive element and not an essential organizing element of the group process across different levels of analysis (project, task, steps).

The activity system model Engestrom (1990) proposed an activity system model that describes collective activities. The model draws from activity theory, outlining the relationships among the major components of an activity (Kaptelinin and Nardi 2006). Engestrom extended the original definition of activity, which was limited to the interaction between subject, object (or objective) and mediating tools of the activity (Figure 6.2, top triangle). He represented the activity as a three-way interaction between subject, object, and a community of reference (Figure 6.3, central triangle). The interactions of subject–object, subject–community, and community–object are associated with three corresponding forms of mediation: tools, rules and division of labor. The model has been used in CSCW to analyze technology-mediated work practices and design collaborative systems.

Activity theory emphasizes that technology should be studied in the context of purposeful work and in a meaningful context. It broadens the scope of the investigation in HCI along three dimensions. The first is context and levels of analysis: the scope of investigation is expanded from the immediate human–computer interaction to the larger context (i.e., activity), including meaningful goals and accounting for different levels of analysis (activity levels: operation, action, activity). Multiple activity systems can also be examined and related. The second is development: users and artifacts are changed through interaction. Third is the individual–social dimension: individual activities always occur in a social context. The social dimension becomes even richer when we consider collective activities, where the subject is a collective agent, a group of individuals pursuing a common goal (Kaptelinin 1996).

These three contributions from activity theory are compatible with the latest model of group process proposed by McGrath and colleagues (McGrath 1991; Arrow *et al.* 2000):

1. levels of analysis are an important aspect of the theory of groups as complex systems (individuals, groups, organizations);
2. the model of group process operationalizes development as transactions between the acting group and standing group (group development) or between the acting group and members' properties (individual development);
3. the transactions between members' properties, the acting and standing group, and the sociocultural environment correspond to the individual–social dimension of activity theory.

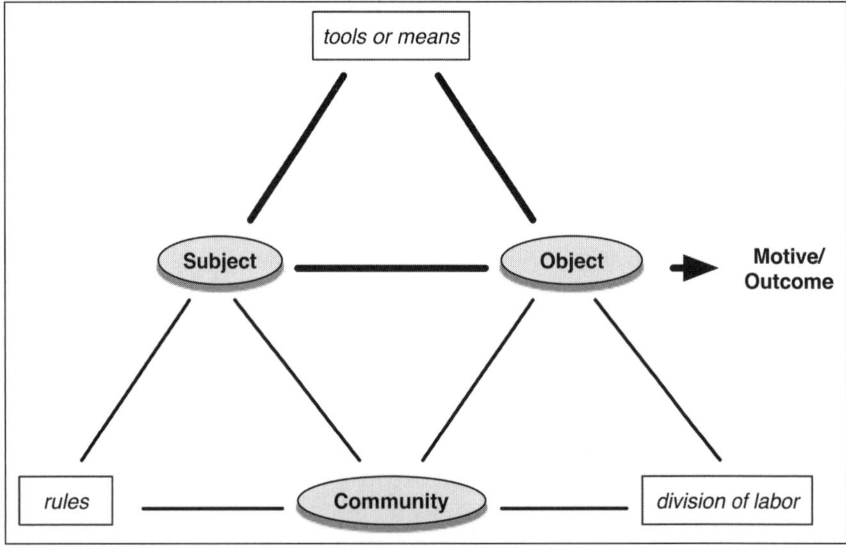

Figure 6.2 Activity system model. Adapted from Engestrom (1990)

Our integrated model

In this section we synthesize elements of the group process and activity system models. We start with a conceptual description; we then describe how the conceptual model has been transformed into an empirical model that we are using to guide our ongoing investigations of common ground in complex technology-mediated group activities.

The conceptual model Our research investigates the formation and use of the group structures that regulate knowledge sharing, coordination and awareness and performance in teams. We examine how these structures affect and are affected by group process in realistic, tool-mediated teamwork (emergency management planning by a small team). In our experimental work on common ground we manipulate or control a range of variables (e.g., collaborative system in use, amount and distribution of shared information), and measure impacts on both group process and products (e.g., amount of shared knowledge, quality of products).

Our model of group process imports the core elements of McGrath's model. We focus on the relationship between the acting group (group process) and the input–output variables forming the standing group (group structures). We assume that when the members become part of a team, they develop the structures needed to regulate the group process (communication, performance, social interactions and power relations). These are reused and adapted as the members keep working together. We measure the changes in, causes of, and consequences of these structures.

Our conceptual model also includes essential elements from the activity system model: the object-focused or purposeful nature of the group work, the mediating role of tools, and the relation to a community of reference. While the last element is peripheral to our current investigations (Figure 6.3, dashed lines), the others are central constructs. The object, or objective, of the group activity is the goal toward which the group process is directed; in our work the team works on the meaningful and motivating activity of emergency management planning. The shared object leads collaborators to organize low-level operations into actions, and actions into an activity. A performance-critical aspect of group process is how well they define and remain aware of the shared object and organize their collaboration toward it. The group process is mediated by the properties of the collaborative tools used by the group. The team's performance on the project will also depend on these properties. Thus, technology and its mediating effects are given a more prominent role in our model than in McGrath's model of group process.

Another aspect imported from activity theory is the materialistic, historical perspective of work. We reconstruct the activity by examining the concrete history of use, and the ways in which tools or artifacts are modified in the activity. This analysis is interpreted with respect to findings from analysis of videos, communication content and structure, and self-reports.

Our conceptual model of group process differs significantly from cognitive models of knowledge sharing and awareness that are based on the metaphor of 'groups as information processors' (e.g., Hinsz 2001). These models assume a symmetrical relationship between humans and computers (following the mind-as-computer metaphor). In this perspective, people's mental functions change their content rather than their structure. We focus instead on human-specific attributes such as intentionality, reflection, development of skills and sharing of strategic knowledge in teams. Information-processing models operationalize work in terms of mental operations. The impacts of technology are analyzed in terms of cognitive functions like perception, attention, encoding, storage/retrieval, response and feedback. Broadening the scope, we attempt to account for the mediation of tools and for how work tools and other artifacts shape (and are changed by) people's work and mental abilities. Finally, we do not analyze the social dimension of work in terms of cognitive primitives (e.g., aggregations of individual cognitive processes). Rather, we investigate systemic structures that emerge at higher levels of social agency. These regulate task performance, social exchange, and group well-being (McGrath 1991; Arrow *et al.* 2000).

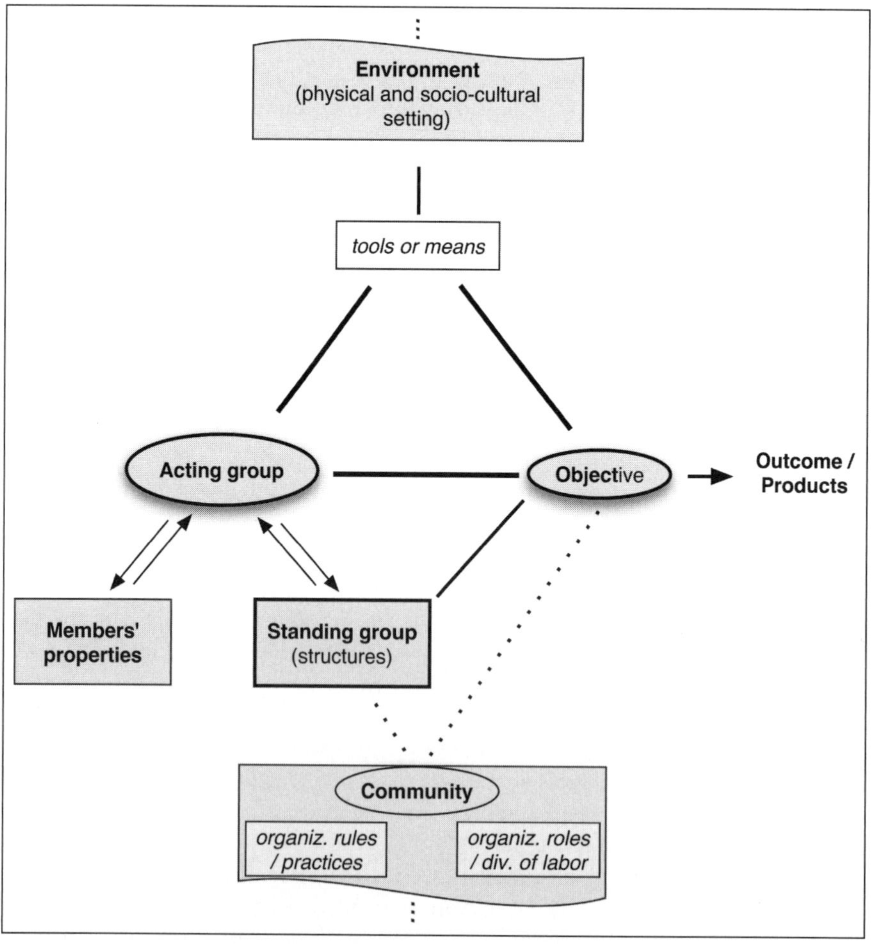

Figure 6.3 Integrated conceptual model

Relation to macro-models of collaboration Warner *et al.* (this volume, Chapter 2) propose a model of macrocognition. Macrocognition consists of internalized and externalized high-level mental processes employed by teams in problem-solving on complex, ill-defined problems. The maintenance of common ground in collaboration is a supporting process that enables the achievement of macro-cognitive functions such as naturalistic decision-making and planning (Klein *et al.* 2003). Note though that our investigations of common ground have a narrower scope than research on macrocognition.

There are theoretical similarities and differences between the two. Both the macrocognitive model and our model of common ground can be seen as a response to the need for realistic and empirically testable models of collaboration. Researchers studying group decision-making and related technology support need frameworks that identify and relate central influencing factors (including context), and that enable them to articulate the internal structure of collaboration. In our conceptual model, we rely on activity theory's representation of 'context in action' – the activity. Motivational, social, material and developmental aspects of cooperative work have more prominence in our model than in the macrocognition model, which is aimed at explaining group-level cognitive phenomena.

More generally, both the macrocognition model and our model of common ground address the need to consider multiple levels of analysis when studying complex systems. The concept of integrative levels of analysis was imported from biology and comparative psychology and reused to explain the behavior of sociotechnical systems (e.g., Korpela *et al.* 2001). Similar to the study of molecules, cells, tissues, organs, systems and organisms, the study of individuals, groups, organizations and society requires concepts and methods that are specific to the unit of analysis considered. Both activity theory and group theory use a multilevel systemic perspective when explaining collaborative phenomena.

Previously, we have described common ground as a subprocess of *activity awareness* for computer-supported teams working on long-term activities (Carroll *et al.* 2006). Building and verifying common ground is a necessary but not sufficient subprocess that enables groups to remain aware of one another's goals, plans and accomplishments so that they can coordinate effectively. Other subprocesses include formation of community practices, establishment of mutual trust and reciprocity (social capital); and human development both as individuals and collectives. These subprocesses co-occur at different levels of analysis and temporal rates.

Empirical model: reference task, manipulations, and measures The CSCW community has a need for empirical models appropriate to the study of collaborative interaction. In this section we operationalize some elements of our conceptual model, drawing from our ongoing studies of common ground. Our empirical setting is small teams performing emergency management planning tasks on shared geographic maps.

An important step in creating the empirical model was developing a reference task. The task is isomorphic to the Noncombatant Evacuation Operation scenario (Warner *et al.* this volume, Introduction), where a multi-expert military team performs a complex problem-solving task. In our task, team members have specific areas of expertise, or roles (public works, environmental, and mass care experts); they manage shared and unshared knowledge in real time, evaluate alternative rescue plans, and choose an optimal plan. The task is designed to support the control, manipulation and measurement of a wide range of variables; some of these relate to individuals (e.g., skills, sociality), others relate to the task (e.g., amount of information provided and processed), and some relate to the group (e.g., group decision quality).

An important characteristic of the reference task is that *team performance depends on members' sharing of role-specific information* in a timely and cooperative fashion (Carroll *et al.* 2007). To facilitate experimental control while at the same time evoking a realistic experience of emergency planning, we modeled the planning task on observations from fieldwork studies of local emergency management teams (Schafer *et al.*, forthcoming). Roles were designed by adapting descriptions of roles by the Federal Emergency Management Agency in the US.

Thus far we have used our conceptual model to plan and interpret the results of two experiments. In the first experiment (completed), 12 three-member teams worked face-to-face with a paper prototype on a table. In the second (ongoing), 12 three-member teams perform the same task, but work remotely using a geocollaborative software prototype. Below, we illustrate how specific measures correspond to variables in the conceptual model that are manipulated, measured or controlled. The mappings constitute our empirical model, summarized in Figure 6.4.

We manipulate three *independent* (input, Figure 6.4, 'In' labels) variables: degree of shared experience with the planning task (Standing Group, In1); degree of shared knowledge about teammate roles (Standing Group, In 2); and the collaboration setting (paper/collocated vs software/remote) (Medium and Environment, In 3).

We measure the effects of these manipulations on group process and performance. At this early stage of model development, we are exploring a wide variety of measures to identify those that are most useful. In Table 6.1 we list variables we have defined to operationalize three aspects of the model: process variables related to Acting Group and Standing Group ('Pr' in Figure 6.4), outcome measures ('Out' in Figure 6.4), and constant or controlled properties of Members, Community, or Environment ('Con' in Figure 6.4).

Model implementation: Example manipulations and effects

To illustrate more concretely the operationalization of our conceptual model, we describe how researchers might draw from the model to manipulate the conditions and effects pertinent of common ground. The examples are drawn from our ongoing investigations as well as other hypothetical empirical settings.

Manipulating the conditions for building common ground

In everyday conversations people develop and share a set of linked assumptions about each other in their current situation (what Clark calls the 'action ladder'). Clark (1996) illustrates with an interaction between a clerk and a customer at a drug store:

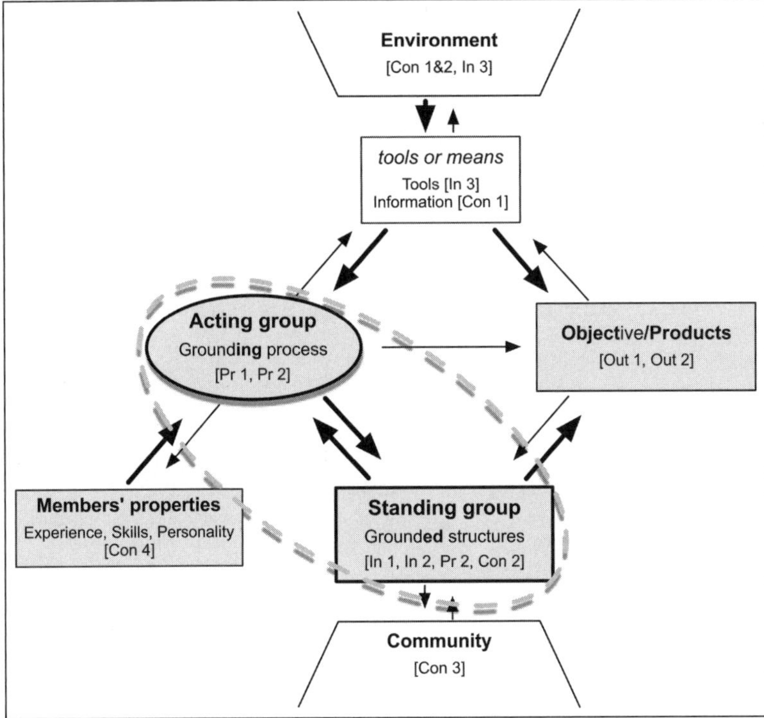

Figure 6.4 Empirical model: variables manipulated, measured, and controlled in our study of common ground in teamwork. This model instantiates the model described in Figure 6.3

> Joe enters the store. Sue (the clerk) is busy checking stock. She says 'I'll be right there' and assumes that: (1) Joe is listening; (2) he will identify the words 'I'll', 'be' and 'there'; (3) he is engaged in recognizing this as a proposition; and (4) he will understand that the proposition is a joint action where his part as customer is to wait and her part as clerk is to finish what she is doing.

This is a stereotypical situation from the point of view of the clerk. The ladder of implicit assumptions that communicators share, know that they share, and verify that they share makes communication more effective. Two common ways to share such assumptions are (1) repeating successful interactions in similar situations; and (2) explicitly communicating about what can be assumed. Similarly, in the context of a complex cooperation and co-production scenario such as our reference task, team members must build regulatory structures that can be later assumed and used to make their group work more effective. These structures regulate various aspects of cooperative work (communication, performance, social and power relations) and are refined as the members keep working together. Over time, these structures act as feedback from the Acting Group into the Standing Group. Within the Standing Group, the structures regulating communication sustain higher-level structures that regulate performance (e.g., decision strategies) and social interactions (e.g., trust relationships).

Table 6.1 Examples of manipulations and measures by model component

Model	Variables	Examples
Process (Acting, Standing group)	[Pr 1] Subjective measures (questionnaire)	• Common ground: perceived gain of shared knowledge, quality of communication, ease of understanding and expression • Awareness: perceived interpersonal awareness, awareness over time
	[Pr 2] Objective measures (recall, video and artifact analysis, communication logs)	• Common ground built: post-task recall • Communication efficiency. For content: queries, breakdowns, ellipsis, references. For structure: turn counts/duration, simultaneous speech • Coordination efficiency: facilitation and decision • Decision process: individual to group judgments
Output (Product)	[Out 1] Subjective measures (questionnaire).	• Performance and satisfaction: perceived quality of the product
	[Out 2] Objective measures (judgments, plan logs)	• Breadth of analysis: coverage of relevant information • Decision optimality: final plan optimality
Constant /Control (Task, Environ. Member)	[Con 1-3] Session variables	• Task [1], Group [2], and Setting [3] properties: e.g., information provided, gender/age mix, lab setup
	[Con 4] Control measures (questionnaires)	• Members' properties: background experience, relevant skills, personality factors

In our experiments we use both repetition and explicit instructions to manipulate degree of shared knowledge. In our experiments, (1) teams enact three similar instances of a planning task; and (2) while all participants learn about their own roles, half are given pre-task briefings about teammate roles and half are not. We assess the effects of these manipulations on both process and outcomes. Other possible manipulations include varying the familiarity of task elements (e.g., local versus non-local geography), training groups in advance on decision-making strategies, or computer-based sharing and reuse of strategies from session to session.

Another class of manipulations of the conditions for building common ground pertains to medium and environment of work. For instance, we use two configurations of medium and environment in our experiments. The three-member groups use either a paper prototype in a face-to-face setting or a software prototype in a remote collaboration setting. The effects of the collaborative technology can then be compared to team work that has minimal tool mediation. Other examples of manipulations of medium and environment include having or not having a shared workspace, ability to view collaborators or visual references in communications, or visual representations designed to support activity awareness in collaborations.

Measuring effects on group process

Questionnaires provide a common and low-cost measurement technique for assessing changes in common ground. Our experiments include a *post-session questionnaire* containing psychometric scales designed to assess common ground and awareness (see http://cscl.ist.psu.edu/public/projects/ONRproject/quest.html for details). In developing these scales, the items were grouped in clusters that focus on conceptually distinct process and outcome variables. Reliability analyses indicate that the items within each cluster are highly interrelated. Thus we compute an average rating for each cluster of items and use these as dependent measures when assessing effects of our experimental manipulations. In one study, statistical analyses revealed that these self-report measures increased over the three sessions, but that the manipulation of pre-session briefing (on roles) had no effect.

An objective measure of common ground built in a collaborative session can be obtained from *post-task recall* data. Following McCarthy *et al.* (1991), we assume that if common ground is successfully built, team members will recall most of the solutions they agree upon and the arguments that specific partners provide. We also assume that team members will agree about their recollections. In one experiment, we analyzed participants' recall of their team's first, second and third best rescue plans (shelters and route chosen). Almost everyone recalled the first plan chosen, suggesting a ceiling effect. However, recall for second and third alternatives increased over repeated sessions, consistent with a gradual build-up of shared knowledge.

Other interesting measures could focus on *how common ground increases* or perhaps *why collaborators choose to share specific pieces of information*. For example, in one experiment we tabulated whether and how often information was used at different stages of collaboration: individual acknowledgment before the group work; team discussion during the session; and individual recall after decisions were made. We used this frequency data to compute frequency ratios as the collaboration progressed through its stages, arguing that this could be seen as related to information retention. We observed that retention of individual pieces of information tended to increase over the three sessions. More qualitatively, we also noted that what was recalled at the end of a session seemed to be at a more abstract level than the specific concerns raised during the collaborative session.

An effective way to initiate the study of group process is to carry out a *qualitative analysis of group interaction videos*. Such an analysis can suggest behavioral patterns and general themes or hypotheses about how teammates are interacting or how they are using features of a tool or the setting. An important goal at this stage is to search for behavioral indicators of the underlying regulatory structures that are (or will become) part of the Standing Group. In the paper prototype study, five coders analyzed the videos of the twelve teams (three runs for each team). This analysis pointed to several interesting patterns:

- The teams used two different regulatory structures, one for facilitating group work and one for managing task decisions. One or two members played the role of the facilitator, decision controller, or both. Over repeated sessions, these role(s) might remain stable, attenuate, or pass from member to member.

- The default strategy used to select among alternatives was simple exclusion. A few teams later moved to a more systematic parallel analysis of alternatives.
- The dominant method for sharing role-specific knowledge was to 'push' information; information 'pull' via role-specific requests was less frequent.
- Common ways for sharing and recording information on the paper maps included were Post-Its and direct annotations. All teams used graphical annotations and deixis, where pointing was used to recap options and decisions. In a few cases, Post-Its or annotation were used at the bottom of the map for more general information.

More systematic (and higher-cost) techniques for studying group process include the analysis of communication structure and content. Such analysis requires verbatim transcripts of conversations with timestamps for each conversation turn. For instance, Sellen (1995) developed a set of quantitative measures to analyze *communication structure*. She was interested in conversations of multi-party meetings in face-to-face versus video-mediated settings. She defined (1) measures of turns (frequency, duration, and distribution of individual and group turns); and (2) measures of simultaneous speech (time, interruptive and non-interruptive, one-person talk time, overlap in turns, pauses between turns).

We are extending these measures of communication efficiency, coding transcripts for higher-level structures of group coordination, decision control and social relationships. This analysis is providing us with indicators of communication efficiency that are helping to clarify our earlier qualitative analysis of the videos. For example, we compared turns and completion time for the first and third sessions of eight teams. We found that individual turns tend to increase by the third run; at the same time task completion time tended to decrease. This may indicate that communication and work efficiency are increasing together across the repeated trials of the planning task. We also found that team facilitators tend to have more individual turns per session. Finally these conversation-based measures can be used to investigate changes in group dynamics, for instance how teams respond to the early emergence of a 'controlling' leader, or an alliance between two team members at the expense of a third. We are using the turn-taking and simultaneous speech measures to distinguish among situations when a peripheral team member seeks to regain some control, when a team member elects to withdraw, and so on.

The analysis of *communication content* is a useful complement of the analysis of structure. To do this, we start with a heuristic analysis of discourse structure to identify emergent patterns, then analyze critical incidents – positive or negative episodes that are suggestive of strategies for building common ground. Discourse analysis classifies utterances according to function. Computational linguists have developed coding schemas that have been used to compare face-to-face communication with computer-based communication. We use a conversation game analysis method to classify and measure the relative proportion of dialogue acts like query, explain, check, align, instruct, acknowledge, reply, clarify, and manage. By analyzing dialogue patterns (e.g., query ⇒ reply ⇒ acknowledge) across conditions we hope to identify markers of the grounding process. For example, we might examine the proportion of different dialogue acts (e.g., query) by session and by role. We can also study qualitatively different types of breakdowns, and the use of deixis, ellipsis, anaphora, and co-referring expressions (e.g., Monk *et al.* 1996).

Finally, we can track and measure the *judgment process*. In our task, participants rate the relevance of each piece of role-specific information, then make individual judgments about the best plan before starting the collaborative decision-making process. In addition to determining the optimality of the final group decision, we tracked how often members replace individual preferences with a novel alternative when working as a team. Looking at a subset of our data, we found that the teams tended to remain 'anchored' on members' initial preferences. Anchoring and confirmation bias limits a team's decision-making potential when solving complex problems (Billman *et al.* 2006). Collaborative systems may help teams become aware of such biases with suitable process visualizations and awareness mechanisms.

Measuring effects on group outcomes

As for effects on group process, a low-cost technique for assessing team performance is a *post-session questionnaire* that contains scales for rating team performance and satisfaction (see http://cscl.ist.psu.edu/public/projects/ONRproject/quest.html). The data indicates that our rating scales of perceived performance are intercorrelated and that aggregated scores increase over the three collaborative sessions, consistent with the expected increase in common ground.

Performance can be explained as a function of input or control variables in the model; it can also be explained as a result of the regularities and history of the process (i.e., the feedback cycle between the Acting and Standing Group). Our reference task allows us to measure *quality of decisions* and the *quality of information-sharing* at the same time. More specifically, the task is designed to make the former depend logically on the latter. The optimal plan can be discovered only if the members effectively pool relevant role-specific knowledge (Carroll *et al.* 2007).

The results of our experiments show that the teams were engaged by the experimental task although it was difficult to complete. In one study, teams recommended the optimal (hidden profile) plan in about one-third of the sessions. Not surprisingly, these outcomes also evinced a learning or warm-up effect, with only a couple teams recommending the optimal alternative in the first session, but about half selecting it in the third session. The modest success rate across teams suggests that performance is far from perfect; from a methodological perspective this is a good quality for a reference task (no ceiling effect). Our attention to ecological validity led to a task that was complex and realistic, leaving us with considerable flexibility for assessing variations in how teams share knowledge while working as a team.

Discussion: Toward a generative model

The study of technology-mediated communication preceded the study of computer-supported work. Beginning in the 1970s, researchers studied communication media to identify, compare, and measure effects, but only recently have differences have been explained through comprehensive models that consider the communication context (Clark and Brennan 1991). A similar transition has yet to occur in CSCW. Our integration and operationalization of established theoretical positions is a move in this direction: we have developed and operationalized a conceptual model of common ground in computer-supported teamwork. The model offers a comprehensive view of the common ground process while also considering how this process takes place in computer-supported teamwork.

Communication consists of not simply overt messages, but also covert elements that together comprise a joint action between communicators. The costs and supports of communication must be explained in terms of the communication medium and participants' purposes (Clark and Brennan 1991). Communication as a goal in itself is qualitatively different from communication as part of group work. This differentiates the study of computer-mediated communication from the study of group work, where the processes and structures of communication must sustain higher-level functions of the group like production (e.g., McGrath 1991). In group work, communication pragmatics matters more than semantics and syntax.

We have also argued that CSCW involves mediated, purposeful activity. Collaborative tools, procedures and roles that mediate activity, but also the goal that directs the activity, are key parameters for understanding the types support required for communication. Moreover, the support for communication may need to change as a function of the activity (e.g., inception vs closure) and the kind of the task supported (e.g., group discussion vs decision-making).

Univocal support for knowledge sharing may result in suboptimal performance of decision-making groups. Instead, optimal decision performance needs balanced support for both communication and decision-making, so that the varying needs of the activity can be met. Consider the case of an expert team whose performance depends on thoughtful sharing of role-specific knowledge along with exhaustive assessment the knowledge shared. A system providing initial high visibility for shared versus role-specific knowledge may exacerbate the group bias for shared information (Stasser and Titus 2003). However, a system that provides insufficient visibility to shared knowledge may delay knowledge integration and consensus.

The original concept of common ground as a synthesis of shared knowledge and communication protocols must be extended when we consider *knowledge-intensive* cooperative work. Consider problem-solving teams of specialized experts who must coordinate large information bases containing varied knowledge. The team members build a transactive memory by sharing knowledge about who knows what. In this case, the ability to share strategic knowledge is more important than pooling detailed knowledge.

Model productivity

In future work we will refine the model through empirical studies. In our ongoing laboratory experiments we are comparing the process of building common ground across media and collaborative conditions (Convertino *et al.* 2005; Carroll *et al.* 2007). We will conduct a field trial using our geocollaborative software prototype in support of regional emergency managers working on planning activities. This field study will enable us to focus on processes that are not easily simulated in a laboratory: skill development, trust relationships and community variables. At the same time, it will allow us to validate and extend both our conceptual and empirical models of the common ground process.

From a methodological standpoint, we will continue to refine our reference task for geocollaborative emergency planning. For example, we are investigating the potential for enhancing the task with a corpus of reference data that can be made available to HCI and human factors researchers. This would be part of a general attempt to help these communities recognize their common focus and to cumulate measures and results around reference tasks and data.

From a design standpoint, we argue that a conceptual model annotated with empirical data from the field and the lab can be useful in generating design ideas as well as guiding evaluation of design solutions. Consider, for example, the case of conflicting effects between an increase in common ground on one hand, and suboptimal group decisions due to biases such as anchoring on the other. Without an empirical model it would be difficult to explain why or predict when an optimal form of support for common ground can lead to suboptimal decisions. It is the conceptual model that provides the interpretive lens for suggesting and assessing trade-offs among the variables affecting group process, the activity objective and mediating technology.

References

Arrow, H. and McGrath, J.E. (1995). 'Membership Dynamics in Groups at Work: A Theoretical Framework', in B.M. Staw and L.L. Cummings, (eds), *Research in Organizational Behavior* 17, 373–411.

Arrow, H., McGrath, J.E. and Berdahl, J.L. (2000). *Small Groups as Complex Systems*.

Benyon, D. and Imaz, M. (1999). 'Metaphors and Models: Conceptual Foundations of Representations for Interactive Systems Design', *Human–Computer Interaction* 14, 159–189.

Billman, D. *et al.* (2006). Collaborative intelligence analysis with CACHE: bias reduction and information coverage. *HCIC 2006 Winter Workshop. Collaboration, Cooperation, Coordination*.

Carroll J.M. (ed.) (2003). *HCI Models, Theories, and Frameworks: Toward a Multidisciplinary Science*. (New York: Morgan Kaufmann Publishers).

Carroll J.M. *et al.* (2007). 'Prototyping Collaborative Geospatial Emergency Planning', *Proceedings of ISCRAM 2007*.

Carroll, J.M. *et al.* (2006) 'Awareness and Teamwork in Computer-Supported Collaborations', *Interacting with Computers* 18(1), 21–46.

Clark H.H. (1996). *Using Language*. (Cambridge, UK: Cambridge University Press).

Clark, H.H. and Brennan, S.E. (1991). 'Grounding in Communication', in L.B. Resnick *et al.* (eds), *Perspectives on Socially Shared Cognition*, pp. 127–149. (Washington, DC: American Psychological Association).

Convertino, G. *et al.* (2005). 'A Multiple View Approach to Support Common Ground in Distributed and Synchronous Geo-Collaboration', in *Proceedings of CMV 2005*, pp. 121–132.

Engeström, Y. (1990). *Learning, Working and Imagining. Twelve Studies in Activity Theory*. (Helsinki: Orienta-Konsultit).

Grudin, J. (1989). 'Why Groupware Applications Fail: Problems in Design and Evaluation', *Office: Technology and People* 4(3), 245–264.

Hinsz, V.B. (2001). 'A Groups-as-Information-Processors Perspective for Technological Support of Intellectual Teamwork', in M.D. McNeese *et al.* (eds), *New Trends in Collaborative Activities: Understanding System Dynamics in Complex Settings*, pp. 22–45. (Santa Monica, CA: Human Factors Ergonomics Society).

Kaptelinin, V. (1996). 'Activity Theory: Implications for Human-Computer Interaction', in B.A. Nardi, (ed.), *Context and Consciousness: Activity Theory and Human–Computer Interaction*. (Cambridge, MA: The MIT Press).

Kaptelinin, V. and Nardi, B. (2006). *Acting with Technology: Activity Theory and Interaction Design.* (Cambridge, MA: MIT Press).

Korpela, M. *et al.* (2001). 'Two Times Four Integrative Levels of Analysis: A Framework', in N.L. Russo *et al.* (eds), *Realigning Research and Practice in Information Systems Development. The Social and Organizational Perspective*, pp. 367–377. IFIP TC8/WG8.2 Working Conference, Boise, Idaho, USA.

Kraut, R.E. (2003). 'Applying Social Psychological Theory to the Problems of Group Work', in J.M. Carroll (ed.), *HCI Models, Theories, and Frameworks: Toward a Multidisciplinary Science*, pp. 325–356. (New York: Morgan Kaufmann Publishers).

Letsky, M. *et al.* (eds) *Macrocognition in Teams: Understanding the Mental Processes that Underlie Collaborative Team Activity.* (New York: Elsevier).

McCarthy, J.C. *et al.* (1991). 'Four Generic Communication Tasks which must be Supported in Electronic Conferencing', *ACM/SIGCHI Bulletin* 23, 41–43.

McGrath, J.E. (1984). *Groups: Interaction and Performance*. (Englewood Cliffs, NJ: Prentice-Hall).

McGrath, J.E. (1991). 'Time, Interaction and Performance (TIP): A Theory of Groups', *Small Group Research* 22, 147–174.

McNeese, M.D. *et al.* (eds) (2001). *New Trends in Collaborative Activities: Understanding System Dynamics in Complex Settings* (Santa Monica, CA: Human Factors Ergonomics Society).

Monk, A. *et al.* (1996) *Measures of Process. CSCW Requirements and Evaluation*, pp. 125–139. (Berlin: Springer-Verlag).

Nardi, B.A. (ed.) (1996) *Context and Consciousness: Activity Theory and Human-Computer Interaction.* (Cambridge, MA: The MIT Press).

Resnick, L.B. *et al.* (eds) (1991) *Perspectives on Socially Shared Cognition.* (Washington, DC: American Psychological Association).

Russo N.L. *et al.* (eds) (2001) *Realigning Research and Practice in Information Systems Development. The Social and Organizational Perspective.* IFIP TC8/WG8.2 Working Conference, Boise, Idaho, USA.

Schafer, *et al.* (forthcoming). Emergency Management Planning as Collaborative Community Work.

Sellen, A.J. (1995). 'Remote Conversations: The Effects of Mediating Talk with Technology', *Human–Computer Interaction* 10, 401–444.

Stasser, G and Titus, W. (2003). 'Hidden Profiles: A Brief History', *Psychological Inquiry* 3–4, 302–311.

Chapter 7

Agents as Collaborating Team Members

Abhijit V. Deshmukh, Sara A. McComb and Christian Wernz

Network-centric warfare has emerged as a key defense doctrine in response to a new geopolitical reality of having to respond to multiple, distributed incidents or threats, with an ill-defined or rapidly changing enemy. Transformational teams are a key element of the network-centric doctrine (United States Navy 2004; United States Army 2006). Such teams are characterized as having an *ad-hoc*, agile membership with rotating, heterogeneous members who have distributed, asynchronous relationships. Another key characteristic of the future combat environment is the pervasiveness of massive amounts of information, enabled by large numbers of field-deployed sensors, satellite-based images, network connectivity and enhanced communication channels. Given the abundance of information, geographic and cultural diversity of team members and the dynamic nature of the team and its tasks, transformational team members need ways to augment their decision-making, development of shared understanding and intelligence analysis capabilities.

Software agents and autonomous systems offer the possibility of enhancing these capabilities because of their ability to plan and adapt appropriately in new contexts, learn, move as needed, and work cooperatively with people and other agents (Campbell *et al.* 2000; Cohen and Levesque 1991; Deshmukh *et al.* 2001; Middelkoop and Deshmukh 1999; Tambe *et al.* 1999). A common example of a simple software agent is the Microsoft Office assistant. This agent can observe how you utilize a Microsoft product and provide suggestions for more effective use.

In this chapter, we focus specifically on augmenting the cognitive capabilities of teams. We approach augmentation from a cognitive perspective, because the increased information-processing requirements and dynamic nature of tasks test the limits of human capabilities. Past research has shown that for individuals such limitations can be reduced through augmentation (Schmorrow and McBride 2004; Schmorrow, Stanney, Wilson and Young 2005). Moreover, cognitive augmentation may enhance macrocognitive processes that are essential drivers of team collaboration and overall performance. Macrocognitive processes, as discussed in by Warner, Letsky and Wroblewski (Chapter 2, this volume), include a range of internal and external cognitive activities, from visualizing complex information to developing individual knowledge that is interoperable across team members to developing a set of alternative solutions. As we will discuss, agents can help team members with their macrocognitive processing by creating visualizations of high-dimensional data, maintaining a knowledge repository, and simulating options under consideration. The augmentation possibilities are virtually endless, but the current state of knowledge regarding human–agent interaction in a team domain is underdeveloped. Thus, our purpose here is to suggest a research agenda for determining optimal configurations of human–agent teams.

The chapter begins with a review of the agent augmentation literature, highlighting the limited examination of team-level human–agent collaboration, the limited use of agents as first-class team members, and the lack of research on augmenting macrocognitive processes. We then introduce a conceptual framework for designing and managing effective human–agent teams, including (1) their sociotechnical structure, (2) the team-level augmentation strategies and corresponding characteristics of agents that may be most effective for human–agent collaboration, and (3) the collaboration performance monitoring required to dynamically aid the cognitive processes of teams and their members. Next, we present examples of the types of agents that could be created. In doing so, we highlight several combinations of augmentation strategies and agent characteristics that could assist teams completing different types of collaboration tasks and delineate how these agents enhance various macrocognitive processes. We conclude by discussing our framework's contribution to the study of human–agent teams.

Agent-based cognitive augmentation

Cognitive augmentation through agents refers to supporting human information-processing by means of a computer-based system that has sensing capabilities to monitor the human and its environment and an 'understanding' of how and when to augment the cognitive capabilities of a human. The concept of computer-assisted, cognitive augmentation dates back to the 1960s when Licklider (1960) and Engelbart (1963) envisioned a human–computer symbiosis. They proposed to extend the capabilities of the human brain by computers to achieve a system with superior information-processing capabilities. This visionary concept inspired research in human factors engineering (Schmorrow, Stanney, Wilson and Young 2005) and computer science, particularly in the area of artificial intelligence (Wooldridge and Jennings 1995) where agents represent the building blocks for augmentation.

Agents, in this context, are defined as autonomous software units that combine inherent knowledge and goals with information about the environment and take action to support other agents or humans (Bradshaw 1997; Negroponte 1997). Middleton (2002) classifies four broad types of agent support systems:

1. character-based agents, that are designed after real-world characters, such as a police dog, and can provide an intuitive interaction right from the start
2. social agents, which acquire their information by interacting with other agents and fusing their information
3. learning agents, which either monitor the human, ask for feedback or are directly re-programmed by the user, and
4. agents with a user model, which is either behavioral or knowledge-based.

The categories are non-exclusive and a particular agent can be a combination of these four categories.

In our review, we focus on the research that includes a human component in conjunction with an agent or system of agents. Much has been written about multi-agent systems that we do not review, but may be of interest as background information, including an introduction to multi-agent systems (Weiss 1999), a comprehensive survey of the theory and practice of agent systems (Wooldridge and Jennings 1995), practical and industrial application examples of such systems (Parunak 1998), and design guidelines for agent systems (Middelkoop and Deshmukh 1999).

Individual human augmentation

To provide humans with the appropriate cognitive augmentation, agents have to know the human's current cognitive state by assessing, for example, workload or comprehension level (Dorneich, Ververs, Mathan and Whitlow 2005). Through neural measures (e.g., EEG), behavioral measures (e.g., reaction time) and physiological measures (e.g., heart rate) a proxy for the cognitive state can be determined. For a comprehensive list of measures see Bruemmer, Marble and Dudenhoeffer (2002) and Lyons (2002). One of the challenges in an augmentation system is to translate the human's actual cognitive state into augmentation requirements. To that end, Plano and Blasch (2003) distinguish between task analysis and cognitive task analysis. Task analysis interprets the physical activities of the user interacting with the system, whereas the *cognitive* task analysis uses a mental model of the operator to interpret the user's action and infer the user's thoughts. Thus, cognitive task analysis assists in determining the expected cognitive load for a given task. This expected load, when compared to actual cognitive load, identifies when and how much augmentation is needed.

Once the augmentation needs are established, agents can augment the human's information-processing capabilities appropriately. The goal is to achieve cognitive congeniality (Kirsh 1996) and not to overload the human. Agents can reduce cognitive load by performing functions such as:

1. presenting task-relevant information to optimize sensory processing
2. sequencing and pacing tasks to minimize cognitive load, and
3. delegating tasks to reduce number and cost of mental computations (Schmorrow, Stanney, Wilson and Young 2005).

Cognitive augmentation of individual humans can be employed in many different general application domains. Schmorrow, Stanney, Wilson and Young (2005) consider operational, educational and clinical domains. In the operational domain, a wide variety of types of agents and tasks have been examined. For example, Plano and Blasch (2003) created a data-fusion system that provides humans with the right information at the right time to enhance their performance on search tasks. Moreover, through their target tracking interface design, which includes displaying information and audio warning, humans can better filter the information provided by the agents. In the education domain, agents can play an integral part in scenario-based training systems. For instance, route planning agents (Payne, Lenox, Hahn, Sycara and Lewis 2000) and information retrieval/look-up agents (Harper,

Gertner and Van Guilder 2004) may be used to vary the training exercises based on decisions made by the trainees. In the clinical domain, patients with attention-deficit hyperactivity disorder (ADHD) can be helped by agents that reinforce good paying-attention behavior (Schmorrow *et al.* 2005). A fourth domain for consideration is the military domain where significant advances in human–agent collaboration have been made. Honeywell's human–agent cognitive system provides an illustrative example (Dorneich *et al.* 2005). This system supports soldiers in fast decision- making situations. Agents monitor the state of humans through sensors and determine the human's engagement, arousal stress and comprehension level. Agents can take proactive actions to mitigate cognitive overloads.

Agents that can decide how and to what degree to augment the human have to be designed. Moreover, they have to be able to decide when and who to augment. Logically, tasks should be assigned to humans and agents based on which entity can best perform the task in question. For instance, agents may be better suited to evaluating different options and searching for an optimal solution because they can search systematically through a solution space whereas humans jump around and are guided by their prior knowledge, experience and intuition (Burstein and McDermott 1996). Thus, agents may be the best entity to conduct such evaluation functions.

Human–agent teams

Research in cognitive augmentation has focused mostly on individual augmentation and on multi-agent teams. Furthermore, to our knowledge, none has focused directly on supporting macrocognition in teams. Burstein and McDermott (1996) have compiled a list of challenges in artificial intelligence, and the necessary research projects corresponding to those challenges, to enable human–agent collaborations. Their list addresses aspects of specialization, information and communication, but it does not articulate a difference in team vs individual human–agent augmentation. Alternatively, Xu and Volz (2003) present a shared mental model for multi-agent systems, but neglect to consider the human role in the developed system. This theme is common in multi-agent system research as numerous examples of these systems exist that consider teamwork without introducing a human component (e.g., Jones *et al.* 1999; Tambe 1997; Tidhar, Heinze and Selvestrel 1999). For example, STEAM – Shell for TEAMwork (Tambe 1997) builds on the *joint intentions* concept (Cohen and Levesque 1991; Levesque, Cohen and Nunes 1990) and hierarchical *Shared-plans* model (Grosz and Kraus 1996) to provide an environment to deploy multi-agent collaborating teams.

A limited amount of research has focused on human–agent collaboration. For example, Payne *et al.* (2000) devised agents that support time-critical team planning tasks. In their application, three military commanders have to lead their platoons such that all platoons meet at one point at a specified time. Both humans and agents can initiate communication, monitor their respective performance, and execute tasks. Sycara and Lewis (2004) provide a second example. They report on two experiments that demonstrate how agents may be useful in assisting their human counterparts' planning and communication activities, particularly selecting and directing communication. Furthermore, they ascertain that humans find agents more useful when they have been designed with the humans' needs (vs the agents' capabilities) in mind. A third, and final, example is the research of Siebra and Tate (2005) examining how human teams use agent assistants during the planning process. In this work, they describe an integrated, constrained-based framework for collaborative activity-oriented planning, where customized agents can be deployed at any decision level to handle activity planning tasks. In addition to these specific examples, several human–agent systems have been advanced that facilitate collaboration such as Brahms (Clancey, Sachs, Sierhuis and van Hoof 1998), KAoS (Bradshaw, Feltovich *et al.* 2004a), and Collaboration Management Agent (Allen and Ferguson 2002). These systems' specific features are summarized in Bradshaw, Acquisti *et al.* (2004b).

The majority of agents that have been developed for interaction with humans play a subordinate role and make team contributions only under direct human supervision (Burstein, Mulvehill and Deutsch 1999). In order to expand the role of agents in mixed teams, agents must be able to understand and predict human behavior (Bruemmer *et al.* 2002). In other words, they need leadership capabilities, autonomy and proactivity. Incorporating these types of agent characteristics are among the greatest challenges facing the field, but they are essential for effective human–agent systems (Dickinson 1998). As mentioned, agents have typically been subordinate. Their role, however, can span a spectrum from subordinate to equal team-player to leader (Bruemmer *et al.* 2002; Sheridan 2006). To determine appropriately which leadership characteristics along this spectrum they should employ to best support the human entities on the team, agents must have autonomy. Moreover, depending on the situation within the team, agents may have to shift roles dynamically. To facilitate such behavior, Scerri, Pynadath and Tambe (2002) developed an adjustable autonomy model that determines an optimal transfer-of-control strategy. Agent proactivity, another necessary characteristic for effective human–agent collaboration, has received attention recently. Yen and colleagues (2001) devised CAST (Collaborative Agents for Simulating Teamwork), which anticipates information needs based on a shared mental model in multi-agent teams. The CAST model has been extended to include the dynamic nature of anticipated information needs (Fan, Wang, Sun, Yen and Volz 2006). Also during this period, Fan, Yen and Volz (2005) developed a framework for proactive information delivery. This framework combines their conceptualization of information needs with axioms for helpful behavior in large teams to produce a model for agent teams that provide information proactively.

To this point, we have focused on the agent side of the human–agent team. One essential element from the human perspective is trust. Some researchers go as far as seeing trust as one of the greatest challenges for the adoption of autonomous agents (e.g., Bruemmer *et al.* 2002; Middleton 2002). Humans must accept agents as team players in order for the human–agent team to become a reality. Indeed, the human must be able to rely on the information presented by the agent so that the agent's contribution can be used as extended sensory information (Plano and Blasch 2003). Robust agent designs and systematic research directed at understanding how best to augment human team behavior are two critical elements needed to help overcome distrust. As the reliability and usability of agents increase, we can anticipate that levels of trust in agent team members should increase accordingly.

This review highlights the need for systematic research on human–agent collaboration that focuses on how to embed agents in teams as fully functional team members. The extant research we have just described provides integral pieces necessary for accomplishing this objective. Several critical gaps, however, must be addressed. First, the research has not focused on designing agents that can function as team members by supporting critical processes such as macrocognitive processes. Second, no design methodology exists for ascertaining optimal human–agent team configurations. Third, research on tracking team progress and dynamically changing agent support based on that progress is needed.

Designing effective human–agent teams

To address the gaps identified in the literature review, we present a conceptual framework for designing effective human–agent teams (see Figure 7.1). Three salient features of this framework deserve mention. First, the framework considers human–agent teams as sociotechnical systems (STS), where humans, agents and the environment are intrinsic resources that can be used to accomplish a complex task (Emery and Trist 1960). Second, the framework is driven by a focus on using agent team members to augment the capabilities of the team as a whole. Third, the components necessary to facilitate enhanced team performance through human–agent team processes are embedded in a collaboration environment that is continuously monitored. This environment constitutes elements such as the collaboration situation parameters (i.e., time pressure, information and knowledge uncertainty, quantity/dynamics of information – see Letsky, Warner and Salas, the introduction to this volume), the team's task, and the human composition of the team (i.e., diversity, co-location), etc.

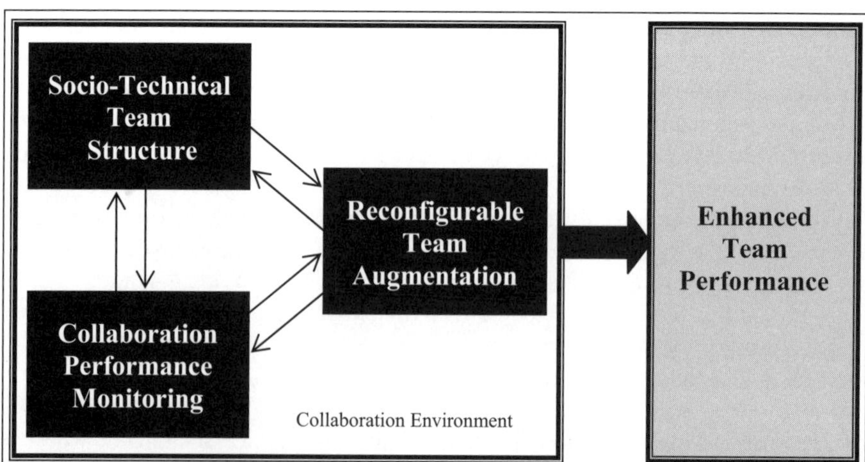

Figure 7.1 Conceptual framework for designing effective human–agent teams

An extensive team literature exists that describes the human dimension of our system. We, therefore, focus on the human–agent interactions in the system by outlining the research needed to help us understand (1) the cognitive and coordination mechanisms used by human–agent teams, composed of rotating, heterogeneous members (humans and software agents) who have distributed, asynchronous relationships, as they collaborate to make decisions, develop a shared understanding, and/or conduct intelligence analysis and (2) ways in which these mechanisms can be augmented dynamically. Once completed, this research will advance a theory of human–agent reconfigurable collaboration and a set of team augmentation guidelines. To further articulate the research agenda, we highlight the basic research required in the following three areas: (1) sociotechnical team structure, (2) reconfigurable team augmentation, and (3) collaboration performance monitoring.

Sociotechnical team structure

Basic research is needed to determine *how* to design and optimize a human–agent STS. STS structure is a function of the assignment given to the team, environmental–domain considerations and the characteristics of the players in the system (both humans and agents). The challenge facing scholars is identifying how these inputs can be transformed into an optimal team structure and a set of recommended processing strategies. Thus, the overall goal of this research area is to develop a set of models and guidelines that can be used to rapidly design network-centric transformational teams of humans and agents for specific tasks.

One possible way to accomplish this goal is to develop models of the human–agent STS by exploiting the extant literature on economic models of teams (Levchuk, Yu, Levchuk and Pattipati 2004; Marschak and Radner 1972; Mesarovic, Mako and Takahara 1970; Rusmevichientong and Van Roy 2003; Schneeweiss 2003). The STS team structure problem can be formulated as an expected utility maximization problem, where the STS structure and coordination rules are decision variables; task coordination and performance requirements represent the constraints; and utility is a function of the efficiency of processing the collaborative task and associated costs (information, delay and decision). Cognitive task analysis can be used to determine the task structure, cognitive requirements of tasks and coordination necessary to perform the tasks. The resulting model should describe the functional representation needed to staff the human side of the team. Moreover, it should identify where agents can most effectively assist their human counterparts. We envision the STS structure to have agents at different levels:

1. supporting the team as a whole
2. providing support for subgroups of humans, and
3. helping individual team members perform their tasks.

To identify the appropriate processing strategies for mixed-initiative systems, the work by Levchuk and colleagues (2004) can be used to formulate strategies for mission monitoring, information-processing and command processing. Their model can be enhanced, for example, by developing multiscale decision-making techniques that can cope with not only organizational, but also geographical or temporal scales needed to model network-centric human–agent teams (Krothapalli and Deshmukh 1999) and by incorporating a corresponding economic incentive scheme for the human–agent interaction that accounts for both rationality and bounded rationality issues (Kahneman 1994; Simon 1997).

Reconfigurable team augmentation

We now address the issues related to developing a fundamental understanding of *what* type of augmentation will most enhance team cognition and collaboration in a manner that improves overall team performance. Several types of augmentation actions have been postulated for mixed-initiative systems, ranging from augmented sensors, to individuals, to teams and organizations (Schmorrow and Krause 2004). Here, we focus on team-level augmentation.

The agents have to decide which action to take based on the state of the system, their own policy demands, costs of the actions, anticipated impact on performance and activities of other agents. Typical agent assistance could be in the form of augmented capacity, knowledge sharing or task reallocation. Augmented capacity refers to enhancing human decision-maker's capacity to communicate, observe/scan, or complete cognitive–motor task processing. Knowledge sharing encompasses many activities such as knowledge–information transfer among STS members that facilitates knowledge interoperability and enables and ensures shared team understanding. Task reallocation refers to realigning task assignments among members to level the workload. We envision that assistance will change dynamically based on workload changes; thereby requiring reconfigurable agents. This research should result in guidelines for supporting key cognitive processes with agents, thereby enhancing team collaboration and, ultimately, performance.

The type of augmentation support that an agent can provide depends on the agent's characteristics. Several dimensions of agent characteristics have been discussed in the literature, such as autonomy, social ability by means of agent-communication language, and proactiveness (Wooldridge and Jennings 1995). These characteristics determine the capabilities the agents bring to the team. In this section, we focus on six characteristics that are relevant for incorporating agents in mixed-initiative teams, where both software agents and humans collaborate to improve the overall team performance.

1. *Autonomy: fixed/adjustable* Autonomy is a fundamental characteristic of an agent that relates to its ability to exhibit goal-directed behavior, where the goals are inherent to the agent itself and not prescribed by outside entities (Bradshaw 1997). A software agent that does not have any autonomy may not be much different than a computer program that is executed by the user when needed. However, by incorporating the ability to pursue its own goals within the constraints placed on its actions, the computer programs start to interact in a more meaningful and richer manner with humans. Hence, by definition agents have some degree of autonomy (Franklin and Grasser 1996). The level of autonomy that an agent has may be fixed throughout its

lifetime or may be adjustable based on the needs of the situation. One can view adjustable autonomy as a mechanism for the appropriate trade-off of authority and responsibility between various team members (human or agent) during the execution of a task. Automatic adjustments of autonomy can be made either by the agents themselves or by the humans on the team. Agent-initiated adjustments in autonomy might typically be required when the current configuration of human–agent team members has led to, or is likely to lead to, failure and when no set of competent, authorized humans are available to make adjustments themselves (Barber and Martin 1999). The mechanisms for adjusting autonomy can be coarse-grained, such as mode setting, or fine-grained, based on specific policies that provide guidelines for limiting agent actions under specific situations.

2. *Reactivity: reactive/proactive* The level of reactivity assigned to the agents determines how the agents initiate an augmentation action or are triggered to make a decision. We can consider agents along a continuum of reactive to proactive response capabilities. Purely reactive agents respond to the current state of the entity they are supporting or to external stimuli, such as a request for assistance from humans or change in environmental variables. Agent reactive capabilities can be based on inputs from the human team members or sensors that provide information about the current state of individual or collaboration activity. Proactivity refers the agent's ability to take initiatives, make conscious decisions, and take positive actions to achieve chosen goals (Fan *et al.* 2005). The proactive capabilities of agents can be based on predictive models of team task performance that the agents use to decide on appropriate type and level of augmentation (Georgeff and Lansky 1987). These predictive models allow the agents to anticipate when individual team members may be expected to exceed their cognitive capacity threshold or communication capabilities or if a task needs to be reassigned to an agent.
3. *Planning horizon: myopic/look-ahead* The agent decision-making planning horizon can be myopic, where an optimal augmentation action is chosen considering only the current state of the system, or look-ahead, where the optimal decision policy is chosen based on implications along the entire task horizon. Single-stage optimal-action selection models may be used by agents to determine the optimal augmentation actions and level in the myopic planning horizon case. In the look-ahead planning models, the agent decision policies can be formulated as Markov Decision Processes (MDP) (Puterman 1994). The reward function of the MDP depends on the efficiency of the overall task completion and the cost of augmentation, and may be additive over the task horizon. The optimal augmentation policy, which belongs to a set of actions representing all possible augmentation choices, minimizes the objective's expected value over the entire task horizon. Given the computational complexity of MDP formulations, approximations and heuristic approaches are often necessary to allow these look-ahead models to be implemented for agent augmentation with real-time performance requirements.
4. *Learning: enabled/not enabled* Agent decision policies can be stationary, wherein they do not vary over time, or non-stationary, where the optimal

augmentation policies evolve over time based on observations of the system. The evolution of decision policies based on observations can also be referred to as learning. In the situation when the decision policies are stationary, we can argue that the agent does not learn from its experience. The evolution of optimal augmentation strategies occurs due to two non-exclusive reasons:

(a) the agents could learn about the impact of their past augmentation actions and adjust their strategies accordingly, and
(b) agents in a loosely-coupled system could anticipate other agents' and humans' actions while formulating their own optimal strategies.

Different learning strategies for agent augmentation policies can be used, ranging from smoothing functions to reinforcement and Q learning methods (Barto and Mahadevan 2003; Greenwald, Shenker and Freidman 2001) that will allow the agents to infer causal relationships between their actions and team performance. A key concern in learning models is the robustness of the resulting strategies.

5. *Position in team hierarchy: leader/follower* In mixed initiative teams, agents can function at various levels of the team hierarchy (Burstein and McDermott 1996). Typically, one envisions agents in the follower role. However, agents can play the role of a team leader when the task to be performed is too complex for a single human being to manage or when the leader of the team is unavailable. Agents can move between these roles based on the state of the team and the task they are performing.
6. *Level of support: team/individual* This aspect refers to the entity that is augmented or supported by the agents. The agents may support individual team members or the team itself. In augmenting individuals, agents can function as personal assistants or monitor the team member's cognitive state. Team-level augmentation support can be provided in terms of enhancing the ability of the group to complete the task in an efficient manner. For example, a team-level support agent can assist in sharing of appropriate knowledge among team members to enable planning tasks. Team-level agents can also monitor overall collaboration activity to decide when augmentation is needed.

Collaboration performance monitoring

Research activity is needed here to develop a fundamental understanding of *when* and *who* among the team would benefit from augmented capabilities. We suggest designing a monitoring and diagnostic system based on individual- and team-level metrics that will indicate when the human decision-maker needs assistance and what types of assistance would be most beneficial for ongoing team processing and overall team performance. Thus, the overall goal of this research activity should focus on creating a collaboration performance monitoring system that can be embedded into the computational architecture. This real-time system needs the ability to facilitate dynamically adaptive augmentation based on the collaboration environment's status.

Several appropriate metrics, including indicators such as biometrics and role overload assessment, have been developed to determine when individuals can benefit from assistance (see for example, Schmorrow and Kruse 2004). Research is needed to establish the appropriate team-level metrics. Here we discuss potential process and outcome measures that may be useful. We begin with examples of direct and surrogate measures of cognitive processes. Mental model convergence can be used as a direct measure of cognitive processes by ascertaining how much shared understanding exists among team members. Using short surveys and calculating a team convergence score and individual member contributions to that score (see for example, McComb and Vozdolska 2007), we can highlight who on the team has dissimilar views. Surrogate measures of team-level cognition can also be examined, such as conflict and resource mapping. Conflict among team members may be beneficial or detrimental, depending on the type and amount of conflict transpiring (Jehn and Mannix 2001). Limited amounts of task and process conflict allow teams to exchange ideas and be creative. Any interpersonal conflict and high amounts of task and process conflict debilitate a team's ability to work on their assignment, thereby adding to the cognitive load associated with teamwork. Resource mapping onto the task demands will identify areas where expertise is appropriately (or inappropriately) assigned to tasks (Blackburn, Furst and Rosen 2003). If the expertise is not assigned effectively, the cognitive demands placed on the team to accomplish the task will be greater than if the appropriate personnel were assigned. We focused our examples on assessing cognitive processes, but a performance monitoring tool could be expanded to include a myriad of team process variables as well (e.g., coordination mechanisms).

While process measures may provide the best opportunities to identify when the team needs support and/or redirection, outcome measures can also assist with this endeavor. Blackburn and colleagues (2003) identified several outcome measures that organizations routinely use to assess the output of the team. They are quality (e.g., the ability of the proposed solution to meet mission objectives), quantity (e.g., the amount of membership turnover), creativity (e.g., the number of novel concepts generated), cost (e.g., the additional monies necessary to implement changes in mission scope), and timeliness of deliverables (e.g., the progress to schedule). Performance assessment may also be focused on perceptions, such as satisfaction with the team and degree of cohesion among team members. Ensuring satisfaction and cohesion will enhance the teams' abilities to collaborate effectively on the current project and on future endeavors (Hackman and Oldham 1980). These outcome measures should be incorporated into the ongoing monitoring of the team so that any necessary midcourse corrections can be implemented to curtail underperformance when it is identified.

Augmenting team macrocognitive processes

Macrocognitive processes are one of the primary means by which transformational teams complete their objectives. Human capacity for these processes is inherently limited. Thus, introducing agents as fully functioning team members may increase a team's cognitive capacity. In Figure 7.2, we provide examples of agent support that may enhance macrocognitive processes. Our examples depict a wide array of agents that showcase the three augmentation strategies (i.e., augmented capacity, task reallocation and knowledge sharing) and embody different combinations of agent characteristics (i.e., reactivity, autonomy, planning horizon, learning, level and hierarchical position). Moreover, we focus on aiding human team members working on three types of collaboration tasks:

1. team decision-making/course of action selection
2. development of shared understanding, and
3. intelligence analysis (team data processing).

These example agents represent a spectrum of complexity. For instance, one of the least complex agents is the agent assisting in simulating outcomes of options available to human decision-makers. This agent's role on the team would be performing what-if analyses to ascertain the probable results of various options being considered by the human members of the team. It augments capacity because it can execute several instances of simulations simultaneously and keep track of the solutions, which would be unfeasible for humans. This particular agent is unsophisticated in that it completes a specific task when it is put into service by a human team member. In other words, it is a follower that is reactive, does not have any autonomy, focuses only on the task given (i.e., it is myopic), and cannot learn as it simulates. This particular agent could be more sophisticated if, for example, it was programmed with learning capabilities. These capabilities may, perhaps, allow it to learn what strategies are best suited for the extant scenario and to use this information to tailor its simulation parameters.

At the other end of the spectrum, an agent designed to enable and enforce the explicit, agreed-upon protocols governing interactions is highly sophisticated. This agent needs to monitor interactions among multiple team members over a period of time to determine if the team members are interacting in an appropriate and meaningful manner. In order to achieve the task of enabling interaction, the agent needs to facilitate knowledge interoperability among team members. As the shared understanding among team members increases, and their corresponding mental models converge, the nature of interactions among them will change. Monitoring and enabling protocols, especially when the content of the interactions is changing over time, requires the agents to be proactive and to look ahead in terms of the expected actions by team members. The agent needs to learn about the team members' common knowledge and shared understanding by observing the interactions among them over time. Furthermore, the agent needs to assume a leadership role in governing the flow of information at the team level.

Our purpose in presenting these examples of agent support is to demonstrate how various agent designs, from simple to complex, can augment macrocognition across collaboration task types. This set of nine agents, presented in Figure 7.2, is by no means an exhaustive list of potential agents that can be introduced into transformational teams. Rather, they exemplify the myriad of ways in which agents can augment macrocognitive processes and illustrate the range of sophistication that can be designed into these agent team members.

Agent Characteristics:
Autonomy: A= Adjustable, FX=Fixed
Reactivity: R=Reactive, P=Proactive
Planning Horizon: MY=Myopic, LA=Look Ahead
Learning: LE=Enabled, LNE=Not Enabled
Hierarchical Position: L=Leader, F=Follower
Level: T=Team Level, I=Individual Level

Team Task Types	Examples of Agent Support	Macrocognitive Processes: Individual mental model construction	Individual task knowledge development	Visualization/representation of meaning	Knowledge interoperability	Iterative information collection	Team shared understanding	Develop solution alternatives	Convergence of mental models	Team agreement on a solution	Team negotiation of solution alternatives	Team pattern recognition
Team Decision Making/ Course of Action Selection	*Simulating options under consideration* Augmentation Type: Augmented Capacity Agent Characteristics: FX, R, MY, LNE, F, T/I		✓		✓	✓		✓				
	Preparing/updating a dynamic schedule Augmentation Type: Task Reallocation Agent Characteristics: A, P, LA, LE, F, T						✓				✓	
	Notifying all team members when goals are achieved Augmentation Type: Knowledge Sharing Agent Characteristics: FX, R, MY, LNE, F, T		✓				✓	✓	✓	✓		
Development of Shared Understanding	*Creating visualizations of high dimensional data* Augmentation Type: Augmented Capacity Agent Characteristics: A, P, MY, LE, F, T	✓	✓	✓	✓	✓	✓	✓	✓	✓	✓	✓
	Maintaining knowledge repository Augmentation Type: Task Reallocation Agent Characteristics: FX, R, MY, LNE, F, T/I	✓	✓		✓	✓	✓		✓			
	Enforcing explicit, agreed upon protocols governing interactions Augmentation Type: Knowledge Sharing Agent Characteristics: A, P, LA, LE, L, T				✓		✓		✓			
Intelligence Analysis (Team Data Processing)	*Mining intelligence data from distributed sources to identify correlations* Augmentation Type: Augmented Capacity Agent Characteristics: A, P, MY, LE, L/F, T/I		✓			✓	✓	✓				✓
	Monitoring cognitive states and (re)negotiating responsibility for pre-selected activities Augmentation Type: Task Reallocation Agent Characteristics: A, P, LA, LE, L, T/I		✓					✓			✓	
	Maintaining up-to-date information in the collaboration space Augmentation Type: Knowledge Sharing Agent Characteristics: FX, R, LA, LNE, L, T	✓	✓	✓	✓	✓	✓	✓	✓			✓

Figure 7.2 Examples of agent support

Benefits of agents as team members

The basis of the approach we have presented is that agents can function as team players and achieve parity with their human counterparts for task planning and execution in transformational teams. Our focus has been on demonstrating how macrocognitive processes can be augmented (see Figure 7.2). Two additional benefits may result from the incorporation of agents into teams. First, these agents may enhance the team's overall performance. These enhancements may be as seemingly simple as utilizing an agent to manage a team's dynamic schedule, thereby helping the team to meet its timing goals. Also, an agent team member may uncover unexpected connections through data mining that may help the team to create unique alternative solutions that would not have been fathomable without the agent's results.

The second benefit is that agents may provide essential elements necessary for team self-management, as we will demonstrate via the example agents already described in Figure 7.2. Hackman (2002) identified five team design/structure elements necessary for effective self-management. The first element is to be a *real team*. Real teams have an articulated task, clear boundaries, explicit authority and stability. Agents functioning as knowledge repositories can help improve stability by keeping information/knowledge available even when team membership changes, which may enhance a team's ability to be a real team.

Second, a *compelling direction* energizes the team, orients their attention/action and engages members' talents appropriately. Agents can help keep the team energized by tracking their forward progress and reporting their goal achievement. Also, they can keep the human team members oriented on their task by creating visualizations of high-dimensional data, thereby keeping them focused on the big picture instead of lost in details that may be cognitively difficult to organize. The third element is *enabling structure*, which includes a clear work design, established norms and appropriate staffing. Agents enforcing the explicit, agreed-upon protocols governing interactions, not only exist as part of the enabling structure, they extend it to include monitoring human team member behavior to ensure that the norms are respected and followed.

Fourth, *supportive contexts* have team reward systems, up-to-date information, educational opportunities and adequate resources. By using agent team members to keep information updated in the collaboration space, all team members have access to current conditions. Finally, *expert coaches* help members focus their efforts equitably, devise appropriate performance strategies, and align individual skills/knowledge with requisite work. Agents monitoring the cognitive load of the human team members can ensure that individual team members do not exceed their maximum cognitive loads. In cases where an individual is approaching their maximum, the agent can reallocate the task(s) to other qualified team members with available cognitive capacity.

Conclusions

Augmenting human cognitive capabilities is not a new phenomenon. Indeed, an example of a commonly seen natural augmentation is a police dog supporting a human partner (Bruemmer *et al.* 2002). The dog augments the human's cognitive capabilities through sniffing. The dog complements its human partner and together they are better at finding drugs and following tracks. The dog also augments the human's physical capabilities through its ability to reach human-inaccessible areas and confront potential threats. Artificial augmentation is also becoming more common place. For instance, robots are routinely used by first responders to enter burning buildings, examine potential bombs, explore unfamiliar terrain, or conduct any other function that may be life-threatening to humans. As our acceptance of these assistants grows and our understanding of how best to assist increases, we envision a day when agents become fully functional members of self-managing teams.

To realize our vision, systematic research is needed to answer questions such as:

- What agent characteristics are most appropriate for various types of augmentation?
- Are different augmentation approaches and agent configurations needed for different collaboration tasks?
- How does the collaboration environment (e.g., distributed vs co-located) in which the team is embedded impact the effectiveness of augmented support?
- How does embedding agents as team players impact team processes and performance?
- Do agents enhance a team's ability to self-manage?

In this chapter, we presented a framework to guide the basic research needed to answer these types of questions. This framework was constructed by organizing and extending the extant literature on human–agent interaction and approaching augmentation from a team cognition perspective. Salient features of this framework are integrating agents as key members of sociotechnical teams, emphasizing team-level augmentation, and incorporating processes to identify and enable different augmentation actions based on the collaboration environment's current state. This holistic view of human–agent collaboration makes our framework a novel approach to this increasingly important body of knowledge.

As the network-centric doctrine is widely deployed in defense and other sectors of society, use of mixed-initiative teams comprised of humans and agents is going to be a critical factor in ensuring successful completion of complex tasks. The framework presented in this chapter provides a foundation for designing and researching such human–agent teams, in which agents play a critical role in augmenting the cognitive capabilities of teams.

References

Allen, J. and Ferguson, G. (2002). 'Human–Machine Collaborative Planning', *Proceedings of the Third International NASA Workshop on Planning and Scheduling for Space*, pp. 1–10. (Washington, DC: NASA).

Barber, K.S. and Martin, C.E. (1999). 'Agent Autonomy: Specification, Measurement, and Dynamic Adjustment', *Proceedings of the Autonomy Control Software Workshop*, pp. 8–15. (New York: ACM).

Barto, A.G. and Mahadevan, S. (2003). 'Recent Advances in Hierarchical Reinforcement Learning', *Discrete Event Dynamic Systems* 13(4), 341–379.

Blackburn, R., Furst, S. and Rosen, B. (2003). 'Building a Winning Virtual Team', in C.B. Gibson and S.G. Cohen, (eds), *Virtual Teams That Work*, pp. 95–120. (San Francisco, CA: Jossey-Bass).

Bradshaw, J.M. (1997). 'An Introduction to Software Agents', in *Software Agents*, pp. 3–46. (Menlo Park, CA: AAAI Press).

Bradshaw, J.M., Acquisti, A., Allen, J., Breedy, M.R., Bunch, L., Chambers, N., *et al.* (2004). 'Teamwork-Centered Autonomy for Extended Human–Agent Interaction in Space Applications', in *AAAI 2004 Spring Symposium*, pp. 1–5. (Menlo Park, CA: AAAI Press).

Bradshaw, J.M., Feltovich, P., Jung, H., Kulkarni, S., Taysom, W. and Uszok, A. (2004). 'Dimensions of Adjustable Autonomy and Mixed-Initiative Interaction', *Agents and Computational Autonomy: Potential, risks, and Solutions. Lecture Notes in Computer Science*, 2969, 17–39.

Bruemmer, D.J., Marble, J.L. and Dudenhoeffer, D.D. (2002). 'Mutual Initiative in Human–Machine Teams', *Proceedings of the 2002 IEEE 7th Conference on Human Factors and Power Plants*, pp. 7–22. (Piscataway, NJ: IEEE).

Burstein, M. and McDermott, D. (1996). 'Issues in the Development of Human–Computer Mixed-Initiative Planning Systems', in B. Gorayska and J.L. Mey, (eds), *Cognitive Technology: In Search of a Human Interface*, pp. 285–303. (New York: Elsevier).

Burstein, M., Mulvehill, A.M. and Deutsch, S. (1999). 'An Approach to Mixed-Initiative Management of Heterogeneous Software Agent Teams', *Proceedings of the 32nd Annual Hawaii International Conference on System Science, HICSS–32*, pp. 1–10. (Piscataway, NJ: IEEE).

Campbell, K.C., Cooper, W.W., Greenbaum, D.P. and Wojcik, L.A. (2000). 'Modeling Distributed Human Decision-Making in Traffic Flow Management Operation', *Proceedings of the 3rd USA/Europe Air Traffic Management RandD Seminar*, pp. 1–9. (Bedford, MA: MITRE).

Clancey, W.J., Sachs, P., Sierhuis, M. and van Hoof, R. (1998). 'Brahms: Simulating Proactive for Work System Design', *International Journal of Human–Computer Studies* 49, 831–865.

Cohen, P.R. and Levesque, H.J. (1991). 'Teamwork', *Noûs* 25(4), 487–512.

Deshmukh, A., Middelkoop, T., Krothapalli, A., Shields, W., Chandra, N. and Smith, C. (2001). 'Multiagent Design Architecture for Intelligent Synthesis Environment', *AIAA Journal of Aircraft* 38(2), 215–223.

Dickinson, I. (1998). *Human Agent Communication, Hewlett-Packard Laboratories Technical Report, HPL-98-130*. Hewlett-Packard Laboratories. (Palo Alto, CA: HP Labs).

Dorneich, M.C., Ververs, P.M., Mathan, S. and Whitlow, S.D. (2005). 'A Joint Human–Automation Cognitive System to Support Rapid Decision-Making in Hostile Environments', *IEEE International Conference on Systems, Man and Cybernetics*, pp. 2390–2395. (Piscataway, NJ: IEEE).

Emery, F. E. and Trist, E. L. (1960). 'Socio-technical Systems', in C.W. Churchman and M. Verhulst, (eds), *Management Science Models and Techniques*, vol. 2, pp. 83–97. (London: Pergamon Press).

Engelbart, D.C. (1963). 'A Conceptual Framework for the Augmentation of Man's Intellect', in P.W. Howerton, (ed.), *Vistas in Information Handling*, pp. 36–65. (London: Spartan Books).

Fan, X., Wang, R., Sun, S., Yen, J. and Volz, R.A. (2006). 'Context-Centric Needs Anticipation Using Information Needs Graphs', *Applied Intelligence* 24(1), 75–89.

Fan, X., Yen, J. and Volz, R.A. (2005). 'A Theoretical Framework on Proactive Information Exchange in Agent Teamwork', *Artificial Intelligence* 169(1), 23–97.

Franklin, S. and Graesser, A. (1996). 'Is it an Agent, or just a Program?: A Taxonomy for Autonomous Agents', *Lecture Notes In Computer Science* 1193, 21–35.

Georgeff, M.P. and Lansky, A.L. (1987). 'Reactive Reasoning and Planning', *Proceedings of the Sixth National Conference on Artificial Intelligence*, pp. 677–682. (Menlo Park, CA: AAAI Press).

Greenwald, A., Shenker, S. and Friedman, E. (2001). 'Learning in Network Contexts: Experimental Results from Simulations', *Journal of Games and Economic Behavior* 35, 80–123.

Grosz, B.J. and Kraus, S. (1996). 'Collaborative Plans for Complex Group Action', *Artificial Intelligence* 86(2), 269–357.

Hackman, J.R. (2002). *Leading Teams: Setting the Stage for Great Performances* (Boston, MA: Harvard Business School Press).

Hackman, J. R. and Oldham, G. R. (1980). *Organization Development* (Reading, MA: Addison-Wesley Publishing).

Harper, L.D., Gertner, A.S. and Van Guilder, J.A. (2004). 'Perceptive Assistive Agents in Team Spaces', *Proceedings of the 9th International Conference on Intelligent User Interface*, pp. 253–255. (New York: ACM Press).

Jehn, K. and Mannix, E. A. (2001). 'The Dynamic Nature of Conflict: A Longitudinal Study of Intragroup Conflict and Group Performance', *Academy of Management Journal* 44(2), 238–251.

Jones, R.M., Laird, J.E., Nielsen, P.E., Coulter, K.J., Kenny, P., Koss, F.V. *et al.* (1999). 'Automated Intelligent Pilots for Combat Flight Simulation', *AI Magazine* 20(1), 27–41.

Kahneman, D. (1994). 'New Challenges to the Rationality Assumption', *Journal of Institutional and Theoretical Economics* 150, 18–36.

Kirsh, D. (1996) 'Adapting the Environment Instead of Oneself', *Adaptive Behavior* 4(3–4), 415–452.

Krothapalli, N.K.C. and Deshmukh, A.V. (1999). 'Design of Negotiation Protocols for Multi-Agent Manufacturing Systems', *International Journal of Production Research* 37(7), 1601–1624.

Levchuk, G.M., Yu, F., Levchuk, Y.N. and Pattipati, K.R. (2004). 'Networks of Decision-Making and Communicating Agents: A New Methodology for Designing Heterarchical Organizational Structures', *9th International Command and Control Research and Technology Symposium*, pp. 1–24. (Arlington, VA: CCRP).

Levesque, H.J., Cohen, P.R. and Nunes, J.H.T. (1990). 'On Acting Together', *Proceedings of the Eighth National Conference on Artificial Intelligence*, pp. 94–99. (Menlo Park, CA: AAAI Press).

Licklider, J.C.R. (1960). 'Man–Computer Symbiosis', *IRE Transactions on Human Factors in Electronics* 1(1), 4–11.

Lyons, D. (2002). 'Cognitive workload assessment', *Proceedings of the 2002 DARPA Augmented Cognition Conference*, p. 64. (Arlington, VA: DARPA).

Marschak, J. and Radner, R. (1972). *Economic Theory of Teams* (New Haven, CT: Yale University Press).

McComb, S.A. and Vozdolska, R.P. (2007). 'Capturing the Convergence of Multiple Mental Models and their Impact on Team Performance'. Paper presented at the Southwestern Academy of Management, San Diego, 14–17 March.

Mesarovic, M.D., Mako, D. and Takahara, Y. (1970). *Theory of Hierarchical Multi-Level Systems.* (New York: Academic Press).

Middelkoop, T. and Deshmukh, A. (1999). 'Caution! Agent-Based Systems in Operation', *InterJournal Complex Systems*, 265, 1–11.

Middleton, S.E. (2002). 'Interface Agents: A Review of the Field', Technical Report arXiv:cs.MA/0203012. Available at http://arxiv.org/abs/cs/0203012, accessed 24 April 2008.

Negroponte, N. (1997). 'Agents: From Direct Manipulation to Delegation', in J.M. Bradshaw (ed.), *Software Agents*, pp. 57–66. (Menlo Park, CA: AAAI Press).

Parunak, H.V.D. (1998). 'Practical and Industrial Applications of Agent-Based Systems', Environmental Research Institute of Michigan (ERIM) Technical Report. Available at http://agents.umbc.edu/papers/apps98.pdf, accessed 24 April 2008.

Payne, T., Lenox, T. L., Hahn, S. K., Sycara, K. and Lewis, M. (2000). 'Agent-based Support for Human–Agent Teams', *CHI 2000 Conference on Human Factors in Computing Systems*, pp. 22–23. (New York: ACM Press).

Plano, S. and Blasch, E.P. (2003). 'User Performance Improvement via Multimodal Interface Fusion Augmentation', *Proceedings of the Sixth International Conference of Information Fusion* 1, pp. 514–521. (Piscataway, NJ: IEEE).

Puterman, M.L. (1994). *Markov Decision Processes: Discrete Stochastic Dynamic Programming* (New York: John Wiley and Sons).

Rusmevichientong, P. and Van Roy, B. (2003) 'Decentralized Decision-Making in a Large Team with Local Information', *Games and Economic Behavior* 43(2), 266–295.

Scerri, P., Pynadath, D.V. and Tambe, M. (2002). 'Towards Adjustable Autonomy for the Real World', *Journal of Artificial Intelligence Research* 17, 171–228.

Schmorrow, D.D. and Kruse, A.A. (2004b). 'Augmented Cognition', in W.S. Bainbridge (ed.), *Berkshire Encyclopedia of Human–Computer Interaction*, pp. 54–59. (Great Barrington, MA: Berkshire Publishing Group).

Schmorrow, D. and McBride, D. (2004a). 'Introduction', *International Journal of Human–Computer Interaction* 17(2), 127–130.

Schmorrow, D., Stanney, K.M., Wilson, G. and Young, P. (2005). 'Augmented Cognition in Human–System Interaction', in G. Salvendy (ed.), *Handbook of Human Factors and Ergonomics*, 3rd edn, pp. 1364–1383. (Hoboken, NJ: Wiley).

Schneeweiss, C. (2003). *Distributed Decision Making*, 2nd edn. (Heidelberg: Springer).

Sheridan, T. B. (2006). 'Supervisory Control', in G. Salvendy, (ed.), *Handbook of Human Factors*, 3rd edn, pp. 1025–1052. (Hoboken, NJ: Wiley).

Siebra, C. and Tate, A. (2005). 'Integrating Collaboration and Activity-Oriented Planning for Coalition Operations Support', *Proceedings of the 9th International Symposium on RoboCup 2005*, pp. 13–19. (Osaka: RoboCup Federation).

Simon, H.A. (1997). *Models of Bounded Rationality*, vol. 3. (Cambridge, MA: The MIT Press).

Sycara, K. and Lewis, M. (2004). 'Integrating Intelligent Agents into Human Teams', in E. Salas and S.M. Fiore, (eds), *Team Cognition*, pp. 203–231. (Washington, DC: American Psychological Association).

Tambe, M. (1997). 'Towards Flexible Teamwork', *Journal of Artificial Intelligence Research* 7, 83–124.

Tambe, M., Shen, W. *et al.* (1999). 'Teamwork in Cyberspace: Using TEAMCORE to make Agents Team-Ready', *Proceedings of the AAAI Spring Symposium on Agents in Cyberspace*, pp. 136–141. (Menlo Park: AAAI Press).

Tidhar, G., Heinze, C. and Selvestrel, M. (1998). 'Flying Together: Modeling Air Mission Teams', *Journal of Applied Intelligence* 8(3), 195–218.

United States Army (2006). 'Future Combat Systems', white paper. Available at http://www.army.mil/fcs/, accessed 6 January 2007.

United States Navy (2004). 'Report of the Secretary of the Navy', http://www.defenselink.mil/execsec/adr2004/pdf_files/0007_navy.pdf, accessed 6 January 2007.

Weiss, G. (1999). *Multiagent Systems: A modern Approach to Distributed Artificial Intelligence*. (Cambridge, MA: MIT Press).

Wooldridge, M. and Jennings, N.R. (1995). 'Intelligent Agents – Theory and Practice', *Knowledge Engineering Review* 10(2), 115–152.

Xu, D. and Volz, R. (2003). 'Modeling and Analyzing Multi-Agent Behaviors Using Predicate/Transition Nets', *International Journal of Software Engineering and Knowledge Engineering* 13(1), 103–124.

Yen, J., Yin, J., Ioerger, T.R., Miller, M., Xu, D. and Volz, R.A. (2001). 'CAST: Collaborative Agents for Simulating Teamwork', *Proceedings of IJCAI*, pp. 1135–1142. (Rochester Hills, MI: IJCAI).

Chapter 8

Transferring Meaning and Developing Cognitive Similarity in Decision-making Teams: Collaboration and Meaning Analysis Process

Joan R. Rentsch, Lisa A. Delise and Scott Hutchison

Introduction

Many military and business teams face extremely challenging and difficult contexts and problems. For example, teams of distributed experts working under severe time constraints must solve complex problems that require the evaluation of uncertain, dynamic and excessive information while communicating using information restricting technology. In these teams, viable solutions typically require the integration of expert information initially held by individual team members. These teams have marked advantages over face-to-face teams, because they are flexible in utilizing diverse expertise and are able to work asynchronously and continuously. However, distributed teams are at a disadvantage relative to face-to-face teams because electronically mediated communication is degraded in terms of social and contextual cues.

Such degradation increases teams' susceptibility to time constraints and poor decision-making processes (Martins *et al.* 2004). For example, distributed team members spend more of their time discussing information held in common by several members relative to the time spent discussing unique, typically expert, information held by individual team members (e.g., Stasser and Titus 1987). This poor within-team information management is exacerbated when teams have temporal pressure, which is associated with inhibited sharing of unique information, decreased likelihood of achieving consensus and reduced decision quality (Campbell and Stasser 2006). In addition to temporal pressures, high cognitive load (e.g., due to uncertain, dynamic and excessive information) may reduce sharing unique information and result in poor task performance (Stasser and Titus 1987).

However, particularly within the military realm, high-quality team decisions are characterized by timeliness and accuracy. Therefore, one challenge for team researchers working with the military is to develop collaborative decision-making aids for teams. Understanding the cognitive processes associated with collaborative decision-making and problem-solving is a foundation for the development of collaborative technologies. For example, because knowledge is the basis for team decision-making, effective aids may focus on helping teams to take full advantage of the knowledge possessed by each team member. Each team member's individually held knowledge must be elaborated and integrated with knowledge possessed by other team members. By elaborating and integrating knowledge, the team creates the potential to construct new task knowledge that may be vital for generating effective solutions. Therefore, team researchers need to understand how individual knowledge is transferred from individually held pieces or chunks of information into knowledge that is mutually held and understood in useful and/or similar ways by all team members. Thus, understanding methods for transferring knowledge, making sense of it and developing cognitive similarity with respect to it are likely keys to understanding how to best aid decision-making teams.

Therefore, one purpose of this chapter is to present several process and cognitive variables associated with team decision-making. Information sharing, knowledge transfer, schema-enriched communication, and knowledge objects are necessary for the development of knowledge interoperability and cognitive similarity among team members. We begin by describing the difficulties faced by distributed teams working in computer-mediated environments.

Distributed teams

A distributed team may be defined as:

> A boundaryless network organization form where a temporary team is assembled on an as-needed basis for the duration of a task and staffed by members who are separated by geographic distance and who use computer-mediated communications as their primary form of communication and interpersonal contact.
>
> Kelsey (1999, p. 104)

Distributed teams may be differentiated on the basis of several characteristics including temporal distribution, boundaries spanned, relational boundaries, locational boundaries, and features of member roles (Bell and Kozlowski 2002; Martins *et al.* 2004).

Distributed teams making high-quality decisions must manage communication and cognitive processes effectively using technologically mediated communication. The nature of the technology and its interaction with the nature of the task must be considered, because some of the difficulties encountered by distributed teams appeared to be lessened when members communicated asynchronously (e.g., Straus and Olivera 2000). However, evidence also suggests that synchronous communication is desirable for complex tasks that require information sharing (Hollingshead *et al.*1993). Regardless of the nature of the technology, technology-mediated communication creates challenges associated with communication and cognitive processes in distributed teams.

Technology mediated communication filters visual and contextual cues, providing a low degree of information richness. Low information richness is associated with deindividuation, lack of accountability, and increased frustration and confusion, which may decrease inhibition, decrease self-regulation, increase self-absorption, and increase counternormative behavior (Savicki *et al.* 1996), thereby ultimately deteriorating the nature and quantity of communication among team members. Although empirical findings are inconsistent, in general, members of distributed teams communicate less frequently and less effectively with each other, and they provide more negative and less positive feedback (e.g., acknowledgment of accepting an idea) compared to members of co-located teams (e.g., Walther 1997). Feedback that is given in distributed teams may not be accepted due to the psychological distance that is likely to exist among team members (Ilgen *et al.* 1979). Psychological distance is likely because low information richness is related to decreased levels of relationship quality, team commitment and trust (e.g., Martins *et al.* 2004). Given these difficulties, it is no surprise that team members communicating through technology have difficulty expressing their thoughts, interpreting communications from others and drawing inferences about and predicting other team members' competencies and responses (e.g., Martins *et al.* 2004).

Attempting to increase information richness simply by increasing the quality of audio and video is not likely to be effective, because intangible and invaluable communicative cues seem to be available in face-to-face interactions (e.g., Sellen 1995). Although understanding these cues and integrating them into distributed teams might be useful, perhaps distributed teams simply require completely different types of cues, processes, or understanding in order to improve their functioning (for example, knowledge that is useful about how to manage information flow and how to format information may be different for distributed versus co-located teams).

The deficiencies of technology currently used to mediate communication will likely limit the development of cognitive similarity (e.g., schema similarity, mutual knowledge, shared understanding, shared goals) in distributed teams. Cognitive similarity is associated with high team functioning (e.g., Rentsch and Klimoski 2001). Indirect evidence of the advantages of cognitive similarity in research on distributed teams is provided by the findings that differences between co-located and distributed teams were diminished when distributed team members knew one another prior to working together (e.g., Straus and Olivera 2000). In addition, technologically mediated communications between people with strong relationships were found to be short but frequent (Hart and McLeod 2002). These findings suggest that team members who knew each other may have developed cognitive similarity with respect to the task, how to communicate, the use of technology and so on. One antecedent of cognitive similarity is interaction and discussion (e.g., Rentsch 1990). However, due to obstacles to effective communication in distributed teams, team members may take longer relative to face-to-face teams to develop cognitive similarity. For example, distributed teams working on ambiguous tasks tend to take longer to develop a common goal than face-to-face teams and may be unable to develop useful goals within a limited timeframe. Although, given enough time distributed teams can develop more focused goals than co-located teams (Straus and McGrath 1994). However, collaborative technologies may aid distributed teams to develop useful goals and cognitive similarity quickly.

It is clear that distributed teams working in time-pressured, information overloaded environments while solving complex problems must overcome obstacles to effective communication and cognitive processes. Although technology is creating many of these obstacles, technology designed with sound science of communication and cognition in teams as its basis may also provide the remedy to the obstacles. Below, we describe the cognitive and process variables.

Cognitive similarity in teams

Cognitive similarity is defined as 'forms of related meanings or understandings attributed to and utilized to make sense of and interpret internal and external events including affect, behavior, and thoughts of self and others that exist amongst individuals' (Rentsch *et al.* 2008). This definition is based on two major assumptions. First, although the probability of individuals developing identical interpretations (i.e., shared) is exceedingly small, some level of cognitive similarity is likely to exist among team members. Second, cognition resides exclusively within individuals.

Cognitive similarity is reciprocally related to communication and social interaction patterns (Rentsch 1990). Assuming functional cognitive content and an optimal degree and form of similarity, teams in which members develop cognitive similarity tend to benefit from efficient team process and high-quality performance (Rentsch and Hall 1994). Empirical evidence supports the positive association between team-member cognitive similarity and team performance (Mathieu *et al.* 2000; Rentsch and Klimoski 2001).

We suggest that distributed teams will benefit from cognitive similarity related to how team members should communicate, which may increase cognitive similarity with respect to the task. Both types of cognitive similarity among team members in combination form a configuration of cognitive similarity that may facilitate the team's ability to make high-quality decisions. Clear articulation of the relevant nature of cognitive similarity will be most useful to the development of interventions designed to promote performance in distributed teams.

A *type* of cognitive similarity is defined by the intersection of the form of cognition, the form of similarity, and the content domain (Rentsch *et al.* 2008). Three forms of cognition have been examined in the literature: structured, perceptual and interpretive. Within team research, cognition has been studied primarily as structured or organized knowledge. Team member schema similarity, shared mental models and strategic consensus are examples of structured knowledge that may be similar among team members. The perceptual form of cognition includes perceptions, expectations, or beliefs about the team or work environment, and the interpretive form of cognition include sense-making and interpretive systems (Rentsch *et al.* 2008).

The various forms of cognition may be similar in different ways. Similarity may exist, for example, in congruent, accurate, complementary and similarity forms derived from the social relations model. Various forms of similarity among cognitions will be related differentially to various outcomes. For example, teams on which members possess highly accurate (form of similarity) beliefs (form of cognition) about a commander's intent will likely aim to meet the commander's expectations (team members will understand what the commander wants). However, teams on which members have highly congruent (form of similarity) beliefs for how to respond to a commander's intent may have effective internal communication (because they will agree among themselves on what the commander's intent is), but they may or may not meet the commander's expectations (because they may or may not understand the commander's intent accurately). Characteristics of forms of similarity include level, variability and stability (Rentsch *et al.* forthcoming).

The cognitive content domain may be anything that team members understand or interpret (including teamwork process, task, technology, or commander). However, it is important to note that content domains may overlap such that relatively arbitrary boundaries must be applied in order to differentiate the cognitive content domain of interest from other related or connected domains (Rentsch *et al.* 2008).

Form of cognition, form of similarity, and content domain intersect to form *types* of cognitive similarity. Combinations of types of cognitive similarity yield cognitive similarity *configurations*. Rentsch *et al.* (forthcoming) argued that teams in which members have various cognitive similarity configurations that contain several content domains may be able to extract and synthesize information, communicate effectively, minimize conflict and foster collaboration among themselves. An example configuration of cognitive similarity is accurate (similarity form) perception (cognitive form) of teammates' areas of expertise (content domain) occurring simultaneously with congruent (similarity form) structured cognition (cognitive form) of the team task (content domain) and congruent (similarity form) structured cognition (cognitive form) of the computer-mediated technology used to communicate in distributed teams.

The development of similar cognitions in teams is beneficial to team performance, but distributed team members face challenges to developing similar cognitions, including self-absorption, increased confusion and few acknowledgements of agreement. However, configurations of similar cognitions are especially important among members of distributed teams because team members are operating in an environment where they must overcome obstacles related to understanding their team members (particularly team member areas of expertise), the team task and the processes through which the team members conduct their discussion. To compound these difficulties, team members must also come to similar understandings of the use and features of the technological media through which they communicate. Without a configuration of cognitive similarity with respect to all of these domains, distributed teams can face substantial problems that will inhibit the ability of team members to share information and transfer knowledge, both of which are essential activities of decision-making teams.

Information-sharing among team members

As noted above, distributed team members working under conditions of high temporal pressure and high cognitive load tend to have difficulty sharing information. Sharing information is an essential activity for such teams, because they must extract and integrate team member expertise (unique knowledge) in order to achieve viable problem solutions.

Researchers have studied information sharing extensively using the hidden profile paradigm (e.g., Stasser and Titus 1987) and have found a clear bias toward team members discussing common information at the expense of discussing unique information. This bias leads teams to make poor decisions, a finding that is robust across situations and manipulations (Campbell and Stasser 2006; Stassser and Titus 1987). Although some research has found that the sampling advantage of common information diminishes over time (Larson *et al.* 1998), the unique information revealed near the end of the discussion has minimal impact (Stasser and Titus 2003).

Some task characteristics, such as team members believing that a correct answer exists (Campbell and Stasser 2006), low information load and a low ratio of common to unique information (Stasser and Titus 1987) promote sharing unique information. Unfortunately, these task characteristics are in contrast to the typical military decision-making team task, for which there is usually no demonstrably correct answer and the information load is exceedingly high.

However, social variables have been found to decrease the advantages of common information during group discussion (e.g., Stasser and Titus 2003) with the most promising interventions centering on identifying team members as sources of credible information. In general, assigning team members expert areas for unique information and forewarning members as to who has which set of expert information increases the amount of unique information mentioned, repeated and recalled, and improves the chances of making correct decisions (Stewart and Stasser 1995).

When team members possess unique knowledge and accurately understand where unique knowledge exists within the team, they may be able to elicit increased unique information from each other during group discussions and to reduce the sampling advantage of common information (Stasser and Titus 2003). However, identification of team member expertise will not necessarily enable team members to understand another expert's information, to identify pieces of their own information they should share, or to share their information in a way that their teammates can understand it. Indeed, sometimes team members ignore unique information that other team members share (Stewart and Stasser 1995).

Effective decision-making requires more than merely the pooling of each team member's unique or unshared information. The shared information must be utilized, explicated and elaborated if it is to contribute to a high-quality decision. Information must be shared in a manner that it is understood by and is meaningful to other team members. In short, shared information should transfer knowledge among individuals. In the case of teams of experts, team members must share information so as to transfer expert knowledge to their teammates, who will likely be relative novices.

Transfer of expert knowledge to increase knowledge interoperability in teams

Transfer of expert knowledge is relevant to imparting expert knowledge among team members who may not need to develop deep expertise, but who need to have enough depth of understanding in order to be able to use the knowledge (that is, integrate it with other knowledge) and to contribute to the team's decision-making process. Transferring and receiving expert knowledge will involve active cognitive manipulation, deconstruction and reconstruction of the domain material at different levels of consciousness (e.g., Bereiter and Scardamalia 1996) as team members develop schemas to organize the information. Schemas are flexible, active mental representations of knowledge that provide interpretation and sense-making of information. Schemas are developed through past experiences, can adapt to accommodate new information (Rumelhart 1984), and provide frameworks for encoding new information. Schema-consistent information tends to be attended to and encoded as part of an existing category within the schema. Schemas also affect what information is recalled from memory (Rumelhart 1984). There is evidence that information is not stored as actual memories, but as representations of memory: when information needs to be recalled for use, the schema assists in reconstructing the information (based on prototypical characteristics of its category) before it is applied to a new situation (Bartlett 1932).

Because experts' schemas are structured differently than those of novices, experts may have difficulty in sharing their expert information so that it transfers useful knowledge to relative novices. Expert schemas tend to contain multiple levels with connections between and within the levels. Experts tend to conceptualize and represent problems in more general and abstract terms than novices, who possess schemas that contain many details and center on concrete features.

These differences may explain why experts and novices transfer knowledge differently. For example, experts explain tasks in broad terms, whereas beginners (relative novices), who have some exposure to a domain but who have not achieved expert status, explain tasks using concrete statements. Hinds *et al.* (2001) found that novices who had no experience in a field and who were instructed by beginners performed better on their introductory task than novices instructed by experts. In addition, their findings suggested that after an initial introduction to new knowledge that focuses on concrete features, the introduction of abstract features may be useful. This implies that novices began to develop expert-like schemas.

Information-sharing and knowledge transfer are different in that information sharing consists merely of statement or recitation of the facts, but knowledge transfer requires active explanation of information and presentation of that knowledge in a way that other team members may access and utilize it. Members of decision-making teams must share unique expert information, but more importantly they must transfer knowledge associated with the information in a manner that enables their teammates to assimilate the knowledge or to adapt their schemas to accommodate the knowledge. Ultimately, information is most useful to the team when it becomes knowledge that each team member organizes in a schema and is thereby capable of operating.

Warner, Letsky, and Wroblewski (Chapter 2 of this volume) defined knowledge interoperability development as the 'act of exchanging and modifying a mental model or knowledge object such that agreement is reached among team members with respect to a common understanding of the topic's meaning'. Warner *et al.* (2005) defined knowledge interoperability as 'the act of exchanging useful, actionable knowledge among team members'. Knowledge interoperability can be developed when team members present and elaborate upon information that others can internalize and utilize to identify connections between their own expert information and the information of others.

In order for team members to effectively share information in a manner that conveys meaning and builds cognitive similarity, team members must become aware of which team members hold unique information by being able to communicate effectively. Effective communication enables team members to externalize their uniquely held information and their understanding of that information (i.e., knowledge) in such a way as to increase the probability that their teammates may be able to use the knowledge conveyed (i.e., increase knowledge interoperability). Next, we discuss strategies for team members to transfer their knowledge.

Externalizing tacit knowledge

Difficulty in transferring expert knowledge is complicated by its implicit nature (Lord and Maher 1991). Nevertheless, each team member must externalize (articulate) his or her implicit expert knowledge. Team members must reveal how they organize or make sense of the information (i.e., reveal their schemas) and do so in a way that transfers that information into useful knowledge for their teammates. The team can increase the interoperability of pooled knowledge by identifying connections among the pieces of shared information. The team may develop aids, such as effective communication patterns and knowledge objects, to assist members in articulating their expert knowledge in a manner that enables other team members to operate the knowledge.

Schema-enriched communication

Interaction and communication among individuals are related to cognitive similarity and to team performance. Research has shown that specific types and structures of communication are related to knowledge transfer (e.g., Smith *et al.* 2005). For example, when team members ensured that teammates understood the team's goals and plans, identified information gaps and incongruent information, shared information and questioned assumptions and expectations, teams performed better than when members did not communicate in these ways (Endsley 1997).

These findings suggested that schema-enriched communication may be a key to knowledge transfer and to the development of cognitive similarity (Rentsch *et al.* 1998). Schema-enriched communication related to volunteering and seeking depth of meaning has been shown to be related to team performance. Depth of meaning, a form of schema-enriched communication, is conveyed when team members reveal the organized nature of their knowledge (their schema). Depth of meaning is transmitted in complex communication forms (e.g., explaining one's logic) that reveal deep information in addition to or rather than surface-level communication forms (e.g., a non-specific request). Communication may be characterized by depth of meaning when team members provide information to their teammates by relating how their information connects *explicitly* to other information in a path of logic. In addition, team members may also seek deep meaning from one another.

For example, they may seek to extract their teammates' tacit knowledge by asking about teammates' interpretations or reasonings, asking why the other is pursuing a line of logic, and/or seeking information about the other's assumptions. Depth of meaning communication is schema-enriched because team members are attempting to externalize not just the information they possess but also their understanding and organization of that information. Similarly, team members communicating depth of meaning may assist another teammate in articulating his or her schema. In a study of problem-solving teams, depth of meaning was related to the degree of problem space identification and the consideration of multiple alternatives, which are indicators of effective problem-solving (Rentsch *et al.* 1998).

Schema-enriched communication should aid team members to share information and externalize knowledge, and thereby transfer knowledge within the team. Schema-enriched communication in combination with the development and use of knowledge objects is hypothesized to facilitate knowledge transfer among team members.

Knowledge objects

Knowledge transfer is facilitated by the establishment of common ground. Common ground reflects the notion of team members working cooperatively and sharing information to achieve a similar understanding. Common ground emerges as team members' schemas become more similar (McNeese and Rentsch 2001). Common ground enables team members to develop new knowledge collaboratively, which is not a simple sum of all team members' knowledge, nor is it knowledge previously possessed by any one team member. Rather, based on team discussions and their own knowledge, team members individually create their own meaning and internalize the new knowledge (Stahl 2005). In this way, new knowledge becomes operable for each team member and facilitates his or her ability to identify information that should be shared, connections among pieces of information, and so on. Knowledge objects are tools that may promote knowledge transfer and the collaboration needed to create new knowledge.

Knowledge objects are 'pictures or icons developed by an individual team member about a problem characteristic'. Knowledge objects are anything observable that can convey meaning to another person, including non-verbal behaviors, documents, or physical objects (Nosek 2004). Carlile (2002) highlighted syntactic, semantic and pragmatic forms of knowledge objects that promote knowledge transfer among individuals. Syntactic objects ensure that individuals are using the same language (e.g., all are using the word 'bolt' to refer to an object holding two pieces of metal together). Semantic objects ensure that individuals using the same language are also using the same meanings and interpreting the language in similar ways (e.g., that all are using the word 'bolt' to refer to a hex bolt rather than referring to a lag bolt). Pragmatic objects ensure that individuals are able to use the knowledge they are sharing in practical ways as long as everyone understands the language similarly (e.g., three-dimensional drawings of machinery that illustrate the placement of 'bolts').

Knowledge objects aid team members in transferring their expert knowledge to the extent that the knowledge object relates to the other team members' schemas.

Therefore, the use of knowledge objects surpasses simple information-sharing methods because they aid team members in developing common ground by illuminating each others' expert knowledge and making it possible for other members to integrate that knowledge into a solution. Knowledge objects are particularly useful in transferring expert knowledge, which is difficult to transfer due to it being highly integrated and embedded within the other knowledge within a domain (Carlile 2002). Knowledge objects are likely to facilitate the development of cognitive similarity among experts. However, the functionality of any given knowledge object depends, in part, on perspective-making and, more importantly, perspective-taking.

Perspective-taking

With respect to distributed decision-making teams composed of experts, perspective-*making* is the process of becoming certain in one's own knowledge and perspective so that it can be shared with others. Perspective-*taking* is the process of understanding another's knowledge and perspective and taking them into account when presenting one's own information (Boland and Tenkasi 1995). Perspective-taking involves efforts to see the world through another's eyes and aiming to understand the other's viewpoint by imagining (and ultimately understanding) how the other feels and thinks (e.g., Davis 1994). Perspective-taking is viewed as a cognitive, affective, and/or behavioral process by which individuals can put themselves in another's position and see a situation from the other's perspective (e.g., Duan and Hill 1996).

Perspective-taking is critical in the development of effective schema-enriched communication and the development and use of knowledge objects. The value of communications or knowledge objects lie in their ability to be useful to other teammates and to be operable from the teammates' points of view. Therefore, team members need to understand not only how an object expresses their own knowledge but also whether the object is appropriate for expressing or relating to the knowledge or schema of the other party. By taking the perspective of another, one can understand the other's information and constraints in sharing that information and the value of information for the other. Perspective-taking should also lead to the development of cognitive similarity in the forms of congruence and accuracy among others on the team.

We propose that effective decision-making teams must develop a configuration of cognitive similarity that includes (1) cognitive similarity with respect to communicating with team members and (2) cognitive similarity related to the task. A reciprocal relationship between the former (communication cognitive similarity) and perspective-taking is proposed. Cognitive similarity among team members with regard to communication methods should lead to increased perspective-taking, which should lead to the development and use of meaningful schema-enriched communications and knowledge objects. Schema-enriched communication and the use of knowledge objects should promote knowledge interoperability and cognitive similarity regarding the task. In addition, schema-enriched communication and knowledge objects are expected to foster perspective-taking, and increased perspective-taking will enhance cognitive similarity with respect to communication.

Summary

In order to increase knowledge transfer that takes place within a team, team members must engage in perspective-making and perspective-taking while externalizing their knowledge through schema-enriched communication in combination with the development and use of knowledge objects. Interventions associated with supporting schema-enriched communication and the use of knowledge objects will likely promote information sharing, knowledge transfer, knowledge interoperability and the development of cognitive similarity, ultimately enhancing team performance.

Collaboration and meaning analysis process

In an effort to integrate the above information, we have developed the collaboration and meaning analysis process (C-MAP), which involves the conscious externalization of knowledge through schema-enriched communication and knowledge objects. The schema-enriched communication is expected to facilitate information-sharing primarily and to support knowledge transfer. The use of knowledge objects is expected to support information sharing and to facilitate knowledge transfer primarily. Together, through their effects on information sharing and knowledge transfer, these interventions are expected to facilitate knowledge interoperability and cognitive similarity with respect to the task. Ultimately, team decision-making is expected to be improved. The components and underlying cognitive processes of the C-MAP are presented in Figure 8.1 and are described in more detail below.

Schema-enriched communication functions to convey information in a meaningful way that expresses or elicits structured cognition (schemas). Communication specifically aimed toward explicitly conveying the essence of expert information can assist technology-mediated distributed teams in overcoming communication obstacles they typically face. Such obstacles include members who are unfamiliar with one another, temporal pressures, and information overload. Schema-enriched communication is expected to support sharing uniquely held information in a manner that conveys deep meaning thereby supporting knowledge transfer.

In time-pressured, high cognitive load environments, team members facing ambiguous decision-making tasks may neglect using communication that can help their team members understand their expert information. Therefore, in stressful situations where member attention is diverted from communication processes, communication-related prompts may serve as helpful reminders to team members to focus on schema-enriched communication. These reminder prompts are expected to cause team members to engage in schema-enriched communication in the instance when the prompt is applied. More importantly, however, prompts are expected to promote the active development of a communication schema within each team member that will ultimately enable the team members to engage in the prompted, schema-enriched behavior not only at the time when a prompt has been presented, but at any time when a communication behavior is necessary to facilitate the discussion. Teams in which members develop cognitive similarity regarding communication methods are expected to share and discuss information competently

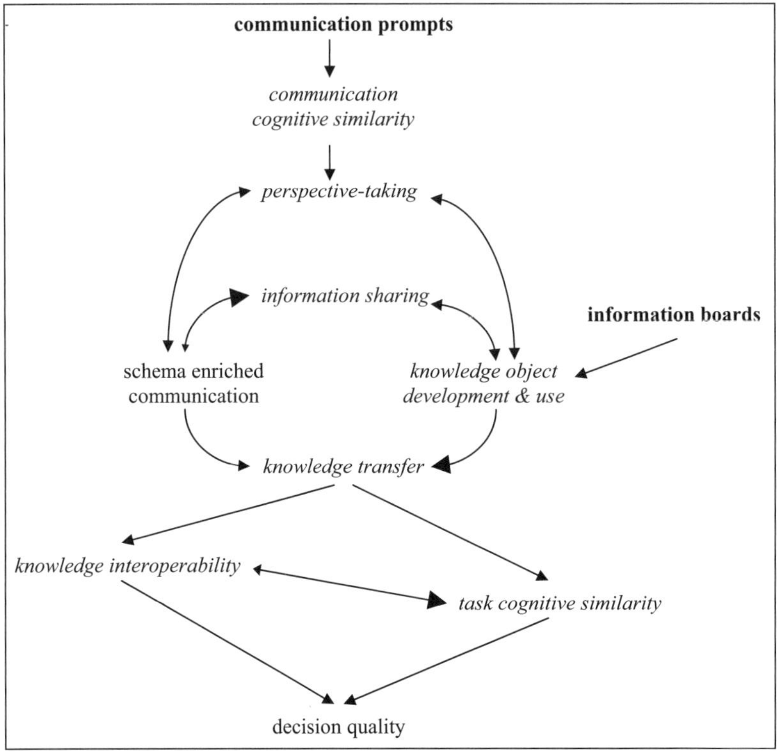

Figure 8.1 Components and cognitive processes of the C-MAP

(Rentsch and Zelno 2003). Thus, one component of the collaboration and meaning analysis process focuses on providing team members with prompts that remind them to engage in schema-enriched communication.

Additionally, because teams of experts may have difficulty communicating their expert knowledge due to its abstract and deeply embedded nature, it is advantageous for distributed experts to externalize the content and the structure of that information. Because externalization of knowledge in this way may not be intuitive for distributed teams, the C-MAP provides training in the development and subsequent use of externalized knowledge objects. The second C-MAP component encourages team members to use a commonly viewable information board on which team members should post important task-related information, particularly expert information. The information board is useful, in part because it enables team members to engage in information sharing by posting facts for everyone to view. Furthermore, the information board is useful because team members may use it to articulate how they structure the information in their minds (i.e., reveal their schemas) by posting related information together and highlighting connections among their own information pieces and the information pieces that others have posted. Because all team members may move any of the posted information, the information board affords opportunities for team members to identify connections among pieces of expert information and facilitates the development of a knowledge object that can assist all members in understanding the nuances of the task information. In this way team members may share information, but more importantly convey the knowledge or meaning associated with the information (i.e., transfer knowledge).

Engaging in schema-enriched communication and using knowledge objects will promote information sharing and will enable team members to make sense of the information such that it becomes meaningful knowledge that is transferred among team members. The transferred knowledge will support the development of knowledge interoperability. That is, the C-MAP will increase the likelihood that most, if not all, of the information and knowledge relevant to the task and possessed by the team will be available and operable for all team members. Knowledge interoperability will develop such that each team member understands the information sufficiently to use it in combination with other available knowledge (i.e., the knowledge is useable for that individual). Through collaborative use of the knowledge, team members are likely to work toward cognitive similarity with respect to how the knowledge is related to other task knowledge. As cognitive similarity increases, knowledge interoperability among team members increases. That is, team members will be able to use that interoperable knowledge in combination with information supplied by other team members because the knowledge is now useful in connecting across team members' expertise areas rather than simply connecting the knowledge within any one team member's cognitive structure. Ultimately, knowledge interoperability and cognitive similarity with regard to the task are expected to enhance team decision quality.

Thus, the C-MAP is intended to increase teams' capacities to exploit the knowledge available to them. It involves the ultimate development of a cognitive similarity configuration that includes schema (form of cognition) congruence (form of similarity) with respect to communication (domain) and schema congruence with respect to the task. Other types of cognitive similarity may also develop, such as schema accuracy with respect to teammates' competencies, preferences and perspectives. This type of cognitive similarity would support perspective-taking, which is important for effective schema-enriched communication and the development of useful knowledge objects. Preliminary evidence collected using face-to-face teams is encouraging and supports the use of the C-MAP. Schema-enriched communications, knowledge interoperability and cognitive similarity were higher for teams using the C-MAP than for control teams (Rentsch *et al.* 2007). Research is currently underway to examine the effects of technology-embedded C-MAP components that are designed to facilitate schema-enriched communication and knowledge externalization in distributed teams. Previous preliminary evidence supports a connection between prompts and team performance in distributed teams (Rentsch and Hutchison 1999).

The italicized processes in Figure 8.1 are those that might be considered macrocognitive processes given Letsky *et al.*'s definition presented in Chapter 2 of this volume. They define macrocognition as 'the internalized and externalized high-level mental processes employed by teams to create new knowledge'. The components in bold type are features of the technology. Future research includes a need to test the entire process presented in Figure 8.1 in different types of teams under a variety of conditions. In addition, process metrics of the cognitive variables need to be developed and validated.

Acknowledgment

Work described in this chapter was supported by Grant Award Number N00014-05-1-0624 from the Office of Naval Research.

Note

The views, opinions, and findings contained in this chapter are the authors' and should not be construed as official or as reflecting the view of the Department of Defense or The University of Tennessee.

References

Bartlett, F.C. (1932). *Remembering.* (London: Cambridge University Press).

Bell, B.S. and Kozlowski, S.W. (2002). 'A Typology of Virtual Teams', *Group and Organizational Management* 27, 14–49.

Bereiter, C. and Scardamalia, M. (1996). 'Rethinking Learning', in D. Olson and N. Torrance, (eds), *Handbook of Education and Human Development: New Models of Learning, Teaching, and Schooling*, pp. 485–513. (Cambridge, MA: Basil Blackwell).

Boland, R.J. and Tenkasi, R.V. (1995). 'Perspective Making and Perspective Taking in Communities of Knowing', *Organization Science* 6(4), 350–372.

Campbell, J. and Stasser, G. (2006). 'The Influence of Time and Task Demonstrability on Decision-making in Computer-mediated and Face-to-Face Groups', *Small Group Research* 37(3), 271–294.

Carlile, P.R. (2002). 'A Pragmatic View of Knowledge and Boundaries: Boundary Objects in New Product Development', *Organization Science* 13(4), 442–455.

Davis, M.H. (1994). *Empathy: A Social Psychological Approach.* (Dubuque, IA: Brown and Benchmark).

Duan, C. and Hill, C.E. (1996). 'The Current State of Empathy Research', *Journal of Counseling Psychology* 43(3), 261–274.

Endsley, M.R. (1997). 'The Role of Situation Awareness in Naturalistic Decision Making', in C. Zsambok and G. Klein, (eds), *Naturalistic Decision Making*, pp. 269–283. (Hillsdale, NJ: Lawrence Erlbaum Associates).

Hart, R.K. and McLeod, P.L. (2002). 'Rethinking Team Building in Geographically Dispersed Teams: One Message at a Time', *Organizational Dynamics* 31, 352–361.

Hinds, P.J. *et al.* (2001). 'Bothered by Abstraction: The Effect of Expertise on Knowledge Transfer and Subsequent Novice Performance', *Journal of Applied Psychology* 86, 1232–1243.

Hollingshead, A.B. *et al.* (1993). 'Group Task Performance and Communication Technology: A Longitudinal Study of Computer-mediated Versus Face-to-face Work Groups', *Small Group Research* 24, 307–333.

Ilgen, D.R. *et al.* (1979). 'Consequences of Individual Feedback on Behavior in Organizations', *Journal of Applied Psychology* 64(4), 349–371.

Kelsey, B.L. (1999). 'Trust as the Foundation of Cooperation and Performance in Virtual Teams' in J. Ajenstat, (ed.), *Proceedings of the 1999 Annual Conference, Information Systems Division vol. 2 part 5*, pp. 103–113. (St. John's, New Brunswick: Administrative Sciences Association of Canada).

Larson, J.R. *et al.* (1998). 'Diagnosing Groups: The Pooling, Management, and Impact of Shared and Unshared Case Information in Team-based Medical Decision Making', *Journal of Personality and Social Psychology* 75, 93–108.

Lord, R.G. and Maher, K.J. (1991). 'Cognitive Theory in Industrial and Organizational Psychology' in M.D. Dunnette and L.M. Hough, (eds), *Handbook of Industrial and Organizational Psychology, vol. 2*, pp. 1–62. (Palo Alto, CA: Consulting Psychologists Press).

Martins, L.L. *et al.* (2004). 'Virtual Teams: What Do We Know and Where Do We Go From Here?', *Journal of Management* 30, 805–835.

Mathieu, J.E. *et al.* (2000). 'The Influence of Shared Mental Models on Team Process and Performance', *Journal of Applied Psychology* 85, 273–282.

McNeese, M.D. and Rentsch, J.R. (2001). 'Identifying the Social and Cognitive Requirements of Teamwork Using Collaborative Task Analysis', in M. McNeese *et al.* (eds), *New Trends in Cooperative Activities: Understanding System Dynamics in Complex Environments*, pp. 96–113. (Santa Monica, CA: Human Factors and Ergonomics Society).

Nosek, J.T. (2005). 'Group Cognition as a Basis for Supporting Group Knowledge Creation and Sharing', *Journal of Knowledge Management* 8(4), 54–64.

Rentsch, J.R. (1990). 'Climate and Culture: Interaction and Qualitative Differences in Organizational Meanings', *Journal of Applied Psychology* 75, 668–681.

Rentsch, J.R. and Hall, R.J. (1994). 'Members of Great Teams Think Alike: A Model of Team Effectiveness and Schema Similarity Among Team Members' in M.M. Beyerlein and D.A. Johnson, (eds), *Advances in Interdisciplinary Studies of Work Teams, vol. 1, Series on Self-managed Work Teams*, pp. 223–262. (Greenwich, CT: JAI Press).

Rentsch, J.R., Small, E.E. and Hanges, P.J. (2007). 'Understanding Team Cognition: The Shift to Cognitive Similarity Configurations', in F. Dansereau and F.J. Yammarino (eds), *Research in Multi-level Issues, vol. 6*, pp. 159–174. (Boston, MA: Elsevier JAI Press).

Rentsch, J.R. and Zelno, J. (2003). 'The Role of Cognition in Managing Conflict to Maximize Team Effectiveness: A Team Member Schema Similarity Approach', in M.A. West *et al.* (eds), *International Handbook of Organization and Cooperative Teamworking*, pp. 131–150. (West Sussex, England: John Wiley and Sons, Ltd.).

Rentsch, J.R. *et al.* (1998). *'Testing the Effects of Team Processes on Team Member Schema Similarity and Task Performance: Examination of the Team Member Schema Similarity Model*', AFRL-TR-98-0070. (Wright-Patterson Air Force Base, OH: Air Force Research Laboratory).

Rentsch, J.R. *et al.* (2008). 'Cognitions in Organizations and Teams: What is the Meaning of Cognitive Similarity?', in B. Smith (ed.), *The People Make the Place*, pp. 127–156. (New York: Lawrence Erlbaum Associates).

Rentsch, J.R. *et al.* (forthcoming). 'Cognitive Similarity Configurations in Teams: In Search of the Team Mindmeld' in E. Salas *et al.* (eds), *Team Effectiveness in Complex Organizations: Cross-disciplinary Perspectives and Approaches.*

Rentsch, J.R. *et al.* (2007). 'Improving Cognitive Congruence and Knowledge Interoperability in Decision Making Teams', Symposium presented at the 2nd Annual INGRoup Conference, Lansing, MI.

Rentsch, J.R. and Hutchison, A.S. (1999). '*Advanced Cognitive Engineered Intervention Technologies (ACE-IT)*'. (Knoxville, TN: Organizational Research Group).

Rentsch, J.R. and Klimoski, R.J. (2001). 'Why Do 'Great Minds" Think Alike?: Antecedents of Team Member Schema Agreement', *Journal of Organizational Behavior* 22, 107–120.

Rumelhart, D.E. (1984). 'Schemata and the Cognitive System', in R.S. Wyer and T.K. Srull, (eds), *Handbook of Social Cognition, vol. 1*, pp. 161–188. (Hillsdale, NJ: Erlbaum).

Savicki, V. *et al.* (1996). 'Gender, Group Composition, and Task Type in Small Task Groups Using Computer-mediated Communication', *Computers in Human Behavior* 12(4), 549–565.

Sellen, A.J. (1995). 'Remote Conversations: The Effects of Mediating Talk With Technology', *Human Computer Interaction* 10, 401–444.

Smith, K.G. *et al.* (2005). 'Existing Knowledge, Knowledge Creation Capacity, and the Rate of New Product Introduction in High-technology Firms', *Academy of Management Journal* 48(2), 346–357.

Stahl, G. (2005). 'Group Cognition in Computer-assisted Learning', *Journal of Computer Assisted Learning* 21, 79–90.

Stasser, G. and Titus, W. (1987). 'Effects of Information Load and Percentage of Shared Information on the Dissemination of Unshared Information During Group Discussion', *Journal of Personality and Social Psychology* 53(1), 81–93.

Stasser, G. and Titus, W. (2003). 'Hidden Profiles: A Brief History', *Psychological Inquiry* 14, 304–313.

Stewart, D.D. and Stasser, G. (1995). 'Expert Role Assignment and Information Sampling During Collective Recall and Decision Making', *Journal of Personality and Social Psychology* 69, 619–628.

Straus, S.G. and McGrath, J.E. (1994). 'Does the Medium Matter: The Interaction of Task Type and Technology on Group Performance and Member Reactions', *Journal of Applied Psychology* 79, 87–97.

Straus, S.G. and Olivera, F. (2000). 'Knowledge Acquisition in Virtual Teams', *Research on Managing Groups and Teams* 3, 257–282.

Walther, J.B. (1997). 'Group and Interpersonal Effects in International Computer-mediated Collaboration', *Human Communication Research* 23, 342–369.

Chapter 9

Processes in Complex Team Problem-solving: Parsing and Defining the Theoretical Problem Space

Stephen M. Fiore, Michael Rosen, Eduardo Salas, Shawn Burke and Florian Jentsch

Overview of approach

Historically, when studying a given phenomenon, scientists typically focus on either a fine-grained level of analysis or a broad conceptualization of the issues associated with their domain of study. The former approach is characterized as a micro-level approach, whereas the latter is described as a macro-level approach. More recently, researchers have suggested that the scientific community needs to attend to inter-relations among these levels. The goal is to better understand the coupling among these levels so as to better clarify their interactions and the impact of each level on the other (Liljenström and Svedin 2005). In this chapter we discuss one such current conceptualization, that of the term macrocognition and its related concepts. This chapter represents an extension of some of the prior theorizing on macrocognition where the term is used to capture cognition in collaborative contexts (Warner, Letsky and Cowen 2005) and encompasses both internalized and externalized processes occurring during team interaction. These processes include not only internalized individual processes such as mental model development, but also externalized processes such as solution alternative negotiation (see Letsky, Warner, Fiore, Rosen and Salas 2007). We take the next step in this definitional exercise and consider a refinement of definitions from the literature. Specifically, we attempt to reify them within the context of what Letksy *et al.* (2007) have described as internalized and externalized cognition. Further, we attempt to describe these not only sequentially, but also dynamically and iteratively, in order to more clearly convey the inter-relations among macrocognitive processes as teams work through the problem-solving process.

Philosophical context

From the start we acknowledge that our approach to this effort is pragmatic. As such, we recognize the dangers inherent in over simplifying complex processes. We take our cautionary approach based upon lessons learned from cognitive engineering research. Using techniques such as 'cognitive task analysis' research has helped to identify misconceptions and also to determine how people respond to dimensions of difficulty. The key finding encompassing this is that people often deal with complexity through oversimplification, which leads to misconceptions and faulty knowledge application. This is labeled the reductive tendency, an inevitable consequence of how people attempt to understand complex tasks (e.g., Spiro, Feltovich, Coulson and Anderson 1989; Spiro and Jehng 1990). For example, dynamic processes are thought to static, continuous processes are viewed as discrete steps, highly interactive factors are thought to be independent, nonlinear systems are misconstrued as linear, and emerging and evolving events are conceptualized as being governed by simple cause-effect relations, etc. Essentially, these complexities are often overlooked because of our tendency to build simplified understanding and explanations leading to misconceptions and possibly errors – what is termed *the reductive bias* (see Feltovich, Hoffman, Woods and Roesler 2004). We suggest that understanding macrocognition is similarly complex and we must recognize the risks inherent when trying to capture this dynamic concept. But, as stated, we must be pragmatic, so we acknowledge that we can not pretend to understand all the complexity inherent in this domain, that is, *truth* with a capital 'T'. Rather, we must work towards understanding *enough* to explain and add predictive utility to collaboration in complex problem-solving, that is, truth with a lower case 't'.

With this being said, we note the following caveats. First, we acknowledge that macrocognition is a dynamic and iterative process. It is essentially a 'moving target' in that processes are parallel and continuous. Nonetheless, in this chapter they are somewhat presented as *sequential* and *discrete.* Our goal is 'freeze' elements of macrocognition in an attempt to better understand certain concepts. In some ways our approach is a conceptual analog to techniques such as those pioneered by Harold Edgerton and stop-motion photography. In that field of inquiry, there was tremendous utility from analyzing phenomena using electronic stroboscopes. Specifically, this facilitated understanding because it allowed enough light to stop objects in motion that the human eye had never seen. So science gained an understanding of everything from the detonation of nuclear bombs to the deceptively sophisticated flow of liquids by freezing the dynamics of complex systems.

Along these lines, we describe the processes involved in macrocognition somewhat in isolation but do so in the context of their broader processes. Our goal is to consider what the literature has presented on related concepts and identify the *constituent elements* across definitions. From these constituent elements we abstract a core definition for a given concept. We have taken this approach in an attempt to bring some order to the terminology and begin the important process of taxonomizing a number of critical theoretical concepts associated with macrocognition. Further, it is these elements that can later become targets for operationalization and measurement.

Research context

First, we wish to delineate teams and the types of teams of interest in this effort. Most succinctly, teams can be considered as 'interdependent collections of individuals who share responsibility for specific outcomes for their organizations' (Sundstrom, De Meuse and Futrell 1990, p. 120), or as 'two or more people who interact dynamically, interdependently and adaptively toward' a shared goal (Salas, Dickinson, Converse and Tannenbaum 1992, p. 4). We embrace these definitions noting the particular importance of interdependence, shared goals, and collective adaptivity. Further, our emphasis is on problem-solving teams. These teams are often ad hoc, and formed to rapidly deal with difficult situations in the short-term. Additionally, given the complexity of the problems these teams face, they are often quite heterogeneous, possessing unique skills and knowledge. Finally, these teams must typically work within environments that are ill-defined and which often have associated with them grave consequences for mistakes.

Second, in consideration of this task context, our objective with this chapter is to codify some of the foundational concepts associated with macrocognition and help the scientific community move forward in its attempts to understand this domain of inquiry. We build upon and refine the framework of collaborative problem-solving being developed as part of the Navy's Collaboration and Knowledge Interoperability program (Letsky, Warner, Fiore, Rosen and Salas 2007; Warner, Letsky and Cowen 2005). Our conceptual work is based upon the Letsky *et al.* definition of macrocognition, that is, 'the internalized and externalized high-level mental processes employed by teams to create new knowledge during complex, one-of-a-kind, collaborative problem-solving.' They describe 'high-level' as cognitive processing involved in the combining, visualizing, and aggregating of information to resolve ambiguity in support of the discovery of new knowledge and relationships. More specifically, internalized processes are not expressed through external means (e.g., writing, speaking, gesture) and they can only be assessed using indirect techniques such as qualitative metrics like cognitive mapping or think aloud protocols or by using surrogate quantitative metrics such as galvanic skin response). Letsky *et al.* (2007) define externalized processes as those higher-level mental processes which occur at the individual or the team level, and which are associated only with actions that are observable and measurable in a consistent, reliable, repeatable manner or explicitly through the conventions of the subject domain having standardized meanings. These are processes utilized by teams in complex environments where collaborative problem-solving focuses on one-of-a-kind situations (e.g., Non-combatant Evacuation Operations).

In the following sections we set out to more fully define the concepts associated with this definition. We present these in such a way that their interactive nature are brought more to the forefront in order for the reader to appreciate the dynamic processes involved in complex problem-solving. In particular, rather than only presenting these concepts taxonomically, we illustrate how they interact with each other in an iterative fashion in support of efforts to comprehend and resolve the problems the team is facing.

Individual and team knowledge-building

Our first major processes involve individual and team knowledge-building. In our presentation of these major processes we suggest a refinement of the current conceptualization as presented in the extant framework (Letsky *et al.* 2007). In particular we suggest that these be described under the general rubric of making ill-defined tasks, well-defined. Specifically, given the task context in which macrocognition emerges, teams must work diligently to resolve ambiguity. As such, much of what is engaged by team members in support of this knowledge-building activity is the reduction of uncertainty. In the context of individual knowledge-building, that is, the *internalization* of data, uncertainty reduction involves a *minimization of uncertainty* related to information in the environment. In the context of Team Knowledge Development, that is, the *externalization* of information, uncertainty reduction describes the use of communication for minimization of teammate uncertainty. We suggest that uncertainty reduction involves both the creation of knowledge as well as the clarification of knowledge. In the former sense, interpretation of environmental cues, for example, would involve knowledge from an individual or from the team adding to data. This productive combination creates new knowledge. In the latter sense, extant knowledge having to do with, for example, multiple and/or ambiguous data points, may be pared down as information is updated. In both instances, processes unfolding within the problem-solving environment serve to reduce the uncertainty that is present.

Although prior conceptualizations of these concepts presented them sequentially, we try to describe them in such a way that we illustrate how the parallel, interdependent, and iterative nature of these processes unfold in the context of collaboration. Figure 9.1 illustrates a conceptual representation of these functions and processes. What this shows is an iterative process engaged by two, four person teams interacting to solve a problem. The upper portion, showing two sets of four circular arrows represents the individuals in the team. The lower portion, showing two larger circular arrows represents the two teams as a whole. The arrows are used to represent the iterative nature of these processes as they unfold individually and collectively. Last, these teams are encompassed by a larger circular arrow to illustrate the overall iterative nature of this knowledge-building effort as it unfolds over individuals, teams, and across teams. We next describe in some detail the sub-processes making up this major macrocognitive process.

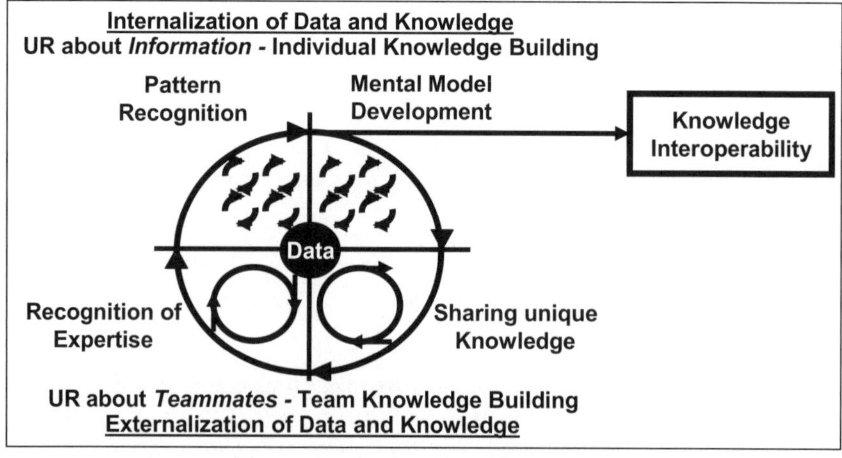

Figure 9.1 Individual and team knowledge-building

UR about information – processes subservient to individual knowledge-building

First, in the context of individual knowledge-building, what we describe as uncertainty reduction about information requires a minimization of sources of uncertainty. This takes place via iterative information collection which involves a retrieving and manipulating of information in such a way that it alters uncertainty. More generally, this involves integrating data to develop new patterns of information. For example, this has been described as a form of conceptual change resulting from a restructuring of data. This restructuring occurs through a variety of processes such as data interrogation, questioning, criticism and evaluation to produce explanations and interpretations (see Brown and Palinscar 1989). More succinctly, individual knowledge-building involves the production and continual improvement of ideas of value to the task context (see Scardamalia and Bereiter 2003).

In this sense, we can more specifically present a *conceptual definition of individual knowledge-building* as, first, requiring the collecting and restructuring of data. This restructuring of data into more meaningful information requires data interrogation. This data interrogation serves to reduce uncertainty associated with that data – an uncertainty arising from either incompleteness, source confusion, and/or a lack of clarity. This interrogation involves processes such as questioning and critiquing in support of explanation and interpretation. This produces a form of conceptual changes whereby the previously experienced uncertainty has been reduced such that the individual is able to more effectively utilize information with less concerns about ambiguity.

Individual knowledge-building is served by a set of processes occurring within and across teams. Essentially, the individuals and their teammates work through an iterative process of internalization and externalization of knowledge as they work to reduce uncertainty associated with data and build knowledge. We turn next to a description of the internalized processes, that is, those processes making up the top portion of the Figure 9.1.

Pattern recognition

At initially an individual level, pattern recognition involves perceptual processes one uses in the identification of individual cues and groups of cues that may be indicative of an important event. Some suggest that pattern recognition requires an awareness of typical world states (Hayes 1989) or an 'ability to discriminate among familiar classes of objects' (Gonzalez and Quesada 2003, p. 287). More generally, Klein and Hoffman (1992) have argued that this process describes ones ability to detect configurations of cues. These configurations can be complex and may not occur simultaneously in the environment. From this need to sometimes detect cue configurations consisting of environmental events potentially separate not only in space, but also in time, we can additionally state that *trend analysis* is a particular facet of pattern recognition. Specifically, the pattern may consist of cues requiring integration across time in order for recognition to occur. For example, Handley (personal communication) views trend analysis as a description of 'long-term overall movement of the time series data'. Similarly, Smallman (personal communication) views trend analysis as 'the identification and monitoring of trends in multi-dimensional evidence evaluation space and the detection of deviations from, and similarity in, those trends over time.'

In consideration of these definitions, we present our *conceptual definition of pattern recognition*. Essentially, detection of the abnormal requires recognition of the normal, and when an anomaly occurs it is recognized as such. This process requires recognition of individual cues as well as constellations of cues. Further, these cues must be discriminated amongst a myriad of other cues in order for idiosyncratic salience to be detected. That is, the relations among patterns of cues may be situation specific such that their criticality changes with particular events. Finally, this must all be recognized as these cues occur separated not only in space, but also in time. It is this detection that sets of a cascading series of individual and collective cognition in service of reducing uncertainty associated with that anomaly, that is, it leads to the development of a mental model associated with this environmental occurrence, a concept we next discuss.

Mental models

From the literature on mental model development, we can see it as the creation of 'psychological representations of real, hypothetical, or imaginary situations' (Johnson-Laird 1999). From a more process oriented perspective, mental models are considered as the 'mechanism whereby humans generate descriptions of system purpose and form, explanations of system functioning and observed system states, and predictions of future system states' (Rouse and Morris 1986, p. 360). More recently, mental model development is seen as the creation of a mental picture by combining, sorting, and filtering information (Warner *et al.* 2005).

In the context of these definitions, we state that our *conceptual definition of mental model development* be construed of as a process incorporating uncertainty reduction by comparing assumptions about cause and effect relations among cues against the external sources of data (e.g., Hoffman *et al.* 2000; Joslyn and Jones 2000; Trafton *et al.* 2000). Individuals engage in this process to develop the knowledge of their present task but contextually embedded within their present problem situation. This may require additional information gathering via sensors or through interaction with teammates. More specifically, the development of this mental model incorporates perceptually and conceptually integrated cause and effect relationships among information culled from the recognition process. From this model one assumes that he is able to describe, explain, and predict a given state of the world.

UR about teammates – processes subservient to team knowledge-building

Next, we suggest that uncertainty reduction in team knowledge-building involves the relationships between uncertainty and communication in teams. More specifically, it involves passive and active strategies that serve to convey and disperse information amongst the team (Fleming, personal communication). For example, team researchers suggest this supports the development of understanding the collective knowledge held by team members. This is an understanding related to task and team related skills and a team's collective understanding of the current situation (Cooke *et al.* 2004). Other team researchers view this analogously but along the lines of a convergence of mental models. This is viewed as an emergent process beginning when team members have unique, independent views of the situation. Through interaction, members' mental model contents become similar (McComb 2007). Last, Stahl (2006) suggests this involves the gradual construction and accumulation of increasingly refined and complex perceptual and conceptual artifacts.

From these views we can glean an abstracted sense of a *conceptual definition of Team Knowledge Development*. First, as teams interact they either implicitly or explicitly engage in processes serving a reduction of uncertainty. This is somewhat akin to individual knowledge-building in that it to involves some form of conceptual change or restructuring. But, it involves more a change in the shared knowledge team members possess. This coalescence and convergence of shared knowledge involves a variety of knowledge types, from the static, (i.e., task and team related skills), to the dynamic, (i.e., a collective understanding of the problem the team is currently addressing). This collectively developed knowledge can encompass shared artifacts produced and refined as teams interact (e.g., graphical representations of task elements conveyed via charts). As such, the conceptual changes experienced at the team level involve a reduction of uncertainty about both the task and the team related artifacts and roles.

More specifically, knowledge-building within a team involves a form of uncertainty reduction having to do with *relationships between uncertainty and communication in teams. These processes may be passive or active and either* convey or disperse information amongst the team (Fleming, personal communication). This is supported by the following processes within the team and these are illustrated in the bottom portion of Figure 9.1.

Sharing unique knowledge

First we can talk about the distribution of knowledge among team members. Social psychology research under the 'hidden knowledge' paradigm has shown in laboratory studies that team discussions tend to converge on common knowledge rather than unique knowledge (see Wittenbaum and Stasser 1996). Specifically, teams often discuss the knowledge that they share rather than the knowledge that is unique to individuals or subsets of individuals within a team. Rather than using the term 'hidden', which implies a purposeful act of explicitly not providing teammates with information, we choose the phrase 'sharing unique knowledge'. This is defined as distributing uniquely-held knowledge – that is, the sharing of task-relevant information held by one or more (but not all) group members. More generally it is an exchange where information is made available to teammates; a communication of unexpressed, unwritten or untaught knowledge of people, artifacts, concepts, and actions related to the task.

In consideration of these definitions, our *conceptual definition of sharing unique knowledge* states that this is a process in which task relevant knowledge held by some fraction of the team is communicated explicitly (e.g., verbal transaction) in such a way that the expression of that knowledge ensures that other members are made aware.

Recognition of expertise

Next is the process 'recognizing expertise' within a team. Here we are discussing the process whereby expressed knowledge is in some way evaluated by team members. This can be seen as an ability of a team to correctly identify and assess the expertise of its members (e.g., Hollenbeck *et al.* 1998; Libby *et al.* 1987; LePine *et al.* 1997; Bonner 2004). Similarly, it occurs when the team considers inputs of individual team members according to the judgment of individual team member's input quality (e.g., Hollenbeck *et al.* 1995).

From this we present our *conceptual definition of recognizing expertise* as involving evaluation of both knowledge and competence. Specifically, in heterogeneous teams, the expertise will be idiosyncratic to the roles held by team members. As the teams interact, they will learn more about the particular type of knowledge each team member holds. This is distinct from recognizing the competence of team members. Competence is independent of type of knowledge and teams engage in interaction processes to understand and sometimes critique the information provided by team members. This leads to a type of dual assessment of the expertise of team members, one involving knowledge and the other involving competence.

Knowledge interoperability

We submit that the functions of individual and team knowledge-building involve both internalized *and* externalized processes employed by the team to create new knowledge. This serves to reduce uncertainty about their situation. We further submit that by iteratively working through the associated processes, and sufficiently resolving uncertainty, the team moves towards *knowledge interoperability*. We *conceptually define knowledge interoperability* as the result of team members learning to represent the gathered information as knowledge objects, icons, or boundary objects (Nosek 2004). The overall effect of this is the creation of mutually specified coordinating representations (Alterman 2001), that is, information that has become interoperable within the teams. Essentially, the outcome of these processes is a construction of a team's preliminary understanding of the situation and their teammates. From this, the teams are able to move to the next major macrococgnitive process, that of more fully conceptualizing the problem they are facing.

Developing shared problem conceptualization

Our next major macrocognitive process is that of *developing a shared problem conceptualization* (Figure 9.2). From the cognitive sciences, this has been said to involve the encoding, representing, and sharing salient aspects of the problem. With respect to this latter aspect, it involves developing overlap between team members' understanding of the essential problem characteristics (e.g., Fiore and Schooler 2004;

Hinsz, Tindale and Vollrath 1997; Orasanu 1994). We more specifically present a *conceptual definition* of this as involving the identification of initial problem states, goals, and operators, that is, the actions that change one problem state into another, along with any restrictions on the operators (Newell and Simon 1972; Hayes 1989).

This function is iteratively supported by the following individual and team macrocognitive processes. What is important to recognize here is that, following the knowledge-building activities in the first major macrocognitive process, individuals within the team use the collaborative activities of this major process to build a perceptual and conceptual understanding of their problem. Specifically, they do this via the externalized processes of visualization of data and through knowledge-sharing and transfer. This enables the team to create a common ground concerning the problem they are facing which, in turn, supports the development of the team's problem model. We discuss each of these in turn.

Visualization of data

The process of 'visualization of data' is considered as a presentation of processed information in a coherent and easily accessible way (Keel, personal communication). The information can be presented in different forms using traditional devices such as pie charts, scatter graphs, line charts etc. (Fayyad *et al.* 2001). Similarly it can be visualization through the use of computer-supported, interactive, visual representations of abstract data to amplify cognition (Card *et al.* 1999). Our *conceptual definition of visualization of data* describes it as the coherent presentation of already processed information using external artifacts that are able to augment cognition.

Knowledge sharing and transfer

The process of 'knowledge sharing and transfer' involves providing and receiving processed information among team members. This may utilize a form of referential communication, that is, communicative acts which are typically spoken but always involve an exchange of information between speakers (Matthias 1999; Yule 1997).

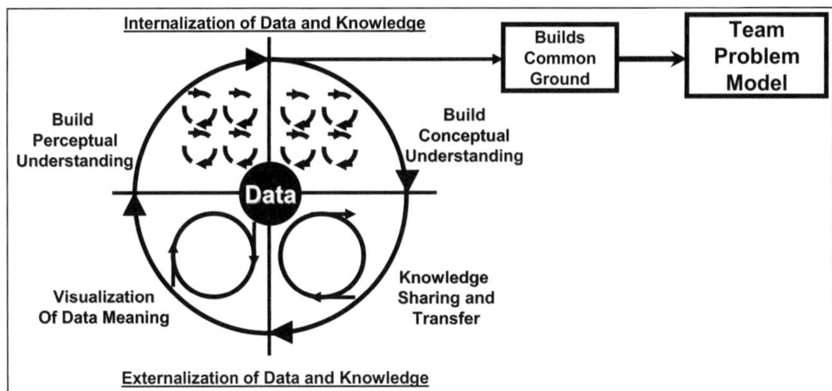

Figure 9.2 Development of shared problem conceptualization

We note that the knowledge that was received cannot be identical to what was transmitted in that there has to be a process of interpretation that is both subjective and framed by the recipient's knowledge. As such, knowledge-sharing implies that there is some generation of knowledge in the recipient whereas we can consider knowledge-transfer as involving information passed to another person in the absence of knowledge generation (e.g., Sharrat and Usoro 2003; Miller 2002). Within this context, our *conceptual definition of knowledge-sharing and transfer* states that it is a process that involves referential communication where, first, there is an exchange of information, and second, comprehension and interpretation of the received knowledge occurs.

Developing common ground

With respect to common ground and its development, when viewed through the lens macrocognition, we consider it as a process of developing shared perceptual and conceptual understanding of the problem situation. In particular, the aforementioned processes of visualization and knowledge sharing, enable the development of common ground – a concept we next describe as consisting of perceptual and linguistic elements.

From the literature on collaboration, researchers suggest more specifically that common ground consists of the pertinent beliefs, knowledge, and assumptions that the involved parties share (Carroll 2006; Klein *et al.* 2004). In the context of macrocognition, we further delineate this concept and suggest that it consists of perceptual and linguistic common ground. Linguistic common ground can be considered a common language for describing tasks (Liang *et al.* 1995), and is sometimes considered as a modification of one's *vocabulary schemas* as referential communication is developed (e.g., Clark and Wilkes-Gibbs 1986). Perceptual common ground concerns shared visualizations, diagrams, or graphical representations that are utilized in collaborative contexts. From this our *conceptual definition of common ground* is that it involves the integration of perceptual and linguistic elements driving the development of conceptual common ground – which is made up of what Clark (1996) describes as joint knowledge, beliefs, and suppositions, and which results from the confluence of mutual visual and linguistic understanding.

Essentially we argue that visualization of data and knowledge sharing and transfer iteratively and interactively support this process and it involves an internalization and an externalization of the knowledge built. As the team moves through these processes they are then able to more specifically focus on the development of the problem model. In particular, once the teams are comfortable with their initial assessment of the data, and have transformed the multitude of data to information and formed a common ground, the team understands that they can then begin to create their model of the problem.

Team problem model

Next is the process of developing the team problem model. Orasanu (1994), in her analysis of problem-solving in teams, notes that it consists of creating a shared understanding of the problem situation. Our *conceptual definition of the team problem model* states that this understanding contains an agreement upon a number of critical problem elements. This includes understanding the specific situation or context in which the problem is emerging, along with some knowledge of the nature and causes of the problem. Concomitant with this is an understanding of the available cues and their meaning. As this builds, it includes a clear and shared understanding of the goals or desired outcomes along with predictions about what is likely to happen in the future with and without action by the team.

Option generation and team consensus development

Last we have *consensus development* (Figure 9.3), a complex major macrocognitive process in that it is served by a variety of processes that involve some manner of manipulating the problem model to generate agreed upon solution options. In particular, we can state that the team is now operating within the problem space, and engaging in a robust interrogation and interpretation of potential solution options in order reach consensus on what to do. Fleming (personal communication), notes that this function 'involves various forms of group judgment in which a group (or panel) of experts interacts in assessing an intervention and formulating findings by vote or other process of reaching general agreement.' This major macrocognitive process is served by the following processes engaged by the team to arrive at the mutual recognition of, and agreement upon, a given proposition, that is, develop consensus. Members within the team work toward agreed upon solutions via internalized processes such as intuitive decision-making and mental simulation and these are evaluated via externalized processes such as negotiation of solution alternatives and storyboarding. We describe each of these in turn.

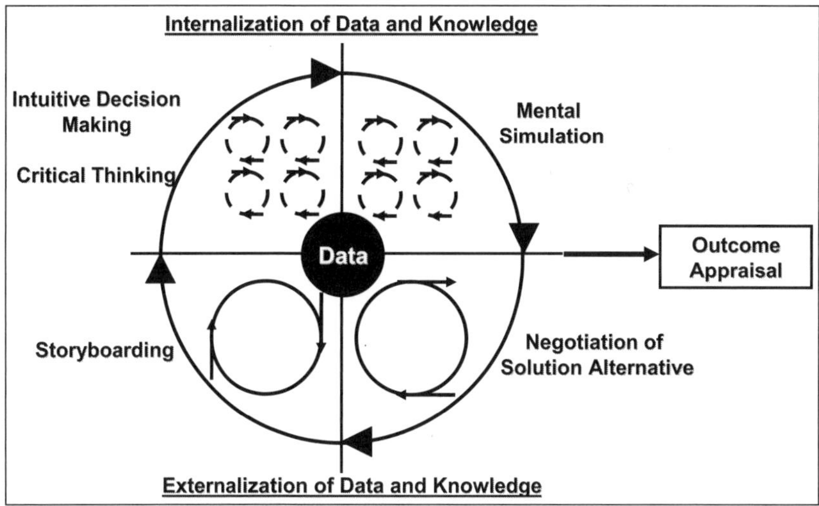

Figure 9.3 Consensus development process

Option generation

At the onset of this major macrognitive process, is that of option generation, where individuals and the team modify and manipulate the problem elements to derive solutions. Here they work to generate alternatives that satisfy some list of attributes and which are reasonable given the constraints of the problem situation (Butler and Scherer 1997; von Winterfeldt and Edwards 1986). Within naturalistic decision-making, researchers argue that option generation does not occur in that experts are believed to retrieve a single option in response to recognition of a problem and that they test that solution through mental simulation and repair it or discard it (Klein 1997). Nonetheless, we suggest that this is still a form of option generation, but it is one that occurs sequentially and in the absence of a more critical comparison of alternatives. Our *conceptual definition states that option generation* involves the identification of suitable decision alternatives or the determination of a course of action among alternatives. This takes place via the following macrocognitive processes.

Intuitive decision-making

First we can consider the internalized processes involving more of a type of 'intuitive decision-making.' Note that in this stage of the problem-solving process the decision in question involves that of generating solution alternatives. In the case of intuitive decisions, little apparent effort is used to reach them, that is, they arise typically without conscious awareness (Hogarth 2001). In support of this, Damasio and colleagues have looked at intuitive decision-making at the neurophysiological level and, within studies of quick and often stressful decision-making have found evidence for a two-step process of decision-making. In the first step an affective or emotional response to the situational context of the decision produces a 'somatic marker' which is thought to be based upon non-declarative or implicit knowledge as well as biases. Only following this might there be a more explicit recall of relevant facts pertaining to prior experience with the decision task. Specifically, 'the hypothesis thus suggests that somatic markers normally help constrain the decision-making space by making that space manageable for logic-based, cost–benefit analyses. In situations in which there is remarkable uncertainty about the future and in which a decision should be influenced by previous individual experience, such constraints permit the organism to decide efficiently within short time intervals' (Damasio 1996, p. 1415). Bierman, Destrebecqz and Cleeremans (2005) assessed this using an artificial grammar decision task and identified what they called a *preconceptual* phase where performance was improved prior to participant's ability to articulate rules they were following. Importantly, this research found that skin conductance measures were able to differentiate correct from incorrect answers prior to response. Interestingly, the galvanic skin response was greater before an incorrect answer was provided. Bierman *et al.* argue that this *somatic marker* may actually function to warn the participant of a possibly incorrect answer. Additionally, some suggest that, when team members are faced with a decision event, they may intuitively compare that event to knowledge potentially held in their shared mental models. If it can be understood by this comparison, Kline (2005) suggests that a team will implicitly know the solution and is able to rapidly reach consensus. From these definitions we derive our *conceptual definition of intuitive decision-making* as that which occurs through internalized processes but in the absence of reportability, that is, the processes preceding a decision can occur without awareness and may be measurable via indirect metrics.

Critical thinking

In addition to processes where team members may *implicitly* know a solution, *critical thinking* may be engaged. This is defined as a deliberate and explicit reasoning process, a process that is to some degree, rule-governed and precise and which may pertain to abstract thought (Hogarth 2005). It essentially entails a determination as to whether or not assessments and/or assumptions have been accurately derived (Ennis 1962; Moore 2004). From this we can note that critical thinking requires an evaluation of the degree to which the team has adequately grasped the meaning of statements and are able to judge ambiguities, assumptions or contradictions in reasoning. From this we derive our *conceptual definition of critical thinking* as that process through which solutions, and the definitions used to generate solutions, are critiqued in an overall assessment of their veracity.

Mental simulation

Following this, individual team members may engage in a more internalized process of 'mental simulation.' In this case one evaluates if an action is satisfactory (Klein 1998). This involves their acting out, in their imagination a mental simulation of the successive steps. This also involves a simulation of the potential outcomes, and problems that are likely to be encountered (Klein 1993). Along these lines, Jones *et al.* (2003) describe this as a process in which mental models are used to run a mental simulation of the situation. From these definitions we derive our *conceptual definition of mental simulation* as the process that enables one to anticipate what might happen over time. This might involve pattern matching between features of the situation and schema of known similar situations in order to base projections on the outcomes of these situations.

Negotiation of solution alternatives

Teams then engage in an externalized process of negotiation in which teams interact to assess a given idea and render a decision though discussion so as to seek resolution to any potential differences of interpretation. This process requires shared recognition of the following points. First, the team must agree that a decision is required on an action related to the interests of many people under time constraints. Second, the team must understand that it is possible to assess and compare the options. Last, the team must acknowledge that only non-coercive means of communication are allowed to reach consensus (Fleming, personal communication). Based upon this our *conceptual definition of negotiation* in the context of macrocognition requires externalized interaction processes to resolve opposing interpretations.

Storyboarding

The externalized analog of mental simulation is the process of 'Storyboarding'. This is used to support collaborative efforts at organizing and communicating knowledge in support of developing a story about the problem. This process takes problem elements and events and conveys and arranges them somewhat sequentially. This involves externalized knowledge using quasi-visual representations for conceptualizing and organizing information. From this we state for our *conceptual definition of storyboarding* that it is an externalized collaborative processes designed to capture what might happen and to try and generate consensus on this story.

What is important to recognize at this point is that, the generated options may still not be agreed upon or be deemed substandard or not plausible after the team engages in the aforementioned processes. In such instances, the team will generate additional options and move through these processes again. If they believe that the conceptualization is adequate, they implement their solution and evaluate its effectiveness in the following major macrocognitive process.

Outcome appraisal

Our final major macrocognitive process in which teams engage is 'Outcome Appraisal' where they evaluate the success of a selected solution option against its effectiveness in meeting the stated goals. Specifically, this requires processes of feedback interpretation, and may involve replanning. This relies upon the interpretation and utilization of 'feedback,' that is, information concerning changes in the environment after the decision has been implemented or information of outcomes or consequences based upon the executed actions. In the context of replanning the team may adapt or completely change its solution if that solution was judged not to meet the goals. This may require that the teams engage in adaptive processes requiring that the team make modifications sometimes on the fly (Klein and Pierce 2001). Within this context our *conceptual definition of outcome appraisal* is use of feedback to evaluate the degree to which an implemented solution met its goals and which may require adaptation and replanning if the solution was not effective.

Processes occurring across functions

Across all of these processes, a set of processes are occurring (see Figure 9.4). Specifically, the following processes form a necessary support and development structure for collaboration to occur across the aforementioned stages. As problem-solving unfolds both individuals and the team must continually coordinate with their teammates, be at least minimally aware of teammate activity, and learn from and reflect upon this activity. First, for effective teamwork, there needs to be the process of 'Coordination'. Most succinctly, this involves 'orchestrating the sequence and timing of interdependent actions' (Marks, Mathieu and Zaccaro 2001, p. 363). We choose this definition because it is most faithful to the etymological origins of the term. In particular, the word coordination is derived from three distinct concepts, those being 'arrange,' 'order,' and 'together.' Second, concomitant with team coordination is the process of maintaining 'Workspace Awareness'. This has been defined as the 'up-to-the-moment understanding of another person's interaction with the shared workspace' (Gutwin and Greenberg 2002, p. 417). Others describe it as a form of Activity Awareness, that is, awareness of project work that supports group performance in complex tasks (Carroll *et al.* 2003). Similarly, this is seen as a form of external cognition in that it involves augmenting cognitive processes with cognitive artifacts, such as external writings, visualizations, and work spaces. What is critical across these conceptualizations of awareness is that it forms a necessary, but not sufficient, component of coordination, that is, in the absence of awareness of collaborative activity, coordination becomes problematic. Finally, we note that there must be a process of 'Team Learning'. This is defined as 'an ongoing process of reflection and action, characterized by asking questions, seeking feedback, experimenting, reflecting on results, and discussing errors or unexpected outcomes of actions' (Edmondson 1999, p. 353).

Conclusions

Summarizing the macrocognitive processes

In short, as teams move through the processes illustrated in Figure 9.1 through 9.3, we can sum up macrocognitive processes as follows. First, as was illustrated in Figure 9.1, individuals within the team, and the teams themselves work to build knowledge. They accomplish this through a process of uncertainty reduction about information and about their team. The individuals work through an iterative process of internalization of knowledge as they engage in pattern recognition and mental model development. Similarly, the teams as a whole work through an iterative process of externalization of knowledge as they share unique knowledge and recognize expertise. These processes serve to produce knowledge interoperability and they launch the team to the next stage.

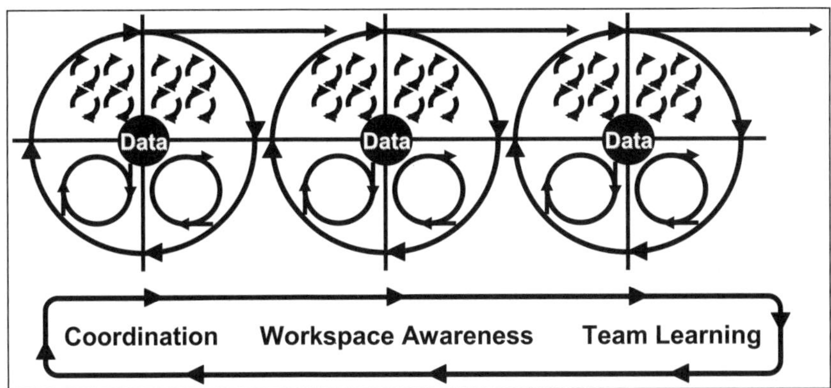

Figure 9.4 Macrocognitive processes occurring across stages

As was shown in Figure 9.2, teams work to develop a shared understanding of their conceptualization of the problem. This involves encoding, representing, and sharing salient aspects of the problem such that the individual team members are able to build perceptual and conceptual understanding. They do this through externalized processes involving an iterative utilization of visualization of data along with knowledge-sharing and transfer. These serve to create perceptual and linguistic common ground through externalization of the knowledge built. These, in turn, coalesce to produce conceptual common ground. All of these drive the development of the team problem model which is then used in the team's attempts at generating solutions.

Finally, as seen in Figure 9.3, teams work to generate options and achieve consensus. They move through a complex process of intuitive decision-making and critical thinking as they examine the veracity of their potential solutions. In service of this, individuals engage in mental simulation to run through their solutions. From the standpoint of externalization of knowledge the teams engage in a negotiation of the solution alternatives as they attempt to explicitly work through the choices. This may involve another externalization of knowledge in the form of storyboarding. These processes are engaged by the team to arrive at the mutual recognition of, and agreement upon, a given proposition, that is, develop consensus.

Recognizing thematic elements across the overall process

When we step back and evaluate what is occurring as collaboration ensues, we can see that as teams move through these stages of problem-solving, there is a continual process of an 'effort after meaning' (cf. Bartlett 1932), that is, an attempt to derive an understanding given a complex set of inputs. Essentially, we see an evolution of understanding within the team. This requires an interplay between perceptual and conceptual processes in that the team must integrate visual and verbal understanding to make meaning. This actually increases in complexity as the team moves through the problem-solving stages. It evolves from a fairly straightforward process of pattern recognition to a more complex visualization of data meaning to storyboarding.

Across these we also see a continual interplay between internalization and externalization of cognition. For example, internal processes such as pattern recognition and mental model development are occurring along with external processes such as sharing unique knowledge and recognizing expertise. Essentially an understanding of constituent elements of a problem is acquired and then integrated for higher level interpretation. This interplay between internalization and externalization of knowledge is at the core of collaborative problem-solving. Individuals within the team, and the team itself, comprehend the elements of their problem situation, by interacting and interpreting the knowledge arising from the environment and that which is held by the team.

We further submit that this evolves from a reduction of uncertainty early in the collaboration process as the team considers both data and their teams mates, to a determination of plausibility. Specifically, first, the teams interact with the environment and each other to better understand incoming data and the competence and knowledge of their team. Second, processes are engaged to act upon acquired knowledge and evaluate the utility and realizability of the identified potential courses of action.

In sum, in this chapter we have attempted to refine the field's conceptual understanding of macrocognition. This was developed within the context of internalized and externalized high-level mental processes. What is next required is a fuller understanding of these processes reified within notions such as combining, visualizing, and aggregating information. Similarly, development of these concepts must be more fully considered through theoretical notions of teams attempting to reduce uncertainty and resolve ambiguity. In addition, these concepts must help us to better understand the discovery of new knowledge and relationships. Finally, from these conceptual definitions, we must develop operational definitions for these terms and move to assess their use in complex, one-of-a-kind, collaborative problem-solving, that is, in macrocognitive problem-solving.

Acknowledgments

The views, opinions, and findings contained in this article are the authors and should not be construed as official or as reflecting the views of the Department of Defense or the University of Central Florida. We thank Nancy Cooke, Bob Fleming, Holley Handley, Paul Keel and Harvey Smallman for their insights on the definitions presented in this chapter as well as the researchers from the Collaboration and Knowledge Interoperability Program for their many helpful discussions and input on the concepts described in this chapter. Writing this chapter was supported by Grant N000140610118 from the Office of Naval Research awarded to S. M., Fiore, S. Burke, F. Jentsch, and E. Salas, University of Central Florida.

References

Alterman, R., Feinman, A., Landsman, S. and Introne, J. (2001). 'Coordinating Representations in Computer-mediated Joint Activities', *Proceedings of the 23rd Annual Conference of the Cognitive Science Society*, pp. 43–48. (Hillsdale, NJ: Erlbaum).

Bartlett, F.C. (1932). *Remembering – A Study in Experimental and Social Psychology.* (Cambridge: Cambridge University Press).

Bierman, D., Destrebecqz, A. and Cleeremans, A. (2005). 'Intuitive Decision Making in Complex Situations: Somatic Markers in an Implicit Artificial Grammar Learning Task', *Cognitive, Affective, and Behavioral Neuroscience.*

Bonner, B.L. (2004). 'Expertise in Group Problem Solving: Recognition, Social Combination, and Performance', *Group Dynamics: Theory, Research, and Practice* 8(4), 277–290.

Brehmer, B. (1990). 'Strategies in Real-time, Dynamic Decision Making', in R.M. Hogarth, (ed.), *Insights in Decision Making: A Tribute to Hillel J. Einhorn*, pp. 262–279. (Chicago, IL: University of Chicago Press).

Butler, A.B. and Scherer, L.L. (1997). 'The Effects of Elicitation Aids, Knowledge, and Problem Content on Option Quantity and Quality', *Organizational Behavior and Human Decision Processes*, 72(2), 184–202.

Card, S.K., MacKinlay, J. and Schneiderman, B. (1999). *Readings in Information Visualization: Using Vision to Think*. (San Diego, CA: Academic Press).

Carroll, J.M. (2003). 'Toward a Multidisciplinary Science of Human–computer Interaction', in J. M. Carroll, (ed.), *HCI models, Theories, and Frameworks: Toward a Multidisciplinary Science*, pp. 1–9. (Amsterdam: Morgan Kaufmann).

Carroll, J.M. (2006). *A Multiple View Approach to Support Common Ground in Geo-Collaboration*, available at http://cscl.ist.psu.edu/public/projects/ONRproject/index.html.

Clark, H. H. (1996). *Using Language*. (Cambridge: Cambridge University Press).

Clark, H.H. and Wilkes-Gibbs, D. (1986). 'Referring as a Collaborative Process', *Cognition* 22(1), 1–39.

Cooke, N.J., Salas, E., Kiekel, P.A. and Bell, B. (2004). 'Advances in Measuring Team Cognition', in E. Salas and S.M. Fiore, (eds), *Team Cognition: Understanding the Factors that Drive Process and Performance*. (Washington, DC: APA).

Damasio, A.R. (1996). 'The Somatic Marker Hypothesis and the Possible Functions of the Prefrontal Cortex', *Philosophical Transactions: Biological Sciences* 351(1346), 1413–1420.

Edmondson, A.C. (1999). 'Psychological Safety and Learning Behavior in Work Teams', *Administrative Science Quarterly* 44, 350–383.

Ennis, R. (1962). 'A Concept of Critical Thinking',. *Harvard Educational Review,* 32, 81–111.

Fayyad, U., Grinstein, G. and Wierse, A. (2001). *Information Visualization in Data Mining and Knowledge Discovery*. (San Francisco, CA: Elsevier).

Feltovich, P.J., Hoffman, R.R., Woods, D. and Roesler, A. (May–June, 2004). 'Keeping it too Simple: How the Reductive Tendency Affects Cognitive Engineering', *IEEE Intelligent Systems* May–June, 90–94.

Fiore, S. M. and Schooler, J.W. (2004). 'Process Mapping and Shared Cognition: Teamwork and the Development of Shared Problem Models', in E. Salas and S.M. Fiore, (eds), *Team Cognition: Understanding the Factors that Drive Process and Performance* (pp. 133–1582). (Washington, DC: APA).

Gonzalez, C. and Quesada, J. (2003). 'Learning in Dynamic Decision Making: The Recognition Process', *Computational and Mathematical Organization Theory* 9, 287–304.

Gutwin, C. and Greenberg, S. (2002). 'Descriptive Framework of Workspace Awareness for Real-time Groupware', *Computer Supported Cooperative Work,* 11(3–4), 411–446.

Hayes, J.R. (1989). *The Complete Problem Solver*. (Hillsdale, NJ: Erlbaum).

Hinsz, V.B., Tindale, R.S. and Vollrath, D.A. (1997). 'The Emerging Conceptualization of Groups as Information Processors', *Psychological Bulletin* 121, 43–64.

Hoffman R.R, Coffey J.W. and Ford, K.M. (2000). *A Case Study in the Research Paradigm of Human-centered Computing: Local Expertise in Weather Forecasting*. Report to the National Technology Alliance, Arlington, VA.

Hogarth, R.M. (2001). *Educating Intuition*. (Chicago, IL: The University of Chicago Press).

Hogarth, R.M. (2005). 'Deciding Analytically or Trusting your Intuition? The Advantages and Disadvantages of Analytic and Intuitive Thought', in T. Betsch and S. Haberstroh, (eds), *The Routines of Decision Making*, (pp. 67–82). (Mahwah, NJ: Erlbaum).

Hollenbeck, J.R., Ilgen, D.R., LePine, J.A., Colquitt, J.A. and Hedlund, J. (1998). 'Extending the Multilevel Theory of Team Decision Making: Effects of Feedback and Experience in Hierarchical Teams', *Academy of Management Journal* 41(3), 269–282.

Hollenbeck, J.R., Ilgen, D.R., Sego, D.J., Hudlund, J., Major, D.A. and Phillips, J. (1995). 'Multilevel Theory of Team Decision Making: Decision Performance in Teams Incorporating Distributed Expertise', *Journal of Applied Psychology* 80, 292–316.

Johnson-Laird, P.N. (1999). 'Mental Models', in R.A. Wilson and F.C. Keil, (eds), *The MIT Encyclopedia of the Cognitive Sciences*, pp. 525–527. (Cambridge, MA: The MIT Press).

Jones, D.G., Quoetone, E.M., Ferree, J.T., Magsig, M.A. and Bunting, W.F. (2003). 'An Initial Investigation into the Cognitive Processes Underlying Mental Projection'. Paper presented at the Human Factors and Ergonomics Society 47th Annual Meeting, Denver, CO.

Joslyn, S. and Jones, D. (2006). 'Strategies in Naturalistic Decision-Making: A Cognitive Task Analysis of Naval Weather Forecasting', in J.M. Schragen, L. Militello, T. Ormerod and R. Lipshitz, (eds), *Naturalistic Decision Making and Macrocognition*. (Aldershot: Ashgate).

Klein, G. (1993). 'A Recognition Primed Decision (RPD) Model of Rapid Decision Making', in G. Klein, J. Orasanu, R. Calderwood and C.E. Zsambok, (eds), *Decision Making in Action*, pp. 138–147. (Norwood, NJ: Ablex).

Klein, G. (1997). 'The Recognition-Primed Decision (RPD) Model: Looking Back, Looking Forward', in C. E. Zsambok and G. Klein, (eds), *Naturalistic Decision Making*, pp. 285–292. (Mahwah, NJ: Erlbaum).

Klein, G. (1998). *Sources of Power: How People Make Decisions*. (Cambridge, MA: MIT Press).

Klein, G. and Pierce, L. (2001). 'Adaptive Teams', *Proceedings of the 6th International Command and Control Research and Technology Symposium*.

Klein, G., Woods, D.D., Bradshaw, J.M., Hoffman, R. and Feltovich, P. (2004). 'Ten Challenges for Making Automation a "Team Player" in Joint Human–agent Activity', *IEEE Intelligent Systems* 9(6), 91–95.

Klein, G. and Hoffman, R. R. (1992). 'Seeing the Invisible: Perceptual-cognitive Aspects of Expertise', in M. Robinowitz (ed.), *Cognitive Science Foundations of Instruction*, pp. 203–226. (Mawah, N.J.: Lawrence Erlbaum).

Kline, D.A. (2005). 'Intuitive Team Decision Making', in H. Montgomery, R. Lipshitz and B. Brehmer, (eds), *How Professionals Make Decisions*, pp. 171–182. (Mahwah, NJ: Erlbaum).

Klusch, M. (Ed.) (1999). *Intelligent Information Agents: Cooperative, Rational and Adaptive Information Gathering on the Internet*. (Berlin: Springer Verlag).

Letsky, M., Warner, N., Fiore, S.M., Rosen, M.A. and Salas, E. (2007). 'Macrocognition in Complex Team Problem Solving', in *Proceedings of the 12th International Command and Control Research and Technology Symposium (12th ICCRTS), Newport, RI, June 2007*. (Washington, DC: U.S. Department of Defense Command and Control Research Program).

Liang, D.W., Moreland, R.L. and Argote, L. (1995). 'Group Versus Individual Training and Group Performance: The Mediating Role of Transactive Memory', *Personality and Social Psychology Bulletin* 21, 384–393.

Libby, R., Trotman, K.T. and Zimmer, I. (1987). 'Member Variation, Recognition of Expertise, and Group Performance', *Journal of Applied Psychology* 72(1), 81–87.

Liljenström, H. and Svedin, U. (2005). 'System Features, Dynamics, and Resilience – Some Introductory Remarks', in H. Liljenström and U. Svedin, (eds), *Micro – Meso – Macro: Addressing Complex Systems Couplings*, pp. 1–16. (London: World Scientific Publications Company).

Marks, M.A., Mathieu, J.E. and Zaccaro, S.J. (2001). 'A Temporally Based Framework and Taxonomy of Team Processes', *Academy of Management Review* 26(3), 356–376.

McComb, S.A. (2007). 'Mental Model Convergence: The Shift from being an Individual to being a Team Member', in F. Dansereau and F.J. Yammarino, (eds), *Multi-level Issues in Organizations and Time*, pp. 95–147. (Amsterdam: Elsevier).

Miller, F.J. (2002). I = 0 (Information has no intrinsic meaning). *Information Research,* 8(1), available at http://InformationR.net/ir/8-1/paper140.html.

Moore, T. (2004). 'The Critical Thinking Debate: How General are General Thinking Skills?', *Higher Education Research and Development*, 23(1), 3–18.

Newell, A. and Simon, H.A. (1972). *Human Problem Solving*. (Englewood Cliffs, NJCA: Prentice-Hall).

Nosek, J.T. (2004). 'Group Cognition as a Basis for Supporting Group Knowledge Creation and Sharing', *Journal of Knowledge Management* 8(4), 54–64.

Orasanu, J. (1994). 'Shared Problem Models and Flight Crew Performance', in N. Johnston, N. McDonald and R. Fuller, (eds), *Aviation Psychology in Practice*, pp. 255–285. (Brookfield, VT: Ashgate).

Orasanu, J. and Salas, E. (1993). 'Team Decision Making in Complex Environments', in G. Klein, J. Orasanu, R. Calderwood. and C.E. Zsambok (eds), *Decision Making in Action: Models and Methods*, pp. 327–345. (Norwood, NJ: Ablex).

Rouse, W.B. and Morris, N.M. (1986). 'On Looking into the Black Box: Prospects and Limits in the Search for Mental Models', *Psychological Bulletin* 100, 349–363.

Salas, E. and Klein, G. (ed.) (2001). *Linking Expertise and Naturalisitc Decision Making*. (Mahwah, NJ: Lawrence Erlbaum Associates).

Salas, E., Dickinson, T., Converse, S. and Tannenbaum, S. (1992). 'Toward an Understanding of Team Performance and Training', in R. Swezey and E. Salas, (eds), *Teams: Their Training and Performance*. (Norwood, NJ: Ablex Publishing).

Scardamalia, M. and Bereiter, C. (2003). 'Knowledge Building Environments: EXTENDING the Limits of the Possible in Education and Knowledge Work', in A. DiStefano, K.E. Rudestam and R. Silverman, (eds), *Encyclopedia of Distributed Learning*, pp. 269–272. (Thousand Oaks, CA: Sage Publications).

Sharratt, M. and Usoro, A. (2003). 'Understanding Knowledge-Sharing in Online Communities of Practice', *Electronic Journal of Knowledge Management* 1(2), 187–197.

Spiro, R.J. and Jehng, J. (1990). 'Cognitive Flexibility and Hypertext: Theory and Technology for the Non-linear and Multidimensional Traversal of Complex Subject Matter', in D. Nix and R.J. Spiro, (eds), *Cognition, Education, and Multimedia*, pp. 163–205. Hillsdale, NJ: Erlbaum.

Spiro, R.J., Feltovich, P.J., Coulson, R.L. and Anderson, D.K. (1989). 'Multiple Analogies for Complex Concepts: Antidotes for Analogy-induced Misconception in Advanced Knowledge Acquisition', in S. Vosniadou and A. Ortony, (eds), *Similarity and Analogical Reasoning*, pp. 498–531. (New York: Cambridge University Press).

Stahl, G. (2006). *Group cognition: Computer support for building collaborative knowledge*. (Cambridge, MA: MIT Press).

Sundstrom, E., de Meuse, K.P. and Futrell, D. (1990). 'Work Teams: Applications and Effectiveness', *American Psychologist* 45(2), 120–133.

Trafton G.J., Kirschenbaum S.S., Tsui T.L., Miyamoto R.T., Ballas J.A. and Raymond P.D. (2000). 'Turning Pictures into Numbers: Extracting and Generating Information from Complex Visualizations', *International Journal of Human–Computer Studies* 53, 827–850.

Von Winterfeldt, D. and Edwards, W. (1986). *Decision Analysis and Behavioral Research*. (Cambridge: Cambridge University Press).

Warner, N., Letsky, M. and Cowen, M. (2005). 'Cognitive Model of Team Collaboration: Macro-cognitive Focus'. Paper presented at the 49th Annual Meeting of the Human Factors and Ergonomics Society, Orlando, FL.

Wittenbaum, G.M. and Stasser, G. (1996). 'Management of Information in Small Groups', in J.L. Nye and A.M. Brower, (eds), *What's so Social about Social Cognition?*, pp. 3–363. (Thousand Oaks, CA: Sage).

Yule, G. (1997). *Referential Communication Tasks*. (Hillsdale, NJ: Erlbaum).

Zsambok, C. E. and Klein, G. (eds) (1997). *Naturalistic Decision Making*. Hillsdale, NJ: Erlbaum.

Chapter 10

Augmenting Video to Share Situation Awareness More Effectively in a Distributed Team

David Kirsh

Team members share and coordinate situation awareness. By talking about what each knows, viewing and handling artifacts together, they become mutually aware of the same states and conditions. The process of working on common questions focuses and coordinates their views and concerns. These standard coordinating and regulating processes in what Letsky *et al.* call 'macrocognition'. When the information gap between team members increases beyond some threshold, however, or the situation awareness maintained by a few diverges from the rest and needs to be disseminated to keep everyone in synch, sharing practices need to change. In such cases, nothing is more effective than giving a presentation face-to-face.

Can we develop multimedia presentations that work equally well whether or not audience and speaker are face-to-face? Are there ways to enhance the cognitive effectiveness of a video of an activity so that distributed team members can share their situation awareness better? These are the general question we set out to explore.

Consider what is special about face-to-face presentations. Presenters stand in front of an audience, usually with visual media such as whiteboard, PowerPoint, maps, perhaps a short video. They reach out to listeners by pointing and gesturing at the media and speaking in a way that explains the visuals. This is all part of the coordination process of macrocognition. The social presence of being face-to-face with their audience helps presenters to capture and manage audience attention. Presenters watch their audience to sense uptake; they gauge their own intonation and emphasis to draw attention to what is most important. They speed or slow their pace to adjust the content bandwidth to what they think the audience can absorb. All these extras come from being face-to-face with an audience, and they are resources that are central to macrocognition, distributed cognition, shared situation awareness, or any of the other expressions used to characterize how groups build common ground, intersubjectivity and group cognition. The speaker and audience observe each other, react to each other and, when permitted, the audience may ask questions and partially drive discourse. Speaker and audience form a closely coupled, coordinated system. Because of the range of these couplings, presenter and audience have a collection of information-bearing channels they can modulate to add to the *cognitive effectiveness* of presentations (gesture, intonation, facial appearance, body language, simultaneous speech and action etc.). Both presenter and audience can use these channels to increase bandwidth and focus.

Presentations are easy to video record so it might seem that sending digitally recorded presentations is a good way for teams distributed in space and time to keep in synch and share situation awareness. Recordings lose some of the spontaneous interactivity between live presenter and live audience; but there remain many channels for information sharing and social engagement. Currently, video summaries are widely used to coordinate global teams in multinational companies that rely on 24-hour work on design projects. As one team finishes its shift, it sends a digital video to the incoming team in some remote time zone documenting what it has done or left undone. In military or police contexts, video is sometimes used to help distributed teams share plans. Videos – especially unmanned autonomous vehicle (UAV) videos that include analysts' comments and annotations – are used to pass on information about strategic, tactical or environmental conditions. The number of post-produced videos, annotated or edited, is growing exponentially; and more recently the trend is to share video that is narrated or commented in real time.

Common sense tells us that a video or graphic that is actively annotated and commented ought to be more effective than one without annotation. For instance, the sports commentator John Madden, typically explains football plays to his television audience by using lines and arrows to show the moves and rationale of the players on the field. This is clearly more effective than simply talking about a play or talking to a static image. But is this the best there is? Should a commentator actively draw over a simple still background image? Should the background be a video, showing players moving? If the background is moving is it best for the annotations to be still? Madden draws his annotations as viewers watch, so they are dynamic over a still background. If the display, by contrast, were a UAV video, should commentators include annotations that present helpful background information? If so, how much? And how should they be laid over the video?

There are few scientific results to motivate guidelines for presenters who want to share situational knowledge among distributed team members. This is surprising given the number of articles on illustration, animation, video and other media. One problem is that there is no adequate conception of cognitive efficiency in this domain. How should we measure the comparative efficiency of video, annotated video, multimedia presentations or even basic non-linguistic representations, such as pictures, illustrations or recorded soundscapes? It is one thing to say of a linguistic item, a sentence for example, that it is cognitively less efficient than a semantically identical one. 'That is something up with which I will not put' seems obviously less efficient than the equivalent: 'I won't put up with that'. We hardly need linguistic and psycholinguistic theories to explain why processing the first sentence is harder. It is quite another thing, however, to say that a picture with one caption is more efficient than another picture and caption, or that one annotated video is more effective than another.

Things get even more complicated when we ask how pointing and gesture, gaze and pace affect cognitive efficiency. These complementary elements modulate and transform content, but do not seem to be channels that carry independent content on their own. They rely on a core channel to carry the basic message and modulate that. Thus, video is a core channel because it carries content by itself, as does audio. But gaze, pointing, pace and most gestures without images, audio or video, carry little content in isolation. There are occasions, to be sure, when specific gestures or facial expressions are sufficient to convey a specific message. However, in general, all such 'extras' are complementary or derivative. This makes analyzing these extras especially difficult.

Passing the bubble of situational awareness

To make this problem more tractable, and more directly relevant to sharing situation awareness of fast-changing situations, we designed experiments to measure how 'passing the bubble' could be enhanced by annotations of different sorts. 'Passing the bubble' refers to transferring situation awareness. Although the notion of situation awareness needs clarification, we use the expression to mean the awareness that a participant engaged in an activity continuously builds up ongoing states, processes and objectives in virtue of their *immersion*. It is knowledge of context. Typically, much of what someone must know about current context is not visible. A newcomer cannot simply look around and figure out what is going on. Situation awareness is much deeper. It refers also to a participant's goals and concerns; to their beliefs about what might happen next, and what they think other participants know and want. The act of transferring to someone else this immersive knowledge of 'what is going on', this knowledge acquired by participating in activity, is the problem of passing the bubble of situation awareness.

Here is a familiar example. When a ship's captain transfers command to another captain he cannot just hand over a set of notes, and say 'here is everything you need to know'. It takes time to acquaint the incoming captain with his concerns, with the things he is monitoring, with what seems to be working well on some problems and what is not working on others. He must share intent, future plans, a sense of the team and so on. Notes alone are insufficient because they need to be situated in real time via annotation and contextual emphasis. As Polanyi famously said in his work on tacit knowledge [the captain] 'knows more than he can say'. He has to *show – not just tel* – his concerns and *share* his immersion. To share this more tacit knowledge the outgoing captain walks around the ship with the incoming captain, gesturing, pointing, and explaining concerns and objectives.

Our approach to studying the efficacy of different video methods for passing the bubble was to create an experimental context where subjects immerse themselves in a computer strategy game, Starcraft™, developing a rich situational awareness of their strategic and tactical position. They know their own objectives and have conjectures and tacit knowledge of where the enemy is and what they might be up to. They must then pass this awareness by means of a presentation to someone who will take over their position and continue the game. We call this new player the 'recipient' or 'bubble receiver'. We tested conditions in which audio and video technology might be used to communicate over distances. To cover both synchronous and asynchronous bubble passing some of our presentations were live and some non-live. *Non-live* presentations were prepared beforehand by the outgoing player and viewed by the recipient in a canned form. To make these presentations we gave presenters as many hours as they wanted to carefully craft presentations. In some conditions they used still images as the background media and annotated them; in others they annotated video snippets they chose from the game; in a third condition, the control case, presenters annotated randomly selected background stills. Our goal in the live and canned presentation experiment was to create enough conditions to titrate out the impact of different components of presentations. We discuss the full experimental design below.

By analyzing these components we have begun to provide an empirical basis for guidelines on how to annotate UAV style videos, and how to design presentations to share situation awareness in distributed teams effectively. These principles will guide presenters in the difficult process of editing, annotating and selecting media from videos for asynchronous sharing; and help guide them in the choice of gesture and annotation for real-time presentations.

Current technology and science

There is an extensive literature on using multimedia to share information. Here we consider two representative studies that are especially relevant. The first, by anthropologist and cognitive ethnologist Charles Goodwin (1994), studies the use of annotated slideshows by expert witnesses in a courtroom. It suggests that to explain how to interpret legal evidence, it is more effective to show annotated still images than video clips of the same evidence. The second, by Mayer *et al.* (2005), is an experimental comparison of the usefulness of animations versus illustrations. In their carefully controlled study, subjects were given different types of presentations on such things as how mechanisms like a flush toilet or car brakes work, and tested for their recall and understanding.

Slideshows are better than video's for presenting to non-experts

In his article 'Professional vision' Goodwin discusses the Rodney King trial in which the Los Angeles Police Department's legal defense team analyzes a citizen's video showing several white policemen kicking and beating an Afro-American male – Rodney King – who was, at the time, lying on the ground. At the end of the trial, the (entirely white) Simi Valley jury found the policemen in the video not guilty of using excessive force – a judgment so at odds with public opinion that it sparked some of the worst race riots and looting in Los Angeles history.

How did the defense succeed in getting the jury to see the episode the way the police did? How did they communicate the situation from the perspective of the policemen who were immersed in the real situation, and who claimed that 'if you were there and involved you would understand that things were not quite what they seem on the video'?

They did what defense teams usually do: they called in an expert witness to explain and interpret the facts. The witness, a specialist in the rule book on police protocol, used a slide show of key frames from the King video to explain to the jury what was important and what was not according to the theory of proper police engagement. During his slideshow, he gestured, pointed, made annotations and of course spoke as engagingly as he could.

Goodwin argues that police, like other professions, have built up practices that depend on shared expertise. In his view:

> profession, a community of competent practitioners, most of whom have never met each other … nonetheless expect each other to be able to see and categorize the world in ways that are relevant to the work, tools, and artifacts that constitute their professions.
>
> Goodwin (1994, p. 606)

The job of the expert witness in the King trial was to get the jury to see the beating the way an expert would, with the categories and interpretations presupposed in professional police work. It was to give the jury enough 'professional vision' that they would override their own prior intuitions about the evidence.

The trouble with just showing a video, even showing it repeatedly, is that to the untrained eye a nasty kick looks just like any other nasty kick, and there is no time to pick up on contextual subtleties. There is no time, as the video plays, for an expert to point out that kick A responds to subtle provocation B, and for the jury to gauge the argument. The *pace* is wrong.

Slide shows are an experts' medium of choice because they control pace: the presenter can take the time needed to complete a point; audience members can be given time to appropriate all the claims. Slides allow an expert to walk step-by-step through still images selected to highlight key features – in this case, body positions or gestures by King and the policemen striking him. Through a narrative overlay, the expert connects the slides explaining how one action meaningfully and reasonably follows another. Each body posture, whether police or King's, is interpreted. Thus one slide, the expert tell us, shows King, a man taller than six feet and well over 200 pounds, in an aggressive posture, getting ready to stand up, and therefore menacing the officers, not submitting. The next slide reveals a policeman kicking King in the gut, but it is in exact conformity with police policy as stated in the police manual. The expert witness points out that it is, in fact, a highly measured escalation to the aggressive act that prompted it. By identifying the key aspects of an image to non-experts, an expert witness slowly teaches jury members how to view the video with professional vision, how to interpret or code the video when they look at it themselves.

It may seem obvious that when an expert is trying to explain to non-experts exactly what is going on, a process that requires micro-analyzing each image, they need more time than a real-time video allows. Attention can be managed quickly only if the features to be identified are familiar and easily recognized. More time is needed for complex structures whose component elements must each be noted and weighed. This is especially so if the grounds for classifying something are tacit, as are most of our perceptual classifications. Even a simple classification – 'this image displays a Delicious apple' – takes time when the basis for the classification needs to be shown. If you must explain that Delicious apples are large and long, with six small knobs opposite the stem, the audience needs time to review the apple. In presentations to other experts this can be truncated; even so, there may be possible confusions, subtle distinctions unnecessary to mention to non-experts that must be ruled out explicitly. For instance, experts know that Delicious apples are usually dark red, with subtly darker stripes. But some Delicious are yellow. If a yellow Delicious is shown, however, it may require discussion to rule out misclassification as Golden Delicious, a related but different variety.

Given the goal of communicating situation awareness, Goodwin's article makes clear that experts and novices have very different needs and that choice of media depends substantially on what is to be learned. Using his article as a springboard, we can distinguish five factors when deciding between using videos versus still images as the background medium; pre-made annotations versus real-time gestures as the foreground attention manager. These factors are:

1. *The expertise of the viewers*. Generally, experts need less time to pick up a bubble whether presented as video or collection of stills. Because they have a deeper understanding of the domain already they are able to register the significance of things pointed out more quickly. Videos and gestures often work best for experts because they are fastest.
2. *The amount of framing or contextualizing language needed* to enable viewers to tune their own sense of what is salient – and whether this can be done offline or needs the presence of imagery for adequate grounding. Some types of situation awareness can be framed quickly when an image is present; others can be adequately framed, or even better framed, without the distraction of images or video, using narrative alone. If little discourse is needed the audience need not be given much in the way of 'situating shots' in either still or video form. Generally, experts get by with considerably less framing than others since they pick up on context quickly. Others need more framing and more language to be brought up to speed. The more contextual framing the better it is to use stills.
3. *How complex the features or structures* are that are to be identified. To show the difference between a 2002 and 2003 BMW 740IL requires looking at many features, some of them detailed. By contrast, few vantage points are needed to distinguish a Granny Smith apple from a Delicious. Again experts need less time than others, though they too need longer the more complex the features being highlighted. At some point a time threshold is reached and only presentations made with stills are practical.
4. *The number of relevant elements* or features in a given scene. If a scene extends over a broad spatial area or a long period of time, or if there are many participants or artifacts involved, it will typically take longer to understand the scene than if there are fewer elements present, or if the process is more localized in time and space. Consider what is involved in identifying the play used in a football or soccer match vs what is involved in identifying an ace in tennis. The one requires tracking the path of several participants while the other requires noting only whether the ball landed within the service marking on the court. The more elements to track the more time it will take.

5. *The time to switch attention* from one feature to another. There are gross psychological limitations on attention switching. The more independent features a subject must attend to, the more time they will need. This is regardless of medium.

The importance of these factors suggests that effective choice of media depends substantially on what is to be learned. In some sense this is obvious. If the material is unfamiliar and subjects must be taught to 'see' what is present, then the media must be present long enough for subjects to study them, and be instructed. Usually this means using static media with text or narration, gesture and annotation. Similarly, the harder or more complex are the things to be identified, learned, or understood, the more time is required and the more useful it is to mark the key elements. Again the need to keep the pace slow enough argues for stills.

All these questions about pace depend substantially on the expertise of the subject. A fly-over showing the location of enemy installations in Starcraft is all an expert needs if they already knows the geography of the area. Somewhat more time may be required for an expert to understand the strategic implications of those assets, but not much. By contrast, less expert subjects need the extra time afforded by stills, to be given a description of the assets, an explanation of their arrangement, and an account of their strategic import.

A further implication of Goodwin's inquiry, for our study, is that it highlights that the goal of an expert witness is often to install *tacit* knowledge in the audience. The elements conveyed in the King presentation were not simply facts about 'what happened when' as much as *ways of seeing*. The jury had to acquire a perceptual skill of interpreting a scene the way a policeman would. When immersion in a situation means *being primed to act appropriately* it is likely to have an implicit component that is hard to impart. To acquire a way of seeing and a *readiness to act* takes more time than just understanding a description. It requires using the appropriate media to prime and train recipients, media that can ready them to react quickly.

Animation vs stills: A brief review

There is a widely held presumption that dynamic media such as animations, video, music, or narration are more effective when the thing they represent is itself dynamic, such as melodies, tempo, driving … or more generally, in domains where spatial or temporal properties of sound, motion or change must be learned. Yet research shows that in most cases dynamic media are *not* more effective than static media for teaching about principles or understanding even when the thing represented is dynamic (Hegarty *et al.* 2003; Narayanan and Hegarty 2002; Palmiter and Elkerton 1993); and that when they are more effective it is usually because the dynamic media carry extra information or offer users interactivity of a sort unavailable in static media (start/stop, zoom, orientation of parts…) (Tversky *et al.* 2002). In their article 'When static media promote active learning' Mayer *et al.* (2005) attempt to meet Tverksy *et al.*'s criticisms concerning the extra information and interactivity of dynamic media by controlling the informational equivalence of media presentations and ensuring that subjects in both animated and static conditions have the same opportunity to interact with their media. They ask whether the cognitive advantages of static illustrations and text outweigh those of animation and narration.

Their article reports on four experiments in which students received a lesson consisting of either computer-based animation and narration, or paper-based static diagrams and text. Both lesson types used the same words and graphics to explain the process of lightning formation (Experiment 1), how a toilet tank works (Experiment 2), how ocean waves work (Experiment 3), and how a car braking system works (Experiment 4). They found that on retention and transfer tests, the static graphic group performed significantly better than the animation group for four of eight comparisons, and that there was no significant difference on the rest. These results support other (methodologically less sound) findings in which static illustrations with printed text outperform narrated animations. The findings suggest that the advantages of the static media presentation are that:

- Learners can control the pace and order of presentation simply through eye movement, whereas in narrated animations pace and order are fixed.
- Learners are saved from wasteful or extraneous processing such as attending to unimportant movements in an animation or having to hold animation frames in working memory to mentally link one to the next. In the static case, all frames are present at once so the subject can compare them at will. Further, only *key* frames are chosen – ones showing the crucial states of the system.
- Deeper more active processing (called germane or generative by Mayer *et al.*) is promoted because learners are encouraged to explain the changes from one static, key frame to the next.

Mayer *et al.*'s results support Goodwin's on the value of still backgrounds for tasks requiring deep processing, but they tell us nothing about the best media for face-to-face and live remote presentations where the pace and order of information is controlled in real time by a presenter or arises from the interactivity between presenter and audience.

The outcome of this brief review of psychological and ethnographic literature is that guidelines for *creating and annotating* presentations are hard to come by. There have been numerous studies exploring the comparative value of stills versus moving media, under a variety of tasks, but few, if any, have focused on the interaction of annotation with background media and few, if any, have been designed to allow the audience to be involved in determining pace and order. Moreover, the type of knowledge transferred has almost always been explicit knowledge – recall of facts or problem-solving knowledge – rather than tacit or implicit knowledge – *the knowledge that may be involved when passing immersive knowledge and situation awareness*. Given current findings, then, it is not possible to plug in:

- a description of the type of information, or situation awareness, to be shared among group members;
- the set of constraints on space, time, interactivity and bandwidth among group members;

- and then calculate a specification of the presentation type that would be most effective.

We turn now to our experimental study which, in part, was designed to bring us closer to this sort of formula.

Conjectures, testbeds and methods

When we began our experimental study of passing the bubble we had a set of conjectures based on intuitions derived from everyday experience and prior studies. Common sense suggests that annotating or gestural pointing to media is bound to be helpful when communicating a complex scene or process, such as a football play, the strategic position of Starcraft™ opponents, the implications of a chess position, or the meaning of a layout of assets displayed on a UAV video. The question of whether the underlying media should be a video or set of selected stills derived from that video, however, is not a matter of common sense but of scientific study.

What sorts of annotation work best? Based on everyday experience again, we predicted an *interaction* among:

- the background media being annotated (video vs still);
- the type of foreground annotation (prepared drawings, real-time improvised drawing, real-time gestures, etc.).[1]

For example, when John Madden draws the path of football players the annotation grows dynamically, but the background is still. The obvious reason the production managers choose to keep the background static is that viewers get overloaded when both foreground (the annotation) and background (the video) move. Movement against movement is too complex; there is too much to attend to, and too much visual distraction.

In the King trial stills were used as the presentation background because video backgrounds move through key scenes too quickly to give the presenter time to get his points across. Their pace is wrong. Effective attention management, here, is a function of how much there is to show and say, and how fast the background is changing. This interaction, though related to the Madden case, seems different than simply minimizing the visual complexity of movement against movement.

To address these and related questions, our first set of experiments was designed to assess the comparative value of different types of annotation on videos and stills.

- With respect to background media (stills vs video) when should annotation be dynamic (moving) or be still?
    - (a) are some content types more reliably acquired from dynamic rather than static annotations?
    - (b) are some types of annotations more effective than others?

---

1 At a more detailed level we expected the form of foreground annotation also to affect performance (e.g., labeling, highlighting, path-drawing, etc.).

- Generalizing across both still and video backgrounds are there dominant rules for annotating presentations?
  - (a) since gesturing in front of presentations is almost always helpful, it is reasonable to expect that annotations that are drawn in front of the viewer (i.e., dynamic annotations) will also always be more helpful than static annotations, whether those annotation are canned or live. Yet canned annotations, unlike live gestures which are ephemeral, persist after they are drawn. Will this persistence become a distraction, raising cognitive load?

Experiment one

Method

Procedure To test how successfully the bubble of situational awareness is passed under various conditions of annotation and media, we recreated a gaming environment in the lab and recruited experienced Starcraft™ players to serve as bubble transmitters, bubble recipients, game opponents and controls. There were ten types of presentation (see Figure 10.1 for the factorial design). Nine types were created offline using three types of background and three types of annotation conditions. A tenth condition was created in real time, immediately after pausing the game. In this real-time condition the impromptu presentation created by the bubble transmitter is given face-to-face to the bubble recipient. We call this last condition 'live'.

To create the game that would form the basis for the presentations, two players of equal strength, A_1 and B_1, played until reaching a natural breaking point about 13 minutes in. A_1 would then take a video recording of the game home and make his or her nine presentations using a video editor. In the live condition, A_1 would give a real-time presentation right after the game was paused. To test the effect of different presentations on transmitting situational awareness, two new players,

Factorial design of Experiment One			
Pre-made Presentations (canned)			
Background	*Foreground*		
	No Annotations	Static Annotations	Dynamic Annotations
Selected Stills	1	2	3
Video Snippets	4	5	6
Control – Random stills	7	8	9
Live			
Real-time selections of video snippets	10 Live gestures		

Figure 10.1 Factorial design of experiment one

A_2 and B_2, would take over the game left by A_1 and B_1 A_2 would take over A_1's seat, and receive a presentation from A_1 B_2 would take over B_1's seat *without* the benefit of a presentation. The game was resumed and played for another 5 minutes. The score was then taken, and we measured the percentage increase that A_2 and B_2 gained over their original inherited position. It was necessary to consider percentage improvement because the presenter, as often as not, created a presentation for the side not in the lead. We were interested in how well the presentation improved the inherited position.

Stimulus To create a presentation the bubble creator (A_1) reviewed the complete video of the game they played in, chose the background media and then annotated it in whatever way they thought would best communicate the strategic position of the game. To choose background media they were told to select:

1. a set of still images that best captured the elements of descriptive and strategic information a player would want to know;
2. a set of video snippets of key activity from which vital situation awareness could be easily acquired;

A third condition – a control – was added to these two:

3. a set of stills randomly selected from the complete video.

The presentation maker annotated these with circles, highlights, arrows and so on. For instance, to help clarify a verbal point he might indicate which part of an enemy installation he is talking about. To explain the path he planned to send his own soldiers on, he might use arrows or draw a route. Dynamic annotations were simply video captures of the physical drawing of arrows, circles and lines on the background media. See Figure 10.2 for an example of an annotated still image.

Live presentations were given in real time by A_1 within 15 to 45 minutes of pausing the game. A_1 was given 5–10 minutes to review the recorded video of the game(s) he just played, and then was sat in front of a monitor, with A_2 to his or her right, and A_1 would choose the parts of the game to jump to and vocally explain what was going on. A_1 was always free to use hand gesture as well as mouse movements to point to elements on the screen. A_2, meanwhile, was given no instructions on how to behave. More often than not they asked A_1 questions.

Figure 10.2 shows a typical annotated Starcraft™ scene with enemy and friendly forces and installations present. Circles identify the locations of each side's assets, colored to clarify identity. The presenters rarely explained the semantics of the annotations they used. Annotations drew attention to activity and helped illuminate the narration. The background in Figure 10.2 may have been a still image or a video snippet showing the forces changing positions over time. If annotations were dynamically drawn in over the video, either indicating where a force would soon go or perhaps tracking the force's movement, then they would constitute a dynamic foreground (i.e., dynamic annotation) over a dynamic background (i.e., a video snippet).

Figure 10.2 An annotated Starcraft game. If annotations are dynamic the evolution of each drawn form is shown in time. If annotations are static they appear all at once fully formed

Subjects We recruited subjects from the UCSD student population. During 2004–06 playing Starcraft™ was popular, though under a revised name of Broodwar, and we easily found 22 experts and 8 intermediates (27 male, 3 female, ages 19–23). All players participated in a round robin so we could determine their relative expertise. We were able to cluster them into smaller subgroups, with enough similarity in expertise in subgroups to enable us to match players of equal ability against each other. Most subjects were given course credit for participating, and cash rewards as a function of their competitive performance against each other. A strong sense of personal competition amongst the players further helped ensure that all players tried their best.

Measurement and metrics To quantify how effective a presentation is, we compared game score improvement, measured as the percentage improvement over the inherited score. The final score a player achieves is a function of several factors other than whether they receive a presentation and its type, however. Game score is also affected by each individual's expertise, the strategic value of each's inherited position and the presentation quality itself. To improve the meaningfulness of our primary measure we introduced a range of controls and corrections to allow us to normalize the score. Our final measure was therefore percentage improvement in normalized score.

For instance, despite our efforts to pair players of equal skill (as shown by their performance in the round robin), there sometimes was a small difference in skill.

To deal with skill difference we introduced a correction factor – a handicap factor – derived by calculating each player's own normative growth factor. We observed that scores grow according to an exponential curve, with players winning more points per minute the longer they are in the game, and at a rate that is proportional both to their inherited score and to their expertise. Using their performance in all the (hundreds of) games they played, whether bubble creator, recipient, opponent or control, in addition to those they played in the round robin, it was possible to assign each player a normative growth curve. The better the player the steeper their growth curve. Accordingly, when players of different skill competed (i.e those with different growth curves) we normalized the better player's percentage improvement by adjusting it downward commensurately to the difference in the two players' growth curves. We calculated and applied a handicap.

To correct for differences in the strategic value of an inherited position we gave the same game paused at 13 minutes to a control team consisting of comparably skilled players neither of whom receives a bubble. We took their final score – normalized for expertise – as an indication of how the game would be played out. There is no other way of control or correcting for the strategic value of a position because the numerical score of a game is determined only by winnings and assets; no points are given for having a good strategic position, a position that should soon deliver points later in the game.

The final factor we controlled for was differences in presentation quality. Since our presentations were made by players chosen for their Starcraft™ skill and not for their presentation-making skills, we expected and found variance in the quality of the presentations they created. To normalize for differences in presentation quality

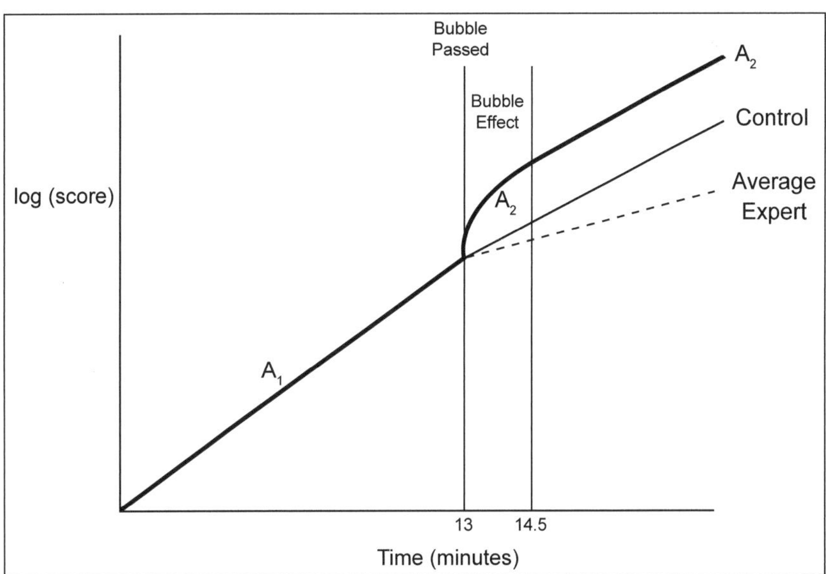

Figure 10.3 Typical growth curves for players before and after receiving a bubble, and for bubble-free controls

we compared the average size of the bubble effect that different presentation makers achieved under each condition and considered it when estimating the average size of the bubble effect per condition. Since we were not confident in our estimate of a presentation maker's skill we only used these measures to see if they significantly altered results. They did not, so we did not in the end normalize scores by this factor.

In Figure 10.3 we distinguish four growth curves. The straight line from the origin to 13 minutes is the rate of growth of A_1, the competitor who plays first and makes the presentation after 13 minutes. The dotted line shows the score we predict an average expert would have at 18 minutes, the time the game ended. By comparing the slope of the average expert's growth curve with the growth curve of A_1 we can tell whether A_1 is better or worse than our average expert. A_1 here is better. The middle line shows the actual score earned by our control – a player of about equal skill who plays the same inherited game as A_2 against another player who, like himself, did not receive a presentation. A_2's curve shows how well A_2 did after being given a presentation. The early steep rise in A_2's growth curve reflects the value of the bubble, and shows that it is most useful in the first minute or two. After that the extra information obtained from the bubble is either stale or insignificant because both players now have their own situation awareness and their own personal immersion.

Results: Experiment one

Bubbles are good. Although we assumed that presentations have a positive effect on performance, it was not a foregone conclusion because of the complexity of metrics and normalizations we used to determine the impact of a bubble on performance. Figure 10.4 shows that receiving a bubble is indeed helpful. Averaging across all presentation conditions, receivers improved their position relative to controls by a sizeable factor, whereas non-receivers playing against receivers did worse than controls.

Dynamic annotation on video is not helpful. When a video presentation is accompanied by narration, as all our presentations are, dynamic annotation is no better than no annotation and tends to be worse than static annotations on video (Figure 10.5). It is not surprising, then, that most annotations overlaid on video or animation during television broadcasts are static, consisting of visual elements such as names, numbers, callouts and arrows that help the narrator refer to elements on the video.

Figure 10.5 shows that although video snippets and stills perform approximately as well when they are not annotated and the presenter simply narrates over them, we found that stills are significantly better than videos as a background medium when they have both annotation and narration. Annotation, whether static or dynamic, on stills significantly improves subjects' performance. On average it does nothing for videos.

Still images are a better background media for passing strategic knowledge than videos. This applies whether the foreground annotation is static or dynamic (Figure 10.5). Our finding is in conformity with much of the literature on illustration vs animation in general, although those studies did not manipulate annotations.

The justification for this claim came from an analysis of what was being annotated: descriptive, pedagogic or strategic information. Experts were asked to review all presentations and classify annotations according to whether they were didactic (describing the meaning of assets), purely descriptive (enumerating the types and position of assets deployed), or strategic (explaining the strategic intent of deployed

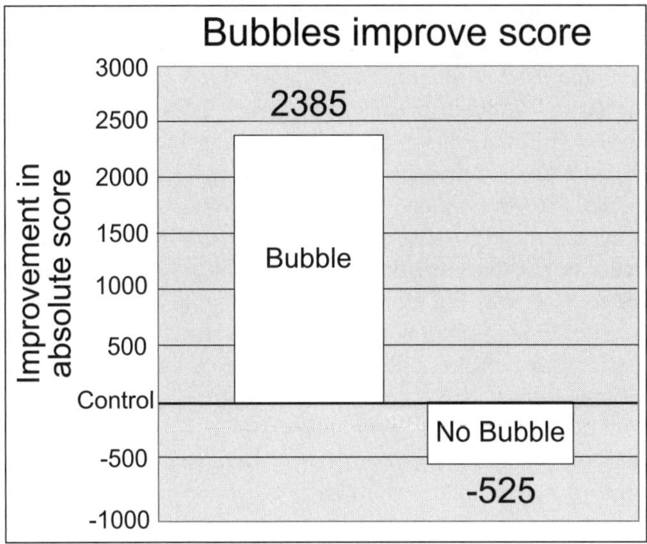

Figure 10.4 Bubble recipients perform better

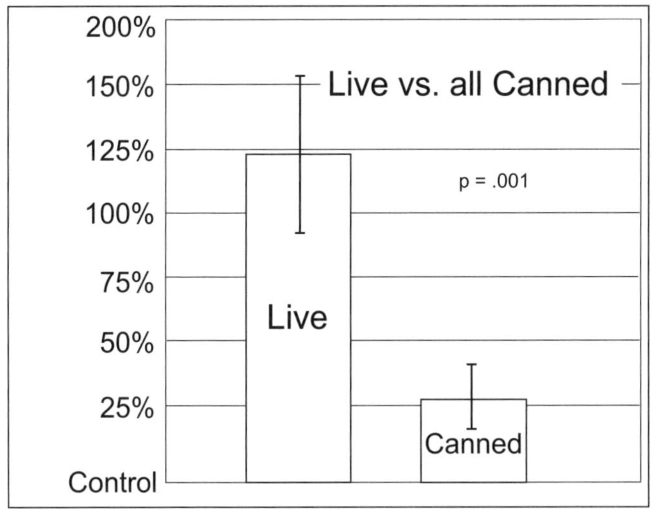

Figure 10.5 Live versus canned. Despite the long hours spent creating canned presentations, live presentations delivery more

assets). Since it takes extra time to explain strategic intent stills were found to be more useful.

Dynamic annotations are not better than static annotations on (selected) stills. This was a genuine surprise (see Figure 10.5). Most of the televised sports programs rely on moving annotations on stills. We found that they may, at times, be worse. This is surprising because dynamic annotations can carry extra information. Based on reactions to John Madden's explanations of football play and from watching people draw on transparencies, we would expect that the pace and activity of drawing seems to have pedagogic value itself. In our controlled experiment we did not find that. Assuming that Madden could simply overlay his drawings on stills, we would predict that he would be as effective, perhaps more so in conveying his ideas if he were to use still annotations on stills. Since background stills already allow him to control the pace of presentation, the pace enforced by showing a drawing being created adds no valuable constraint. It is really just eye candy. If, by contrast, the speed of movement in moving annotations encodes dynamical information that would otherwise be unexpressed, there would, indeed, be a reason to expect an effect from dynamic foregrounds. The majority of our dynamic annotations did not encode information unavailable in static form. Our findings, therefore, suggest that Madden should use static annotations on stills, and that presenters who use PowerPoint slides annotated with highlights, markers, arrows etc., should give up fancy transitions and animations and use annotations that are fully formed.

Live, minimally prepared face-to-face presentations are better than all forms of canned presentations. This holds regardless of how long preparers take to create their canned presentations (Figure 10.6). Live presentations contain gestures and mouse pointing. These function like dynamic annotations on still or video background. Unlike our experience with canned presentations moving annotations, when live, do not seem to overwhelm viewers.

Each line in Figure 10.6 marks a 25 percent increase in score above the control. Receivers of live presentations, on average, scored 125 percent higher than opponents who received no presentation (one of the controls). This compares with a 25 percent

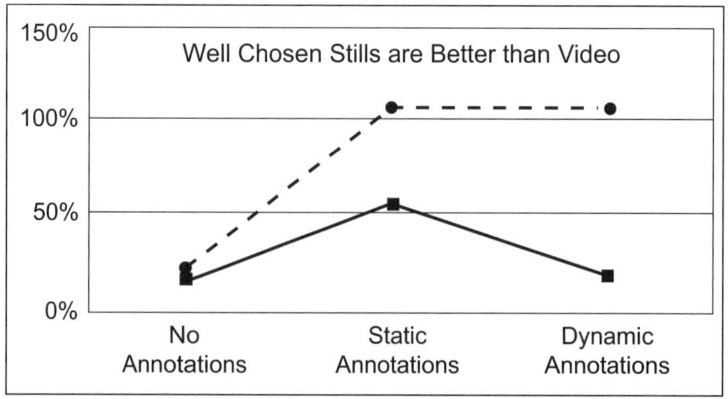

Figure 10.6 Well chosen stills are better than video

average improvement by players who received a canned presentation. (Figure 10.6 shows scores, not growth curves.)

When we surveyed our subjects in an effort to explain this finding, responses suggested that experts find live briefings more valuable than highly prepared canned presentations because:

- Experts like to drive information transfer, and hence strongly prefer interactive (live) transfers, responding less well to forced attention management.
- Canned presentations were too often found to be boring and didactic. Experts don't like to hear things they already know;
- Experts strongly prefer strategic information – which is more abundant when the presenter feels the social presence of the audience. The term 'strategic' refers to information that explains goals, rationale and game objectives, as opposed to descriptive *data*, which experts prefer to gather visually.

Our follow-up analysis of the canned presentations vs the live ones revealed significant differences in the amounts of didactic information presented in each. Evidently what experts like correlates with presentation impact. On closer inspection we came up with four conjectures – aspects of face-to-face interaction – to explain the performance boost found in live presentations.

- Verbal interactivity via recipient questions and oral acknowledgement (I see, uh huh),
- Adaptation of pace to oral and subtle interpersonal cues apparent in face-to-face interactions,
- Social pressure for relevance conveyed through face and body language,
- Presenter's gestures increase precision and bandwidth.

To help explore the relative value of these elements we designed a second set of experiments, concerned solely with comparing different types of live, face-to-face conditions.

Experiment two

The high performance of live presentations in our first experiments surprised us because it showed that impromptu presentations are more effective than carefully crafted canned presentations. Although there were important differences in static vs dynamic annotations, nothing was more effective at passing the bubble than a face-to-face impromptu presentation when presenters were given just a few minutes to prepare. Which aspects of the face-to-face context were responsible for this high performance?

To isolate the active factors and determine their relative importance we manipulated three central aspects:

- Verbal interactivity: the opportunity to ask the presenter questions in real time;

- Co-location: the opportunity to use body language, facial expression and the brute social presence of being in the same room to communicate boredom or interest;
- Synchrony: the time when the presentation is heard, live or after the fact. If live there is the chance that the presenter, even if remotely located, will be affected by knowing another person is listening right now.

It was necessary to leave gestures present in all live conditions because gesture and pointing are an essential part of the way the presenters ensure verbal reference.

Experimental testbed. To give us the flexibility to manipulate synchrony (time), co-location (space), and interactivity between team members we created an experimental testbed that supported high bandwidth connectivity between several rooms. Because we were concerned that annotations and gestures might have different effects if they were done face-to-face, rather than captured and sent by media to an audience somewhere else, or to an audience later in time, we had to develop a testbed that supported audio/video capture and presentation under many different conditions.

Figure 10.7 illustrates the conditions for our live presentation experiments. Note that face-to-face presentations differ from live remote presentations only with respect to co-location. Not indicated as an attribute, however, are the social factors present in co-location that are not present in live remote, such as the co-coordinated, reciprocal body language of the presenter and listener, eye contact and the fact that the presenter has a greater sense of the listener's responsiveness.

In Figure 10.8, two rooms are shown connected through audio and video. The arrangement allowed the video capture of the presenter's pointing and gesturing at the screen in room A to be relayed to a recipient in room B who had good audio contact with the presenter and could freely ask questions or audibly show interest. The presenter in room A could not see the recipient in B and thus had no access to his or her responsive body language or facial expression. Not shown in the figure is room C, much like room B but without two-way audio, where another recipient could sit to watch and listen to A and B but not question.

Full face-to-face interaction was used as a control condition. The recipient sat beside the presenter in room A, the two having the full range of interaction typical of mutual social presence.

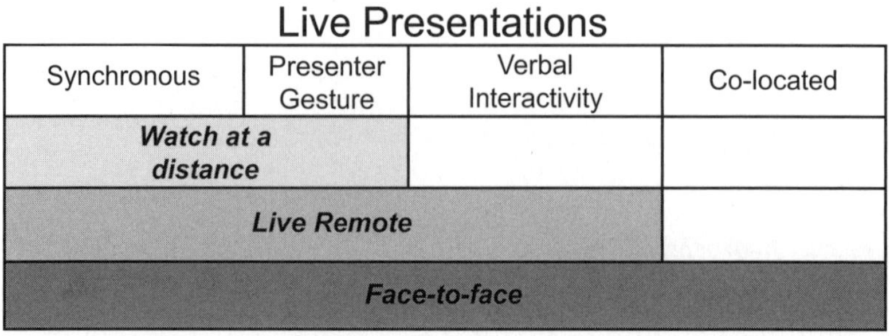

Live Presentations

<table>
<tr><td>Synchronous</td><td>Presenter Gesture</td><td>Verbal Interactivity</td><td>Co-located</td></tr>
<tr><td colspan="2">Watch at a distance</td><td></td><td></td></tr>
<tr><td colspan="3">Live Remote</td><td></td></tr>
<tr><td colspan="4">Face-to-face</td></tr>
</table>

Figure 10.7 Attributes found in different forms of live presentation

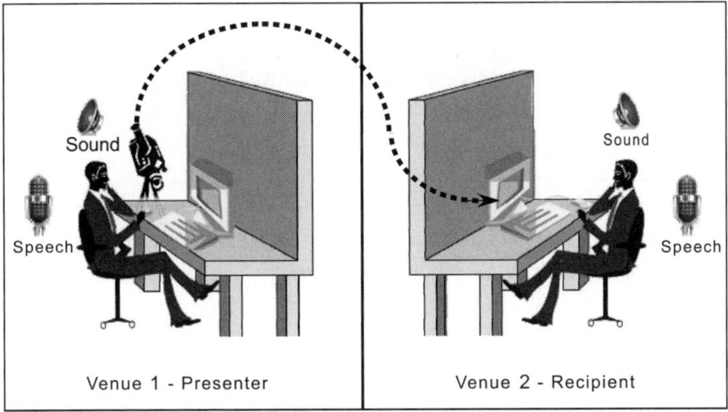

Figure 10.8 Venue 1 – presenter and venue 2 – recipient

Results: Experiment two

The primary goal of experiment two was to determine whether the benefits of face-to-face presentations could be substantially matched by non-face-to-face presentations, and if not, why. The results are mixed.

Live remote presentations are about as effective as typical face-to-face presentations

In the live remote condition, as shown as Figure 10.8, presenter and audience interact in real time from different rooms. Recipient and presenter are in good audio contact throughout, with a good sense of engagement. Gesture and other proxies for annotation, such as mouse movements, are shared one way because the video feed the audience is watching has a wide enough viewing angle to include both the presenter's screen and hands. Equivalent performance with face-to-face is not a big surprise given the positive experiences of collaborators and lecturers who rely on an open audio channel and presenter controlled monitor. However, it is still worthwhile noting that two-way video is not necessary for near face-to-face performance as long as there is high quality two-way audio.

One implication of our result is that real-time video of gesture and mouse movement when accompanied by duplex conversation yields enough mutual awareness that the benefits of social presence found in co-located, face-to-face interactions are duplicated remotely. This hypothesis is further supported by our examination of the content of face-to-face vs live remote presentations that reveals no discernable differences in the number of didactic events or the relevancy level of the content. This suggests that the impatience of expert listeners was ably perceived and accommodated by presenters.

Audio interactivity is necessary for near face-to-face performance

In our *watch at a distance* condition, performance after receiving the bubble was significantly reduced. In this condition recipients had video display of the presenter's screen and gestures, but their one-way audio did not allow them to ask questions; instead they heard the exchange between presenter and recipient in the live-remote condition, occurring at the same time. We duplicated these results when we ran a pilot study where recipients heard a presentation after the fact.

In both cases the effect of denying verbal interactivity between listener and speaker was to *significantly reduce performance.* There seems to be an important difference between watching a presentation over someone's shoulder, and being actively involved in dialogue. Having a sense of being socially present with the presenter, and being an active agent in deciding the type and pace of presented content, makes an important difference in how effectively a bubble is passed.

The implications for collaboration are easy to infer. *Groups need to be in active and synchronous contact, with at least two-way audio and one-way video*, to reap the full benefits of presentations. Since there are only so many audio exchanges that can take place synchronously, the number of distant sites and participants that can be linked together is small. A valuable next step would be to determine how performance degrades as the number of sites increases.

Conclusions

Presentations are a key resource in team coordination. A presenter is not only a communicator of ideas and views, they help to coalesce team conceptions, share situation awareness and helps to shape the way different players on a team understand what is going on and what their role is in joint activity. This view is apparent in one-on-one dialogue when an active listener challenges the presenter to justify their position. The two communicants reach a joint understanding via the mediating artifact of the presentation. It is also apparent when a presenter, in front of a group, responds in real time to the variety of social cues the audience generates as they struggle to assimilate the presenter's content. Presentations are a classic instance of distributed, or macrocognition, as the presentation and the dialectic between presenter and recipient become mediating artifacts through which the group crystallizes ideas, coordinates activity and manages focus.

The challenge for research in this area is to understand the factors influencing the cognitive efficiency of presentations as mediating, coordinating and attention-focusing artifacts. The first study reported here manipulated conditions under which presentations are created, presented and distributed to team members later in time.

We found that when presentations are created without the social pressure that comes from the presences of an impatient audience those presentations tend to contain content that is unnecessarily didactic. This lessens the informational impact of those presentations. We also found that adding graphical annotations, or including gestures, in presentations enhances their effectiveness in sharing situation awareness. Gestures are analogous to moving annotations. But we found that when graphical annotations are all a subject has then static annotations are best. This holds whether the media being annotated is video or stills. In fact, if the background media is a video then adding moving annotations to the video is distracting and is significantly less effective than when the video is annotated with static annotations.

Presentations overall are most effective when given either face-to-face or live over an audio–video connection that allows remote viewers to easily talk with the speaker, and see his or her gestures on the presentation media. When remote audiences lack two-way audio they get less from a presentation, though we cannot say whether it is because they are unable to negotiate with the speaker over the content being presented or because the speaker lacks adequate social awareness of them.

The implication for sharing situation awareness among distributed members is that transfer is best when teammates have two-way audio, when they ask questions, and when they have clear view of gestures and annotation. It is not necessary for the presenter to mark-up video or stills ahead of time; real time presentations, even impromptu ones, work better than pre-made ones, providing the presenter has a few minutes to collect his or her thoughts. Finally, video is not better than still images as the background medium.

Acknowledgments

The author gratefully acknowledges the support of Mike Letsky and ONR grants N00014-02-1-0457 and N000140110551, Joachim Lyon for thoughtful conversations, Joy de Beyer for textual advice, and Thomas Rebotier for help with data mining and statistical analysis.

References

Goodwin, C. (1994). 'Professional Vision', *American Anthropologist* 96(3), 606–633.

Hegarty, M., Kriz, S. and Cate, C. (2003). 'The Roles of Mental Animations and External Animations in Understanding Mechanical Systems', *Cognition and Instruction* 21, 325–360.

Narayanan, N.H. and Hegarty, M. (2002). 'Multimedia Design for Communication of Dynamic Information', *International Journal of Human–Computer Studies* 57, 279–315.

Mayer, R.E., Hegarty, M., Mayer, S.Y. and Campbell, J. (2005). 'When Passive Media Promote Active Learning: Static Diagrams Versus Animation in Multimedia Instruction', *Journal of Experimental Psychology: Applied* 11, 256–265.

Palmiter, S.L. and Elkerton, J. (1993). 'Animated Demonstrations for Learning Procedural Computer-based Tasks', *Human–Computer Interaction* 8, 193–196.

Tversky, B., Morrison, J.B. and Betrancourt, M. (2002). 'Animation: Can it Facilitate?', *International Journal of Human–Computer Studies* 57, 247–262.

Chapter 11

EWall: A Computational System for Investigating and Supporting Cognitive and Collaborative Sense-making Processes

Paul Keel, William Porter, Mathew Sither and Patrick Winston

We introduce EWall, a computational system for users to conduct, for researchers to investigate, and for computational systems to support individual and collaborative sense-making activities. The EWall system engages users in the visual organization of information through the spatial arrangement and modification of graphical objects. Computational agents infer from spatial information arrangements and the collaborative use of information as a basis for directing the flow of information among collaborating users and computational systems. The goal of the agent system is to direct the distribution of information in ways that brings together people with complementary backgrounds, expertise, interests and objectives and that improves team shared understanding. Individual agents represent unique cognitive and collaborative concepts, combine their analyses and autonomously adapt to particular users, tasks and circumstances. The primary contribution of this work lies in the design of concepts and mechanics that enable a non-interruptive interchange of contextual discoveries between humans and computational systems. In this chapter, we discuss objectives underling the design of the EWall system, explain the user interface and the agent system, and provide several examples of agents and how agents can support particular sense-making processes.

Introduction

The creative and incremental discovery of relationships among pieces of data, information and knowledge is commonly referred to as sense-making. Typical sense-making tasks such as brainstorming, decision-making and problem-solving include the collection and visualization of task-relevant information, the analysis and comprehension of information and the subsequent investigation of possible solutions and strategies. Sense-making produces new knowledge through the application of existing knowledge and information to unique situations, tasks and circumstances. The successful execution of sense-making tasks not only depends on the creative interpretation and application but also the availability, fast accessibility and diversity of knowledge and information. Thus, sense-making tasks are commonly conducted in groups of people with different backgrounds and expertise as well as through the use of computational tools for the search, visualization and organization of information. The challenge is to coordinate available knowledge and information effectively as well as to bring together the human capacity for intuitive problem-solving and the computer's capability for processing and visualizing large amounts of information.

The key problem in regards to sense-making activities is that a lot of knowledge is processed in the minds of humans and thus remains invisible to collaborators, observers and computational systems. The externalization of knowledge and exchange of information among humans and computational systems is often time-intensive and interruptive. This is particularly true for the early stages of sense-making processes, where individuals explore the availability and location of task-relevant knowledge and information. Commonly, sense-makers bypass this problem by collaborating with people they previously collaborated with and by accessing information from known sources. This approach has become ineffective within the highly dynamic, de-centralized, globally distributed and information-loaded environments enabled by information technologies. Individuals increasingly collaborate asynchronously and remotely, contribute to multiple tasks and work groups simultaneously, dynamically regroup based on their availability and expertise and deal with large amounts of heterogeneous, distributed and dynamically changing information. To deal with these challenges it is necessary to explore new means for helping sense-makers to effectively coordinate and harvest the knowledge and experience of humans and computational systems.

In this chapter we explain EWall, a unified computational infrastructure for helping humans and computational systems to retrieve and synthesize task-relevant information and knowledge during individual and collaborative sense-making activities. The EWall system is designed for users to conduct sense-making activities, for researchers to monitor and analyze sense-making activities, and for computational systems to support sense-making activities.

- The EWall system provides *users* with a variety of interfaces to easily collect, organize, analyze, discuss and exchange information. A standardized information format referred to as Card is used to represent information in different formats and locations. The EWall interfaces and Cards are aimed at reducing the cognitive burden by enabling the external representation and management of task-relevant knowledge.
- The inclusive representation of information as Cards allows *researches* to observe and investigate the use, evolution and exchange of knowledge and information during collaborative sense-making activities. The EWall system also provides researchers with opportunities to convert discoveries of cognitive and collaborative processes into computational agents that can be used to investigate and visualize particular sense-making activities.
- The agents enhance *computational systems* with the ability to monitor and support human sense-making activities. The agents infer from the spatial and temporal organization and collaborative use of information as a basis for helping users locate and exchange task-relevant information as well as to

learn about and incrementally adapt to particular user preferences, tasks and circumstances.

Research observations and experiments conducted within the framework of our computational environment inform the direction of additional agent implementations that in turn support sense-making processes and subsequently allow for more advanced research experiments. This iterative cycle of operations, experiments and agent implementations incrementally advances the EWall system towards a comprehensive sense-making environment that equally benefits users, researchers and computational systems.

The following three sections introduce objectives, concepts and implementations relevant for understanding the EWall system as a means to explore and support cognitive and collaborative sense-making activities. The section on the EWall user interface introduces technologies for users to conduct and for researchers and computational agents to monitor sense-making activities. The section on the EWall agent system details the utilization of computational agents for representing and supporting particular sense-making processes. The section Discussion examines the use of EWall technologies in relationship to micro-, macro- and metacognitive processes.

EWall user interface

Sense-making activities commonly accumulate large amounts of data, information and knowledge, a lot of which is represented through external visual representations such as notes, graphics, diagrams, sketches, and documents. Sense-makers develop a contextual understanding through the decomposition, combination, comparison, organization and analysis of such representations. The external representation and visual organization of information reduces the cognitive burden by helping sense-makers to comprehend, evaluate and reflect on what they know, to quickly access specific information when needed, to remember relevant thoughts and ideas, as well as to exchange knowledge and develop a shared understanding with other people. The primary problem with the external representation and visual organization of information is that relevant information is commonly extracted from different sources, represented in different formats and stored in different locations (Marshall and Bly 2005). For example, an idea may be written down on a Post-It, a hyperlink may be stored in the bookmark section of a web browser, an e-mail may be copied into a mailbox of a mail handler, and a digital text document may be stored in a shared file system on a server. Effective sense-making requires people to quickly compare and relate task-relevant information items uninfluenced and unhindered by information formats and locations.

EWall Cards

The EWall user interface allows for the external representation of data, information and knowledge in a standardized and visually abstract format referred to as Cards (Figure 11.1, Layout). Cards help users to quickly overview, compare, organize and contextualize information (Robertson *et al.* 1998) as well as to memorize and recall associated contents. Cards are intended to remind of information rather that to present information in detail. In other words, Cards are supposed to omit all information that is unnecessary to compare Cards or to remember the contents associated with Cards. The object-like nature of Cards is intended to engage a human's cognition in ways that allows for the processing of large amounts of information (Byrne 1993; Marcus 2003) and that enables people to think of information as something more tangible and personal that they can possess, understand, collect, trade and exchange (Winnicott 1982).

Card layout The Card layout consists of four components (see Figure 11.1):

1. The Icon area allows for the placement of a representative picture and/or descriptive text to support the easy memorization and recollection of the information referenced by a Card (Lewis *et al.* 2004; Pena and Parshall 2001).
2. The Function area contains the Card's control panel, allowing users to monitor Card states and to operate Card functions such as instant messaging and voting.
3. The Information area is used to display secondary Card information such as document type, modification date, author name and geographic location.
4. The Heading area allows for the placement of a keyword or a heading which provides an additional and complementary option for the access, memorization and recollection of Card contents (Haramundanis 1996).

Card functions Cards combine a large amount of functionality with intuitive operations. The following list outlines s selection of essential Card features (Figure 11.1). Note: [1] indicates planned or partially implemented functionality; [2] indicates terms and concepts that are explained in Chapter 2.

1. The Card layout and functions are customizable, allowing individual users to view shared information in a familiar format as well as to complement Cards with custom functionality for particular work tasks. For example, information

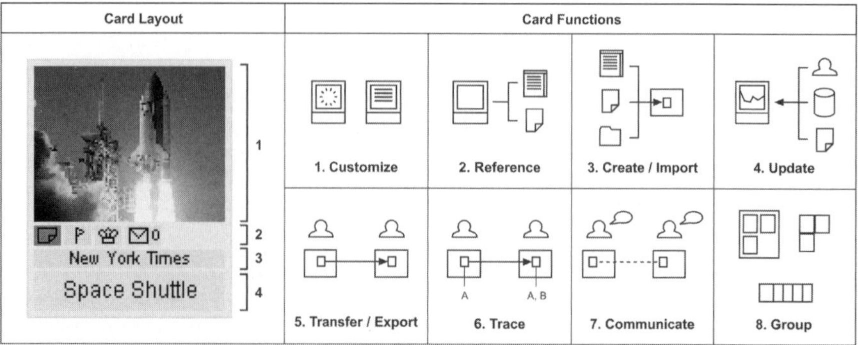

Figure 11.1 Card layout and functions

may be represented with very small Cards that only display an Icon and a Function area and that include a custom function for the categorization of Card contents.

2. Users can hyperlink or attach[1] information to Cards. A mouse-click on a Card can open a web page, a file, a folder, or an executable. This particular option allows users to easily manage information in different formats and locations.
3. Cards can easily be created and imported. Information (such as websites or files) to be displayed, embedded, or hyperlinked can simply be dropped onto Cards or the EWall interface. Cards can also be copied from other users and EWall interfaces. The small effort involved in creating and importing Cards encourages and simplifies the external representation of knowledge and allows users to combine and compare their contributions with information from other sources.
4. Cards can be static or dynamically adapt modifications from remote information sources. For example, a Card may synchronize itself with a Card on another user's EWall interface, a Card in a database, or an Internet-based resource such as a security camera, weather forecast, stock quote or news update. The dynamic nature of Cards is essential for the user to deal with the decreasing permanency of information as well as to remain responsive to changing circumstances.
5. Cards are not confined to one particular location but can be transferred among different user interfaces and computer applications. For example, users can collaborate remotely in the development of a Card arrangement by copying Cards between their individual or shared interfaces. Cards can also be converted into files and transferred through traditional means such as file sharing and email.
6. Cards maintain a detailed log about where they have been, by whom they have been modified, and what other Cards they encountered on their journeys. This information helps researchers and computational systems analyze the history and collaborative use of information.
7. Users can discuss Card contents remotely through a Card's instant messaging feature. A Card indicator informs of instant messaging activity, thus allowing users to easily monitor and simultaneously participate in a large number of discussions on different subjects and with different groups of users.
8. Cards can be grouped and hierarchically structured by placing Cards within the spatial boundaries of bigger Cards. Furthermore, a magnet function[1] automatically joins together Cards that overlap or are being positioned in close proximity. This particular functionality allows users to group Cards in absence of visual indicators. Cards can also be linked[1] to entire Card arrangements. Cards that are grouped or hierarchically structured automatically attach to each other so they can be moved and manipulated[1] as a unit.

EWall views

The EWall user interface consists of two interdependent computer windows, the Workspace view and the Kiosk view.

Workspace view The Workspace view (Figures 11.2 and 11.3) is a canvas for the creation and spatial organization of Cards. The spatial positioning of Cards on the Workspace view suggests a dependable reference system that helps users distinguish, relate and establish the relative importance of information items as well as to develop a contextual understanding of an information space (Kirsh 1995; Pena and Parshall 2001). A collaboratively developed Card arrangement additionally serves as a shared representation or group memory that can enhance group interaction and shared understanding (Pena and Parshall 2001). The Workspace view reduces the cognitive burden by enabling users to consolidate, spatially arrange and comprehend task-relevant information externally (Figure 11.2, Arrow 1). The use of the Workspace view also presents an opportunity for observers, researchers and computational systems to harvest valuable clues about a user's mental consideration, knowledge, expertise and foci during sense-making activities without establishing a direct and potentially distracting and time-consuming dialogue with the users (Figure 11.2, Arrow 2). For example, recently modified Cards might indicate the current focus of attention, while Cards in close spatial proximity might hint discovered relationships. Furthermore, the transfer of Cards between users and user interfaces provides hints about the evolution and collaborative use of information, which allows for various investigations such as the determination of shared understanding, mutual interests and distributed cognition.

The Workspace view supports various macrocognitive processes[2] by enabling users to visually manipulate task-relevant information and explore it creatively:

1. The Workspace view bridges internalized and externalized processes associated with individual mental model construction[2]. More specifically, the Workspace view allows users to externalize some of their mental considerations while simultaneously producing a physical artifact that can be used to communicate particular considerations with other team members.
2. The Workspace view enables individual task knowledge development[2]. Users can explore possible relationships among information items through the spatial arrangement of Cards. In other words, the Workspace view not only provides an interface for collecting and organizing information but also for synthesizing information into new knowledge.

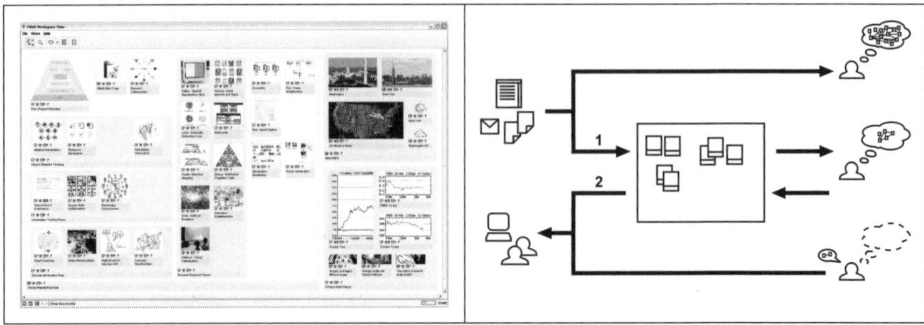

Figure 11.2 Workspace view

3. The Workspace view supports individual knowledge object development[2]. The external representation of data, information and knowledge with Cards translates unique mental interpretations into standardized and sharable information items. Thus, Cards present an effective means for team members to communicate and to consider task-relevant information collaboratively.
4. The Workspace view may be used as a tool for the individual visualization and representation of meaning[2]. Individual team members can use the Workspace view to combine data, information and knowledge into chunks (Cards), to organize and group these chunks in meaningful ways (Card arrangements) and to subsequently promote and explain their considerations to other team members.
5. The Workspace view provides a space for iterative information collection and analysis[2]. More specifically, the Workspace view is a dynamically changing information space that emerges through the continuous addition, modification and removal of Cards and that reflects the emerging considerations of individuals and teams throughout a sense-making process.

Kiosk view The Kiosk view (Figure 11.3) provides users access to information from internal and external information sources such as information available on websites, in databases or the Workspace views of other users. The Kiosk view represents all information as Cards and offers a varicty of visualization options that help users estimate the relevance and discover relationships among information items, learn about other users that may be able to provide expertise on subjects of current consideration, or explore the shared use of information during collaborative sense-making activities. Cards of interest can be copied from the Kiosk view to the Workspace view for further consideration and comparison with previously accumulated information.

Related work

The concept of representing and managing information and knowledge throughout the spatial arrangement of information in an abstract and standardize format offers similarities with various other research efforts and software applications. For example, Pena and Parshall's problem seeking methodology (2001) introduces a sense-making methodology that suggests the use of physical paper cards for collaborative brainstorming activities. The NoteCards (Halasz 2001) developed at Xerox PARC provide an early example of representing and interconnecting thoughts and ideas computationally. Everitt *et al.* (2003) investigated computational opportunities to keep track of and relate physical Post-It notes during collaborative brainstorming sessions. Greenberg and Rounding (2001) focused on the development of real-time collaborative surfaces that allowed individual contributors to post and arrange information in a Card-like format. Wright *et al.* (2006) introduced computational means and procedures to support analytical thinking processes through the creation, manipulation and interrelation of Card-like objects. Several commercial brainstorming and content management applications (such as Inspiration, SmartDraw, QuestMap, VisualLinks, MindManager, Knowledge Forum, Analyst's Notebook and Tinderbox) also offer interfaces for representing and relating ideas through the creation and spatial organization of Card-like objects. The communality and primary application of these tools is to support mental explorations of information spaces during individual and collaborative sense-making activities. Functional similarities among these tools include the representation of information in an abstract format as well as the means to visually represent relationships among information items.

The EWall user interface expands upon previous efforts through the design and realization of an inclusive computational framework for the fast and uncomplicated representation, collection, analysis, management and exchange of information. The EWall Cards combine a customizable and abstract layout with a large number of administrative and collaborative functions. The primary design goal was to introduce a customizable information format that allows for the inclusive representation of information in different formats and locations, that enables a flexible exchange of information among users and across different user interfaces, that dynamically adapts modifications from external information sources, and that provides users with the means to remotely discuss the information contents associated with Cards. The EWall user interface also promotes an environment that minimizes the need for the creation of explicit relationships among Cards so as to accelerate sense-making processes and to allow for more creative interpretations of information spaces. Furthermore, the EWall user interface provides the foundation for the EWall agent system to infer from and support human sense-making activities.

EWall agent system

Sense-making depends on an effective exchange of knowledge and information among humans and computational systems. The goal of the EWall agent system is to help individual users navigate vast amounts of shared information effectively, while remotely dispersed team members combine their contributions, work independently without diverting from common objectives and minimize the necessary amount of

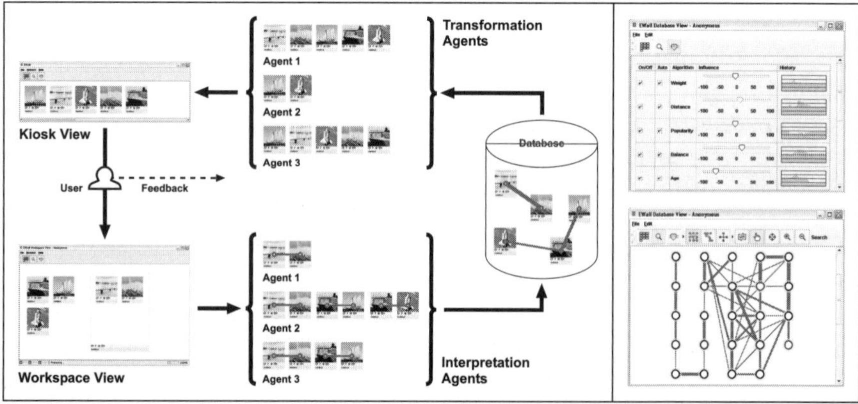

Figure 11.3 Agent system

verbal communication. The key challenge for the agent system lies in determining who needs to know what and from where or how this information may be acquired. More specifically, the agent system must learn about the expertise, knowledge and task foci of individuals as a basis for effectively managing the flow of knowledge and information among users as well as users and computational systems.

The agent system is managed by two types of computational agents: Interpretation agents and Transformation agents (Figure 11.3). Interpretation agents capture some of the users' mental considerations unobtrusively by inferring from the spatial and temporal organization and collaborative use of information on the users' Workspace views. Interpretation agents also synchronize and combine the contributions of multiple users in a shared database. Transformation agents analyze the database contents as a basis for directing the flow of information among collaborating users and for bringing together people with complementary backgrounds, expertise, interests and objectives. Transformation agents communicate their conclusions by providing individual users with a customized selection of potentially task-relevant information in their Kiosk views. In other words, transformation agents do not directly communicate their analyses with users but only adjust the flow of information so as to influence the sense-making process in ways that facilitates collaboration and information sharing. Users may copy information from their Kiosk views to their Workspace views, thus providing the agent system with feedback about its performance and subsequently triggering additional suggestions based on similar considerations. Agents responsible for correctly determining information of interest for particular users become more influential during subsequent iterations. The dynamically changing influence of individual agents allows the agent system to improve the accuracy of its suggestions as well as to incrementally adapt to particular tasks, users and circumstances. The influence of individual agents is temporary, meaning that past successes of agents are only recognized for a certain amount of time. This is to ensure that the system remains adaptive to new users and agents as well as to allow for a more active participation of agents that represent less commonly used concepts.

The agent system is analogous to, and proceeds in parallel with, human observers that review and respond to a user's Card arrangement. Human observers (collaborators, colleagues) can review a user's Card arrangement and respond with suggestions and comments that may or may not influence, inspire or stimulate future additions and modifications to a Card arrangement. Similarly, the agents analyze a user's Workspace view and respond with suggestions for potentially relevant information. The primary difference between human observers and the agents is that the agents do not engage users in a direct dialogue and that the agents offer a complementary computational perspective that draws from an extensive search and analysis of information on the Internet, shared databases and the Workspace views of other users.

Every agent emulates one unique intuitive or explicit human sense-making activity. For example, one agent might focus on collecting contributions from people that have made effective contributions in the past. This particular activity reflects a typical human response in dealing with information uncertainty yet might contradict with various other considerations (represented by other agents) such as, for example, whether the contributor's expertise corresponds with a particular problem domain. The collective application of multiple agents may lead to incompatible results that subsequently require a conflict-resolution strategy. The agent system offers several conflict resolutions options: by default, the agent system favors complementary conclusions shared by the majority of agents. System operators may manually adjust the influence of particular agents over other agents. Alternatively, the agent system can automatically adjust the influence of individual agents based on indirect user feedback (the adaptation of Cards from the Kiosk views). This process offers similarities with the human mind, where different evaluation methods are considered based on past experiences.

Design criteria The agent system is designed based several criteria that ensure the effective integration of computer technologies into human sense-making processes (Figure 11.4):

1. Many traditional computational solutions for the search and retrieval of information are unidirectional, meaning that information and computational analyses primarily flow from computational systems to the users while human discoveries remain inaccessible to computational systems. The EWall agent system enables a *bidirectional* flow of information in which users and computational systems benefit from each others contributions. The agents evaluate the users' Workspace views and the users assess the agent suggestions in the Kiosk views.
2. The bi-directional information exchange is also an *iterative* process that produces an ongoing and evolving flow of information between users and agents. Sense-making activities experience frequent changes in foci, directions and objectives. The continuous exchange of information ensures that users and agents maintain a shared focus and adapt jointly to changing tasks and circumstances.
3. The exchange of information between users and agents is *indirect*, meaning that agents do not disrupt the work of users with distracting inquires about their needs and discoveries but estimate human considerations and requirements by monitoring the handling of information. In other words, users are not required to consider or specifically request agent suggestions and agents do not directly interact with users to find out about their preferences and interests. Users gather knowledge through the casual and optional browsing, evaluation and

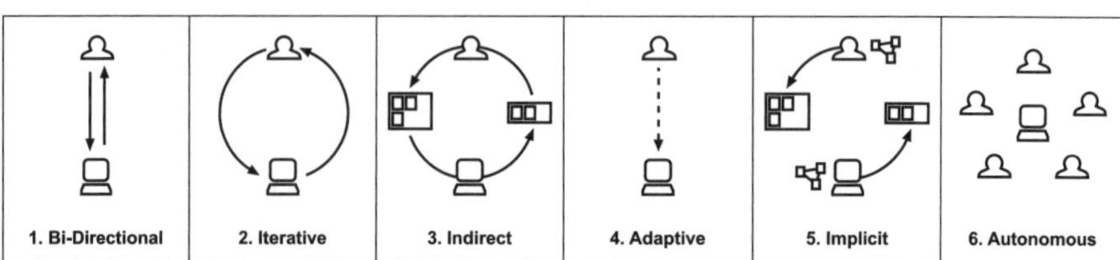

Figure 11.4 Design criteria

acquisition of suggested information by agents, and agents gather knowledge through the analysis of the users' evolving information arrangements and the collaborative use of information.

4. Information copied from the Kiosk view to the Workspace view provides the agent system with feedback about its performance and allows agents to *adapt* to particular user preferences, tasks and circumstances. Agent adaptation eliminates the need for users to adjust software parameters and allows the system performance to improve over time. Agent adaptation ensures that users remain focused on sense-making rather than potentially interruptive administrative tasks.
5. Users and agents communicate their considerations *implicitly*. Users are not required to explicitly visualize discovered relationships among pieces of information for the convenience of computational systems. Instead, agents estimate possible user discoveries of relationships through the analysis of the spatial organization and collaborative use of information. Similarly, agents do not communicate discovered relationships with users but only adjust the flow of information by presenting users with a selection of potentially task-relevant information. The intentional concealing of contextual discoveries ensures that neither users nor agents are limited in the creative exploration and interpretation of information spaces.
6. Agents act as *autonomous* players in a collaborative setting of multiple users and agents in which agents participate and support rather than direct sense-making processes. Agents automatically adapt to users and other agents and can easily be integrated and removed from the system. This particular criterion ensures that existing sense-making settings and processes are not altered or interrupted through the integration of computer technologies.

Benefits The use of the agent system as a computational environment for research investigations and experimentations offers several benefits:

1. The agent system provides researchers with the means to translate observations of cognitive and collaborative processes into computational agents that can be tested individually or collectively. Because every agent only represents one particular concept, the complexity remains minimal and individual operations can easily be monitored and analyzed.
2. The agent system automatically adjusts the influence of individual agents, thus allowing researchers to easily add, modify and remove individual agents without considering the presence and operations of other agents and users.
3. The agent system allows for the easy consolidation of computational options for the analysis and retrieval of information. The agents we introduce in this chapter can easily coexist with, or be replaced by, alternative information analysis and retrieval mechanisms.
4. The agent system provides two comprehensive user interfaces that enable researchers to monitor and control the influence of individual agents as well as to examine the evolving database contents (see Figure 11.3). The user interfaces provide a convenient set-up for researchers to investigate the

impact and dependence among various cognitive and collaborative concepts as well as to adjust the flow of information for particular tasks and in different collaborative settings.

Interpretation agents

The task of the Interpretation agents is to collect Cards from the Workspace views of users, create relationship among Cards, and save the Cards and relationships to a database (Figure 11.3, interpretation agents). Relationships among Cards reflect human mental considerations that are inferred from the users' spatial and temporal organization and collaborative use of Cards. Every individual interpretation agent specializes in creating one particular type of relationship. For example, one interpretation agent might establish a relationship between two Cards because the two Cards have been positioned in close proximity on a user's Workspace view. Another might create a relationship between two Cards that were created during similar time frames. Yet another interpretation agent might create a relationship between a Card pair that is used by multiple users. The number of relationships between two Cards can increase over time, consequently producing varying weights (strength, importance) among different Card pairs. Thus, the addition of every new relationship affects the network balance in part or as a whole. Corresponding relationships conceived by different interpretation agents are likely to become more influential within the database. Contradictory relationships are not resolved but are added to the database, assuming that at a later point in time the accumulation of additional relationships will put an end to the conflict. Consequently, the database is an assembly of both conceivably correct and conceivably incorrect relationships. Interpretation agents can construct either dynamic or static relationships. Dynamic relationships reflect the current state of the database and are reevaluated after every event. Static relationships accumulate over time and never change or expire. The sum of all static relationships represents a database's history or long-term memory while the sum of all dynamic relationships represents a database's current state or short-term memory. The relationships created by the interpretation agents are not meant to be seen by the users but exclusively used to structure the database contents.

We introduce as selection of interpretation agents that primarily infer from the organization of information. These interpretation agents are complementary with other methods for creating relationships among information items such as, for example, the semantic comparison of information contents (Foltz 1996) or the analysis of information access patterns (Sarukkai 2000).

Group 1 A first group of interpretation agents focuses on detecting implicit relationships among spatially arranged Cards and is inspired by Wertheimer's laws of organization in perceptual forms (Wertheimer 2000). Wertheimer suggests a set of eight factors that influence the ways in which humans perceive relationships among spatially arranged objects: The Factor of Proximity suggests the emergence of relationships among objects in close spatial proximity. The Factor of Direction proposes relationships among a sequentially aligned sequence of objects such as, for example, an alignment of objects in rows or columns. The Factor of Objective

Set considers the human ability to remember incremental changes to an arrangement of objects and suggests that previously perceived relationships among objects may dominate even after the emergence of contradictory perceptual concepts. The Factor of Closure focuses on the visual recognition of object clusters. Object clusters commonly emerge through the spatial grouping of information based on specific criteria. The Factor of Similarity refers to the emergence of relationships among similar-looking objects. Color, shape and iconographic representations are among the most influential aspects in regards to this particular perceptual concept. The Factor of Common Fate deals with the effects and the resolution of conflicting perceptual concepts. For example, two objects might be in close proximity yet offer distinct differences in their visual representation. The Factor of Past Experience elaborates on how a human's background might influence their visual perception. For example, people educated in Western cultures develop a tendency to start exploring object arrangements from left to right and top to bottom. Finally, the Factor of the Good Curve discusses the effects of visual ambiguities that may emerge through the intersection of object clusters or the shapes of empty spaces between object clusters. Our current collection of interpretation agents in this group establishes relationships between Cards based on Wertheimer's factors of proximity, direction, objective set, closure and common fate. We plan on implementing additional interpretation agents so as to be able to demonstrate and test all of Wertheimer's principles within the context of our computational environment. Subsequently we will consider complementing this group with interpretation agents that represent other research findings. For example, the work Albert Zobrist and William Thompson (1975) focuses on the recognition of object clusters and introduces concepts such as curvilinear continuity, good closure and overall goodness grouping. Ullman (2000) investigated the cognitive perception of spatially arranged objects based on shapes and colors. Johnson (2001) introduced the concept of 'habituation', the tendency to devote more attention to novel stimuli such as, for example, objects that have not been noticed previously, objects that appear suddenly, or objects that are located in odd or wrong locations.

Group 2 A second group of interpretation agents focuses on recognizing explicit relationships among spatially arranged Cards. Explicit relationships may be assumed among objects that are grouped with a bounding box, that are interconnected with rubber lines, that are stacked, overlap or touch each other, or that are hierarchically organized in a file system. For example, one of the interpretation agents in this group creates a hierarchy of relationships that reflect the spatial nesting of Cards. Another interpretation agent in this group creates relationships between Card pairs connected through the magnet functionality.[1]

Group 3 A third and more experimental group of interpretation agents creates relationships based on the analysis of previously created relationships (independent of external activity) and is primarily active during system idle times[1]. For example, one of these interpretation agents randomly explores the network structure for weakly related Cards (Cards that only maintain few relationships with other Cards). These Cards are temporarily related with more recent additions so as to stimulate their discovery. The goal of this particular interpretation agent is to ensure that past knowledge is not prematurely discarded as well as to instigate a more explorative discovery of information during sense-making activities. Another interpretation agent strengthens interconnections among heavily related network segments in the database. Imagine, for example, a heavily interconnected group of Cards that were created in relation to Apple stocks and another heavily interconnected group of Cards related to Microsoft stocks. If the interpretation agent were to detect a significant number of relationships that interconnect Cards in both groups then it would add one strong relationship (a so-called Highway) between the most heavily related Cards in each group. Subsequent database retrievals of Apple related information would subsequently become more likely to carry along information about Microsoft and vice versa. We also work on an interpretation agent that creates relationships among Cards in the database based on Card activity. Card activity is registered whenever a user selects a Card on their Workspace view, whenever a Card is added or modified in the database, or whenever a Card in the database is accessed, traversed or retrieved. Temporary relationships are created among Cards that experience activity during similar time frames. The goal of this interpretation agent is to foster a distribution of information that reflects and potentially intersects individual user foci. Yet another interpretation agent is specialized on creating shortcuts among indirectly related Cards in the database. For example, a sequential connection of Cards A, B, C and D could produce a relationship between Cards A and D. With this particular agent we expect to foster a more explorative retrieval of database contents that is less focused on strong or direct relationships among information items.

Transformation agents

The task of the transformation agents is to adjust the flow of information among humans and computational systems by providing individual users with a customized selection of potentially relevant information in their individual Kiosk views (Figure 11.3, transformation agents). The relevance of information is determined based on an analysis of the contents in the database (created and managed by the interpretation agents). Because the database reflects and interrelates current and past contributions from different users it provides a rich source of information for transformation agents to learn about the knowledge and expertise of users, the current foci and interests of users, the interactions among users, as well as the collaborative use of information. Every individual transformation agent specializes on one particular aspect when determining the relevance of information for a particular user. For example, one transformation agent might search the database for Cards that maintain relationships with Cards on the user's Workspace view. Another might increase the relevance of Cards that were created by authors from whom the user previously copied Cards. The relevance of information does not necessarily satisfy the interests of one particular user but a community of collaborating users as a whole. For example, if a Card were to be used by all but one collaborating user then a particular transformation agent might increase the relevance of this Card for the one user who is not yet using it. The selection and prioritized sequences of Cards from different transformation agents are combined and displayed on the user's Kiosk view.

Group 1 A first group of transformation agents is focused on facilitating team building as well as converging individual user investigations and foci. These agents select and prioritize Cards based on a comparison of the contents and evolution of individual users' Card arrangements as well as the collaborative exchange of information. A few examples of transformation agents in this group are outlined below.

The *interaction agent*[1] accelerates the information exchange among users that previously benefited from each others contributions. This particular agent prioritizes Cards based on user interactions. An interaction is registered whenever a user copies or views a Card from another user, whenever two users communicate through a Card's instant messaging feature, or whenever an interpretation agent generates a relationship between two Cards created by different authors. Cards are prioritized based on how many interactions the Card recipients have with the Card authors. The interaction agent enables an environment in which users might implicitly and unknowingly start collaborating in teams with unique task foci. The selective exchange of information among team participants also allows for a less interruptive and more efficient development of team knowledge[2].

The *balance agent*[1] prioritizes Cards based on the orderliness and modification frequency of Card arrangements. An unstructured arrangement of Cards is more typical during the early stages of sense-making sessions when people primarily focus on exploring and collecting information. A structured arrangement of Cards is more typical during the end of sense-making sessions when people conclude their analyses and organization of information. The goal of this agent is to encourage individual users to be broad and explorative at the beginning of sense-making sessions and increasingly focused and organized during the end of sense-making sessions. For example, if, at the beginning of a sense-making session, the Card arrangement of a particular user is very organized and only slowly receives new additions then this user is presented with contributions by other users whose Card arrangements are less organized or rapidly accumulate new information. Thus, the balance agent contributes to the development of team knowledge[2] by intersecting the unique sense-making methods of individual team participants.

The *relevance agent* helps users detect potentially relevant information based on the contextualization of Cards by different people. The relevance agent prioritizes Cards by the number of relationships that exist between a Card in the database and a Card on a user's Workspace view. In other words, the Card relevance is determined by the number of relationships between a Card that the user is not using and a Card the user is using. For example, user A might align Card 1 and 2 on his Workspace in ways that prompts one of the interpretation agents to create a relationship between the two Cards. The adaptation of Card 1 by user B would subsequently trigger the promotion of Card 2. The distribution of information by this agent also makes users aware of other users that may have similar interests, that may be experts on subjects of current consideration, or that might benefit from a direct collaboration. The estimation and distribution of relevant information provides an effective means for the relevance agent to foster the development of knowledge interoperability[2], the development of team shared understanding[2], as well as the sharing of hidden knowledge[2].

The *popularity agent* prioritizes Cards by their number of relationships with other Cards in the database as well as by their current and past number of users. Cards that maintain many relationships with other Cards and that are used by many users are considered central to the efforts and considerations of collaborating users. The goal of this agent is to help users converge their individual views and opinions by accelerating the shared use of information. The danger of the popularity agent is that it promotes information that already dominates the collaborative environment, thus decreasing the chance for less influential yet potentially crucial information to enter the users' discourse. Thus, the popularity agent is also equipped with mechanics to explore the database contents in ways that allows for incidental discoveries through the occasional retrieval of less dominant information. The primary objective of the popularity agent is to support the convergence of individual mental models to team mental models[2] by accelerating the utilization of commonly used information.

The *focus agent*[1] attempts to intersect the foci of individual users by promoting Cards that have recently been created, modified or repositioned on a user's Workspace view as well as Cards that are related to Cards with recent activity. During the early stages of collaborative sense-making activities this agent helps users to maintain some awareness of what other team participants are currently working on before engaging in a direct dialogue. During the later stages of sense-making activities this agent helps user to converge their individual investigations and to maintain a shared focus on issues. The focus agent can accelerate team negotiation[2] and team pattern recognition[2] by promoting information in ways that increases joint attention to particular issues.

The *pattern agent*[1] distributes information in ways that fosters shared understanding by helping users establish and recognize similar relationships among pieces of information. This agent compares the Card arrangements of individual users and modifies the flow of information in ways that might increase the chance for users to produce similar Card groupings. For example, if user A would group together Cards 1–4 and user B would group together Cards 1–3 then the pattern agent would promote Card 4 to user B. Helping individuals detect and consider similar relationships among information items can greatly benefit the development of knowledge interoperability[2] and team knowledge[2].

The *context agent*[1] examines the shared use of information in different contexts. Although different users may be using the same information they may use or interpret information differently (Reddy 1979). The context agent distributes information in ways that helps users become aware of and intersect different information contexts. For example, if user A places Card 1 close to Card 2 and 3 and user B places Card 1 close to Card 4 and 5 then the context agent would recognize an incompatibility in the contextual use of Card 1 by user A and B. The context agent would subsequently promote Cards 2 and 3 to user B and Cards 4 and 5 to user A in an attempt to inspire both users with information that emerged through alternate readings of shared information. Bringing about a shared contextual use of information among individuals is likely to benefit the development of team shared understanding[2].

Group 2 A second group of transformation agents is designed to minimize the promotion of Cards that are same or similar to Cards a particular user is already working with. These agents compare the history of Cards in the database with the history of Cards on a user's Workspace view to detect whether a particular Card was created or previously adapted by the user or whether a Card is a modified version of a Card the user is currently working with.

Group 3 A third group of transformation agents is concerned with the retrieval of Cards with same or similar properties as Cards on a user's Workspace view.[1] Examples of Card properties may include categories, geographic locations, authors, ownerships, votes, priorities, comments, references, modification dates and times, associated documents, as well as Card appearances such as sizes, shapes and colors. This group may also include agents that conduct semantic comparisons of text in Card headings and associated documents.

Discussion

The EWall system was designed as an enabling technology for monitoring, measuring, analyzing and computationally supporting micro-, macro- and metacognitive processes.

- Microcognitive processes[2] are low-level mental processes that are intuitively deployed throughout sense-making activities. Typical examples of macrocognitive processes include aspects of attention and perception. The EWall interpretation agents are designed to recognize microcognitive processes by inferring from how humans collect, spatially arrange and exchange information. The determination and analysis of these microcognitive processes provides a basis for humans and computational systems to construct higher-level mental processes.
- Macrocognitive processes[2] are higher-level mental processes that are primarily deployed by individuals and teams when engaging in complex sense-making activities. Typical examples of macrocognitive processes include the development of knowledge interoperability, team shared understanding and convergence of individual team mental models. The EWall transformation agents are designed to detect and monitor occurrences of macrocognitive processes as a basis for supporting the information exchange in ways that allows sense-makers to combine their individual views and opinions more efficiently.
- Metacognitive processes[2] help individuals and teams to reflect on and develop an overall awareness of collaborative sense-making activities, settings and objectives such as, for example, the development of team agreement and shared knowledge. Various EWall information visualizations (such as the one illustrated in Figure 11.3, right-hand side) help sense-makers observe and analyze dynamically changing networks of people and information thus allowing for a more comprehensive understanding of particular considerations within the context of emerging team and knowledge structures.

The research challenge is to find effective means for monitoring and measuring micro-, macro- and metacognitive processes as well as to determine which microcognitive processes play a part in macrocognitive processes and which macrocognitive processes are used in metacognitive processes. Any of these mental processes can be internalized or externalized[2]. Internalized mental processes remain invisible in the minds of people while externalized mental processes can be inferred from the communication and physical representation of information. While there are various quantitative and qualitative techniques for evaluating externalized mental processes, only a few qualitative techniques are available for investigating internal mental processes. The qualitative evaluation of internalized mental processes depends mostly on user feedback (e.g., on questionnaires) whose accuracy is often distorted by the users' subjective reflection of previous activities, thoughts and events.

The EWall system can provide researchers with opportunities to monitor and explore internalized and externalized mental processes more easily and accurately. Most essential is the ability for researchers to investigate internalized mental processes without engaging users in districting and non-task relevant dialogues. For example, a spatial arrangement of Cards on a Workspace view allows researchers and computational systems to estimate possible relationships between information items that are not explicitly externalized. In other words, the use of the Workspace view encourages users to intuitively externalize some of their mental considerations (think aloud) in ways that is meaningful to human observers and computational systems. Thus, the EWall system provides a unique framework for naturalistic decision-making (NDN) (Zsambok and Klein 1997) by enabling researchers to monitor and analyze internalized and externalized mental processes during real-life activities.

The EWall system can also help researchers to establish metrics for measuring individual and collaborative sense-making activities by deploying individual agents that count occurrences and monitor changes to particular cognitive processes. For example, one agent may measure similarities among individual Card arrangements while another agent may count interactions among users. A third and higher-level agent could leverage the conclusions of the previous two agents by investigating possible connections between users with similar Card arrangements and users that frequently interact. The primary benefit of the agent system is that multiple cognitive processes may be investigated simultaneously.

The EWall system consists of modular components that can be combined and used for a various different sense-making applications and settings. For example, the Non-combatant Evacuation Operation (NEO) scenario[2] may be conducted with the use of particular EWall technologies. In the NEO scenario, a diverse team of specialists (such as a weapons specialist, a political analysts, and an environmental expert) must decide collaboratively on a strategy for a military operation focused on rescuing a group of civilian support personnel caught up in a battle situation. During the knowledge construction stage[2] the individual specialist may use their personal Workspace views to collect and organize task-relevant information in their unique areas of expertise. The Kiosk views provide the specialists with a preview of what information their colleagues are currently considering. The specialists may copy Cards from their Kiosk views to their Workspace views, thus producing an initial set of shared information. During the team problem-solving and consensus stages[2] the specialists may communicate with each other through the instant messaging system built into Cards, view the Card arrangements of their colleagues, and combine their individual contributions on a shared Workspace view. This particular approach allows locally and remotely collocated specialists to incrementally transition their individual investigations into a collaborative effort as well as to implicitly contribute to and benefit from the development of team shared knowledge, understanding and consensus.[2]

Conclusion and future work

In this chapter we introduced EWall, a computational environment for people to engage in the visual organization of information through the spatial arrangement and modification of graphical objects. A system of computational agents infers from the spatial and temporal organization and collaborative use of information and automates the exchange of information among collaborating users. We believe that the EWall system can impact future research and development of enabling technologies for collaboration and knowledge management by:

- Supporting the asynchronous and intermittent participation of remotely dispersed collaborators that have varying levels of involvement, may not know each other, may not work on a common task, may operate under different circumstances, or may have different foci, interests and objectives.
- Enabling team members to work independently without diverting from common objectives, increase shared awareness and understanding and encourage anonymous contributions that are less influenced by social factors such as reputation, prestige and organizational status.
- Assisting team management by connecting together users with similar or complementary responsibilities, interests and backgrounds, by dynamically regrouping users based on their availability and expertise and by enabling simultaneous contributions to multiple tasks and work groups.
- Accelerating information sharing by helping users navigate large amounts of information, determine information relevant to their work, investigate the certainty associated with information and minimize the necessary amount of verbal communication during the acquisition and exchange of information.

The EWall concept has evolved over many years in direct response to particular military needs and applications with primary focus on command and control collaboration. Our research objectives are currently shifting from the development of our computational environment to the investigation of cognitive and collaborative sense-making processes with the goal of conceiving, implementing and testing additional computational agents. We are also evaluating a variety of alternative options for balancing the influence of individual agents through the use of genetic algorithms, neural networks and autonomous agent negotiation techniques.

Acknowledgments

Portions of this publication are reprinted, with permission, from Paul Keel, *EWall: A Visual Analytics Environment for Collaborative Sense-Making*. Information Visualization 2007; 6. ©2007 Palgrave Macmillan. This research was sponsored in part by the Human Systems Department of the Office of Naval Research (Grant N000140410569 and N0001407M0072). Current and past research contributors include Edith Ackermann, John Bevilacqua, Jeffrey Huang, Mike Kahan, Paul Keel, Yao Li, William Porter, Mathew Sither and Patrick Winston.

References

Byrne, M.D. (1993). 'Using Icons to Find Documents: Simplicity is Critical', in *SIGCHI Conference on Human Factors in Computing Systems*, pp. 446–453. (Amsterdam, The Netherlands).

Everitt, K.M. Klemmer, S.R. Lee, R. and Landay, J.A. (2003). 'Two worlds Apart: Bridging the Gap Between Physical and Virtual Media for Distributed Design Collaboration', in *SIGCHI Conference on Human Factors in Computing Systems*, pp. 553–560. (Ft. Lauderdale, Florida, USA).

Foltz, P.W. (1996). 'Latent Semantic Analysis for Text-based Research', *Behavior Research Methods, Instruments and Computers* 28, 197–202.

Greenberg, S. and Rounding, M. (2001). 'The Notification Collage: Posting Information to Public and Personal Displays. *SIGCHI Conference on Human Factors in Computing Systems*, pp. 514–521. (New York: ACM).

Halasz, F.G. (2001). 'Reflections on NoteCards: Seven Issues for the Next Generation of Hypermedia Systems', *Computer Documentation* 25, 71–87.

Haramundanis, K. (1996). 'Why Icons Cannot Stand Alone', *ACM SIGDOC Asterisk Journal of Computer Documentation* 20.

Johnson, M. (2001). 'Functional Brain Development in Humans', *Nature Reviews/Neuroscience* 2, 475–483.

Kirsh, D. (1995). 'Complementary Strategies: Why We use our Hands when we Think', in *Proceedings of the Conference of the Cognitive Science Society*, pp. 212–217. (Hillsdale, NJ: Lawrence Erlbaum).

Lewis, J.P. Rosenholtz, R. Fong, N. and Neumann, U. (2004). 'Identifying and Sketching the Future: VisualIDs: Automatic Distinctive Icons for Desktop Interfaces', in *ACM Transactions on Graphics*, pp. 416–423. (New York: ACM).

Marcus, A. (2003). 'Fast Forward: Icons, Symbols, and Signs: Visible Languages to Facilitate Communication', *Interactions* 10, 37–43.

Marshall, C.C. and Bly, S. (2005). 'Saving and Using Encountered Information: Implications for Electronic Periodicals', in *SIGCHI Conference on Human Factors in Computing Systems*, pp. 111–120. (New York: ACM).

Pena, W. and Parshall, S. (2001). *Problem Seeking: An Architectural Programming Primer*. (New York: John Wiley & Sons).

Reddy, M.J. (1979). 'The Conduit Metaphor: A Case of Frame Conflict in Our Language about Language', in A. Ortony, (ed.), *Metaphor and Thought*, pp. 284–324. (Cambridge: Cambridge University Press).

Robertson, G. Czerwinski, M. Larson, K. Robbins, D.C. Thiel, D. and van Dantzich, M. (1998). 'Data Mountain: Using Spatial Memory for Document Management', in *UIST Symposium on User Interface Software and Technology*, pp. 1153–162. (New York: ACM).

Sarukkai, R. (2000). 'Link Prediction and Path Analysis Using Markov Chains', in *International World Wide Web Conference*, pp. 377–386. (Amsterdam, The Netherlands: North Holland Publishing).

Ullman, S. (2000). 'Visual Cognition and Visual Routines', in S. Ullman, *High Level Vision*, pp. 263–315. (Cambridge: MIT Press).

Wertheimer, M. (2000). 'Laws of Organization in Perceptual Forms', in S Yantis, (ed.), *Visual Perception: Essential Readings*, Key Readings in Cognition. (Philadelphia, PA: Psychology Press).

Winnicott, D.W. (1982). *Playing and Reality*. (London: Routledge).

Wright, W. Schroh, D. Proulx, P. Skaburskis, A. and Cort, B. (2006). 'The Sandbox for Analysis: Concepts and Methods', in *SIGCHI Conference on Human Factors in Computing Systems*, pp. 801–810. (New York: ACM).

Zsambok, C. and Klein, G. (1997). *Naturalistic Decision Making*. (Mahwah, NJ: Lawrence Erlbaum Associates).

Zobrist, A. and Thompson, W. (1975). 'Building a Distance Function for Gestalt Grouping', *IEEE Transactions on Computers* C-24, 718–728.

Chapter 12

DCODE: A Tool for Knowledge Transfer, Conflict Resolution and Consensus-building in Teams

Robert A. Fleming

Introduction

The Decision-Making Concepts in a Distributed Environment (DCODE) program is a prototype team decision support concept initially developed at the SPAWAR Systems Center in San Diego. It is designed to work within the framework of MIT's Electronic Card Wall software (Keel 2005).

DCODE supports two phases of collaborative team problem-solving: an individual team member's knowledge construction/analysis phase as well as the team's collaborative process of conflict resolution and consensus-building. DCODE takes the internalized macrocognition assessments used by an individual team member for course of action (COA) analysis and externalizes them into coded iconic displays. The graphical representation of the grouped icons assists the individual team member in forming an overall assessment of his information 'pool', and simplifies the task of COA comparison and selection. These coded individual member displays are then used by the entire team to assist in the collaborative macrocognitive processes of conflict resolution and consensus-building needed to produce the overall team COA recommendation.

Prior to the advent of high speed and broadband communication technology, one of the primary causes for a lack of consensus in a command and control distributed decision-making team was the distribution of the information itself. Critical information was often simply not available to all participants, and the differing information pools resulted in differing recommendations. Improved communication technology has significantly reduced the impact of this problem on team performance and decision-making capabilities. The DCODE design is based on the assumption that the lack of a team consensus is now more often due to the differing interpretations and weightings that the various team members assign when viewing the exact same pool of retained information items. It is these individual and unique cognitive assessments that result in a lack of team consensus on any decision option recommendation. Examples of differing cognitive assessments would include determining the impact of an item on a particular decision option, assessing the relative importance of that item to other retained items or ranking the quality of the information source.

Several recent changes in the general composition of command and control collaborative groups have also exacerbated the effect of individual differences. Diverse team components (e.g., legal, medical, first responders, security) may have markedly different perceptions as to the impact and importance of a particular information item, and differing culture backgrounds (as found in multinational coalition operations) have been shown to have unique behavioral characteristics that influence cognitions and decision-making (Hofstede 2001).

In the team process of conflict resolution and consensus-building, a critical first task is determining which of the myriad decision-relevant items reflect the key assessments differences causing the lack of consensus. These conflicting items are often small in number but are difficult to isolate because there is usually no physical display of the actual individual cognitive assessments. While current electronic transfer protocols usually have a number of 'meta tags' (such size of document, format, date of origination, URL, author, keywords, copyright restrictions, abstract, etc.) attached to an information item, they rarely display, or 'tag' the critical information about an individual's subjective assessment of the impact of that information on a particular decision option.

DCODE proposes that these internalized cognitive assessments be externalized through a knowledge elicitation template, and then encoded into simple icons which are physically attached as Cognitive Assessment Tags (CATs) to each retained item. The DCODE concept includes the development of (1) an assessment template for capturing individual assessments and (2) a color-coded system for attaching and displaying these CATs on each decision-relevant information item. A conceptual overview of the CATS concept is shown in Figure 12.1. It shows an analyst who has just received some incoming information and determines that it is of very high importance and has a negative impact on the decision option of using a SEAL team. These represent his CATS for this item, and the DCODE template translates them into an easily understood size and color-coded icon which is then attached to the retained information item.

While meta tags are very useful in finding content information, CATs are very useful in searching for those *differences* in the cognitive information assessments that the various team members have assigned to that specific item. Since it is these differences that are contributing to a lack of consensus, the display and review of these cognitive tags can be a critical factor in reducing the time a group spends on conflict resolution. The DCODE technical approach emphasizes that interoperability of collaborative knowledge management requires the display, exchange and integration of both information object (IOB) information content as well as the cognitive assessments and interpretations that each individual has assigned to that content. The DCODE concept is most applicable to a distributed group decision-making task where:

1. time is not a highly critical issue, i.e., a minimum of a few hours are available for the decision process
2. there are a limited set of viable options
3. team members are collecting, analyzing and weighting incoming information items

4. each team member forms their own recommendation based on their cognitive assessment of the items in their information pool
5. information and recommendations are freely exchanged within the group, and if there are conflicts, the group attempts to resolve these and reach consensus on an overall group recommendation, and
6. if consensus cannot be reached, the recommendation typically follows the majority opinion, or is determined by the senior participant.

This structure is very similar to the Operational Planning Team (OPT) described in Chapter 2 and discussed in detail by Jensen (2002). Additionally, wherever possible, the current chapter will use examples from the Non-combatant Evacuation Operation (NEO) scenario from Chapter 2 of this volume.

The OPT decision process, as well as the DCODE concept, is most representative of analytic/rational decision-making, where incoming information is compared and evaluated against each set of COAs. This is in contrast with Naturalistic Decision-Making (NDM) process proposed by Klein (Klein *et al.* 1993). NDM studies how people actually make decisions and perform cognitively complex functions in demanding high-stakes situations, usually with severe time constraints. It assumes that decision-makers in these situations do not have the time luxury of comparing and contrasting incoming information against each possible decision option. They often use a process call 'satisficing' (Simon 1957), where they do not test each possible option but rather select the first option that reaches a minimum set of acceptable criteria. Obviously, NDM and rational decision-making are not dichotomies but

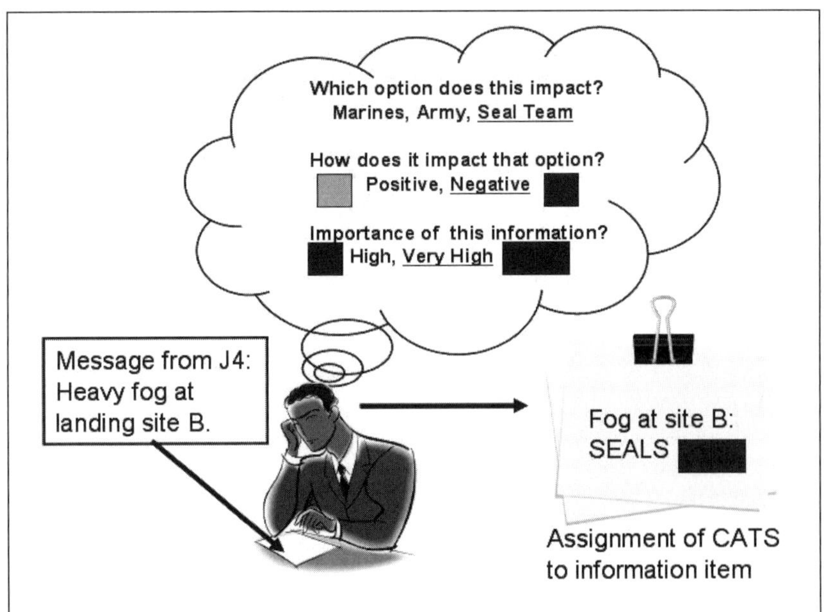

Figure 12.1 Elicitation and encoding of Cognitive Assessment Tags (CATS)

rather points along a continuum that are determined by a variety of factors, including time availability and the expertise of the decision-maker.

This chapter discusses five critical components of CATs:

1. What are the minimum essential cognitive assessment parameters that need to be attached to the item? e.g., importance, impact, information quality, etc.;
2. What are the user interface design issues related to the structure of the knowledge elicitation template needed to extract these internalized cognitions?
3. What type of iconic display representations are best used in converting these extracted macrocognitions and attaching them to each information item?
4. How are the displayed CATs best used by the analyst to reach his specific recommendation? and
5. How does the team use the CATs to improve the speed and efficiency with which they address the issues of conflict resolution and consensus-building?

Preliminary research on early versions of the DCODE concept (Fleming 2003; Fleming and Cowen 2004; Cowen and Fleming 2004) has indicated that the display of quantified subjective information assessments does improve the quality of the decisions and is rated as a strong preference by users.

Basic cognitive assessment tags (CATs)

An analyst in an operational planning COA selection environment is usually exposed to constrained time pressures and an ongoing heavy cognitive load. If we are to ask the analyst to perform an additional task (i.e., 'tagging' each retained information item with their subjective assessments), this new requirement must not unduly increase their cognitive workload. Therefore, it is important to determine the minimum number of CATs that need to be elicited and attached to each retained item. There appear to be four minimally essential CATs:

1. which options or COAs are impacted by the item
2. what is the nature of the impact on that COA, i.e., supportive or non-supportive
3. how important is this item relative to the other retained items, and
4. what is our assessment of the quality of this information.

Course of action CAT

Any information item that is retained is retained because it has some impact on one or more of the options/COAs under consideration. For example, a new information item which reports 'Heavy fog at proposed AH-1 landing site' would probably only affect the COA(s) involved with that landing site, and the analyst would attach the appropriate COA CAT to that item. The overall evaluation of any COA would be then be based on the pool of information items that have been assigned to it by the analyst. An item may also be retained because the item appears relevant but cannot yet be assigned to a specific COA.

Impact CAT

An item is retained because it has something to say about the viability of particular COA, i.e., it either supports the COA (positive impact) or is non-supportive (negative impact). Impact is obviously not a dichotomy but rather a continuum which may range cognitively from something like 'fatal flaw' to 'slam dunk'. The analyst attaches their assessment of the impact of this item as a CAT and later uses all the pooled impacts to make their COA recommendation. Clearly, analysts can differ in their assessment of the same information item such as the landing site example above, where one analyst may consider it positive because it helps conceal the landing while another team member may assess it as negative because it makes the landing more risky.

Importance CAT

By definition, all retained items will be perceived to have some level of importance to the decision-making task, but they can differ significantly relative to one another. Importance can be viewed as the weighting given to each impact assessment. In the example above, fog at the site may have a fixed impact effect on some COAs, but the importance of the information would vary as a function of whether, say, a COA proposed the site as the primary landing site vice a back-up landing site.

It is worth noting that impact and importance are not completely independent cognitive parameters. It is difficult to imagine an item that has a very high importance assessment but only a mild positive or negative impact. There is, however, enough independence at the intermediate ranges of each to make this a useful distinction.

Information quality CAT

The impact and importance CATs are designed to take the information item at face value, i.e., given this is valid information, what is its impact and importance? However, information can differ significantly in quality, and the analyst needs to add an information quality CAT to any retained item. Issues could revolve around the credibility of the information source, the level of uncertainty associated with the information specifics and the timeliness of the information. Because most military analysts typically deal with relatively high-quality information, this could probably be an 'exception only' type of CAT where the analyst need do nothing if the information is, as expected, of acceptable quality, and only adds this quality CAT to the item when there is some specific concern about its quality.

While there are a number of other CATS that can be beneficially added to an information item, the four CATs above are considered here to be the basic essentials. Thus in the example about the landing site, the basic assigned CATS might be: Related to Marine force option, supports that option, is of medium importance, No issues with quality of information.

Knowledge elicitation approach

Another interface issue is how to efficiently elicit these macrocognitive CAT assessments and convert them into externalized representations that can viewed, stored and exchanged between team members. Again, the primary consideration is to minimize the cognitive load placed on the contributing analyst. The analyst has incoming relevant information from a wide variety of source types, e.g., e-mail, computer message traffic and displays, hard copy reports, phone and face-to-face conversations, personal experiences, etc. The DCODE concept is that for any retained item, the analyst creates a small computer-generated information card, referred to here as an information object (IOB). Each IOB contains a number of administrative information items (tracking number, originator, time tag, etc.) that are automatically generated, as well as a 'content' area where the analyst can enter or paste keywords and a very brief description of the item. The IOB also has an area for activating a knowledge elicitation template which the analyst would use to externalize the CATs he associates with the enclosed information content. The design of this template needs to capture the four CATs described above and involves a trade-off between level of granularity and ease of use. Many opinion surveys often use a seven-point scale to span an assessment dimension, e.g., seven options available from 'Strongly oppose' to 'Strongly support'. While this level of granularity captures a more precise indication of the internalized macrocognition, its increased complexity causes a greater cognitive burden than a shorter or binary yes–no option, and this negative effect can be cumulative when a large number of assessments are required. The DCODE approach is to use a minimum number of intradimensional segments and to have the displayed template include all possible selections as simple 'radio' button choices. The goal is to have the analyst complete all the CATS within a few seconds by using a simple 'point and click' mode.

Most military decision-making tasks involve only a limited number of options or COAs. In the template, these can be listed next to radio buttons and the user would have the option of selecting more that one button as the information may be relevant to several COAs. For a larger number of COA options, a drop-down menu might be utilized.

Impact is a critical CAT and it is proposed that this be a five-segment selection, two levels for negative impact (e.g., 'Negative', 'Very negative'), two for positive impact and a mid-dimension 'Neutral' choice for information that is relevant but for which the analyst has not yet classified the impact. Only one impact choice can be selected for this CAT.

Importance is viewed as a two-dimensional selection, i.e., an 'Average' category and a 'High' category. 'Average' does not indicate neutral or marginal, but is assigned to those items that have sufficient impact/importance to meet the criterion agreed upon for item retention. 'High' is for those items which clearly surpass that criterion. Only one importance choice can be selected for this CAT.

As noted above, quality of information has a default preselection of 'Acceptable', so that the user need make no input for the large majority of items that are assumed to have no real quality issues. If the analyst perceives a quality issue he can select any or all of the subcategory options of Source credibility, Timeliness or Uncertainty.

With this knowledge elicitation template structure, almost all items retained by the user can, with three point and click selections, complete all the required externalizations of the CATs.

Display of CATs

Once the required CATS have been elicited by the template, the interface issue of information display needs to be addressed. Since the individual analyst will use a pooled perspective of the IOBs to search for an overall assessment of the viability of any particular COA, the information coding system for CATS should complement this cognitive search strategy. Size and color coding have been shown to be particularly effective in search tasks (Tufte 1989; Galitz 1996) and it is proposed that Impact CATs be color coded and Importance CATS be size coded. For the impact dimension it is recommended that the five selections be color coded as follows:

- Very negative – Red;
- Negative – Yellow;
- Neutral – Gray;
- Positive – Pale green;
- Very positive – Bright green.

For the Importance dimension it is recommended that the two selection options be coded as follows: Average – one size unit; High – two size units. Thus, a single icon, varying in color and size, can represent any one of the ten impact/importance combinations (five for impact, two for importance). The CAT reflecting the impacted COA would be in text format. In order to declutter the IOB format, the quality of information CAT would only be displayed when the default 'Acceptable' option was overridden. When any other quality category is selected, a 'flag' would be displayed on the IOB, and the user could then click on it to drill down to the amplifying information.

The basic IOB structure is shown in Figure 12.2. The left-hand box is the basic IOB framework, and is blank, with the internal text describing what information should be entered in each area. The right-hand box shows the user had added the basic information content to the IOB, but this is unevaluated information, i.e., does not contain and CATs. When the user wishes to append his CATs, he can click on the 'CATS' box and the knowledge elicitation template will become available for completion.

An example of two completed IOBs is shown in Figure 12.3. On the left, team member Collins has evaluated the mud slide information item (#117) as having a Negative (yellow) effect upon the Army option and is of average importance (one box filled). He also has some question about the credibility of the information source (bottom of the IOB). One the right, team member Baxter has evaluated the fog information item (#123) as having a Positive (pale green) effect upon the Marine option and is of high importance (both boxes filled).

It is assumed that when each member creates an IOB it is also simultaneously posted to a community workspace where it becomes available to all the other team

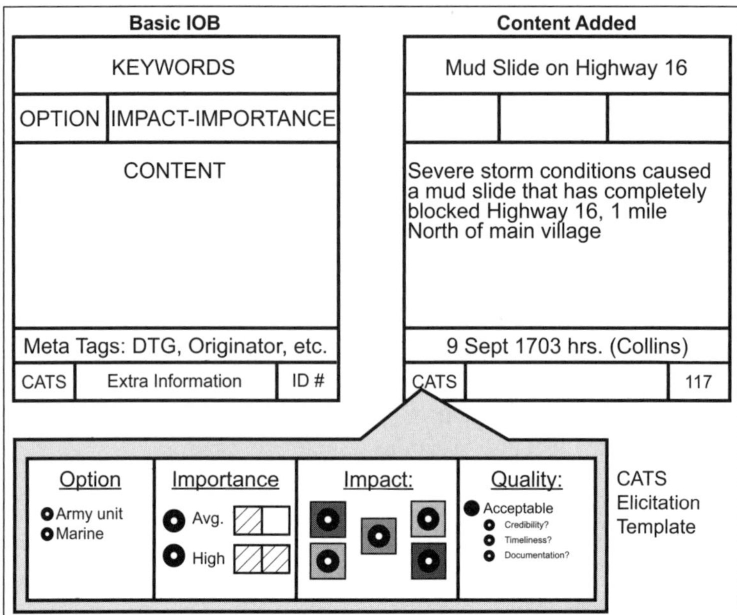

Figure 12.2 Basic IOB structure, and the CAT template

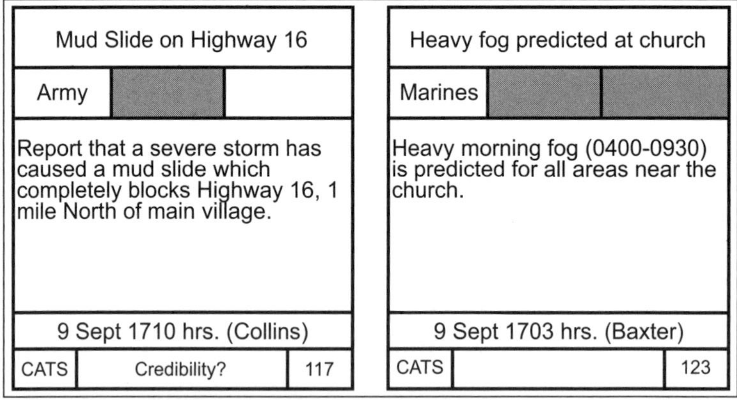

Figure 12.3 Two completed IOBs

members. The information content is shared but each team member can 'pull' any IOB into his own workspace and assign his own CATs. When a team member feels they have collected enough IOBs, they would use this pooled information to make their individual option recommendation. A hypothetical distribution of pooled IOBs related to two viable decision options is shown in Figure 12.4 (textual content has been removed for the sake of brevity and clarity). It can be seen that he has collected eleven IOBs related to the Army option and ten IOBs for the Marine option.

A critical individual macrocognitive process is how an analyst goes about making an overall recommendation from his pool of IOBs. They can attempt to cognitively pool and integrate all previously elicited CATs, but this can be a difficult and error-prone process that relies heavily on short-term memory. A proposed DCODE tool helps in this process by assigning numeric values to each particular impact–importance CAT pair, and presenting summary scores. This changes the qualitative macrocognition of 'my overall assessment' into a quantitative score that can directly compare all COAs for individual option recommendations, and it available to other participants for analysis and discussion. It is assumed that at the beginning of any assigned decision-making task, the team has agreed to certain standards regarding the weightings and integrations of CATs. For example they may specify that a 'Very positive' CAT would be assigned a value of +2, 'Positive' a +1, 'Negative' a -1 and 'Very negative' a value of -2. Similarly they may dictate that an 'Average' importance rating weights the impact score by a factor of ×1 while a 'High' importance weights that score by a factor of ×2. Thus, an impact of 'Very negative' with 'High' importance would be assigned a -4 score (-2 × 2). DCODE would automatically generate a small bar chart to show the algebraic summary of all the IOBs the participant is using. A summary chart for the IOBs in Figure 12.3 is shown in the lower right corner and indicates that, for this collection of IOBs, the Army option has an overall summary score of +3 while the Marine option has a summary score of -6. On the basis of his pooled IOBs and the summary chart information, this participant would recommend the Army option.

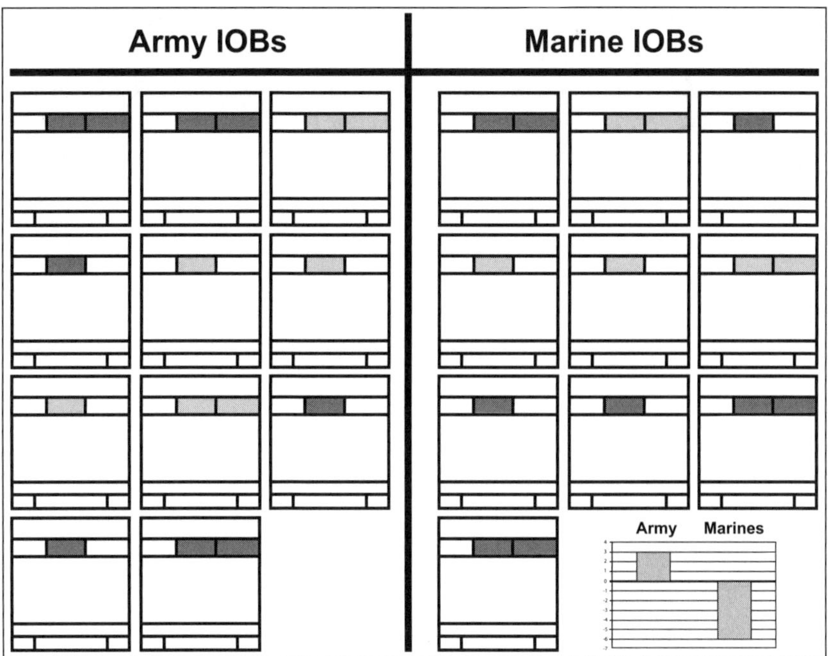

Figure 12.4 A sample individual workspace and summary chart

Consensus

A critical team macrocognitive process is conflict resolution and consensus-building. In any decision-making task there is often a lack of consensus among the participants regarding the group's COA recommendation. Because of the pooled and shared IOBs, each participant has the same information pool available, but has come to different conclusions. Their interpretation of the information, as reflected in their CATs, is the critical determiner in their recommendation. It is not unusual to find the situation shown in Figure 12.5, where three participants have made their summary recommendation, two supporting one option, the other supporting a different option. This particular display reflects the NEO decision task (Warner, Wroblewski and Shuck 2003), and shows Baxter strongly recommends the Army option, Collins gives a mild endorsement to the Marine option and Malone gives a stronger endorsement to the Marine option. Since all participants have the same pool of IOBs, it can be assumed that the differences in the summary scores are due to differing CATs. The team needs to isolate those critical CAT differences that most contribute to these different recommendations.

Many IOBs were used in the analysis – which ones truly reflect the critical CAT differences causing conflict and lack of consensus? Locating these key differences is a basic requirement in conflict resolution and consensus-building and is often the major contributor to the excessive amount of time the team devotes to this macrocognition task. DCODE has a proposed methodology to quickly display those conflict items (i.e., the IOBs) that are most contributing to team conflict, significantly reducing the energy and time the team spends on this task

The proposed DCODE process would look at each unique IOB (based on ID#) and compare the summary score assigned to it by a team member with the summary score assigned to it by all the other team members. All difference scores would be summed (based on absolute value, not algebraic sum, because magnitude of the differences is the critical issue) and those IOBs with the highest discrepancy score would be displayed to the team.

Figure 12.6 shows a hypothetical sample of the three IOBs that have the largest disparity scores from our team of three decision-makers. Malone and Collins both recommend the Marine force option for the NEO, while Baxter recommends the Army option. The display shows the three summary discrepancy scores, displayed on the right side, which most contribute to this difference of opinion. Baxter feels

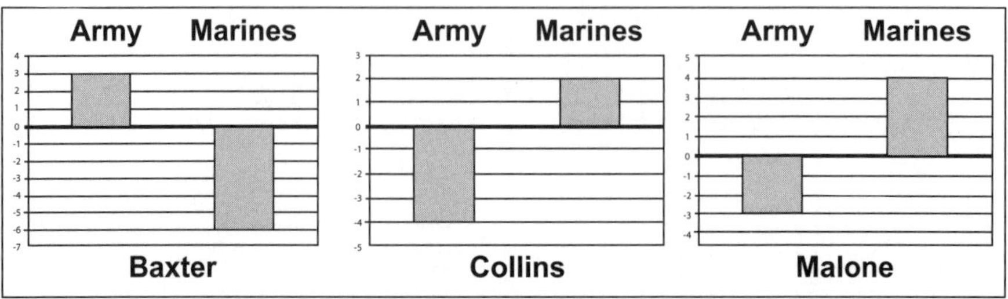

Figure 12.5 Summary option recommendations from three team participants

item #130 supports the Army option (a score of +2), while both Collins and Malone feel it has a very negative impact on the Army option (scores of –4). Baxter's score is 6 units from Collins and 6 units from Malone, giving a total discrepancy score of 12. For IOB #132, Baxter assesses it as a Negative impact on the Marine option, while Collins and Malone assign a positive impact. This figure shows that the team should focus discussion on determining the causes for the differences found in these IOBs, as they contain the most diverse opinions. For example, in IOB #130, Collins and Malone may have given a negative rating because they believe riots in the city will put the island's defensive forces on heightened alert, making detection of the Army units more likely. Baxter, on the other hand, perceives it as a positive factor in favor of the Army option, as he believes it will actually distract the local defense forces and make them much less effective. The team would then use their most accessible communication system (face-to-face, phone, chat box, etc.) to discuss, persuade, negotiate and resolve the key discrepancies and reach consensus.

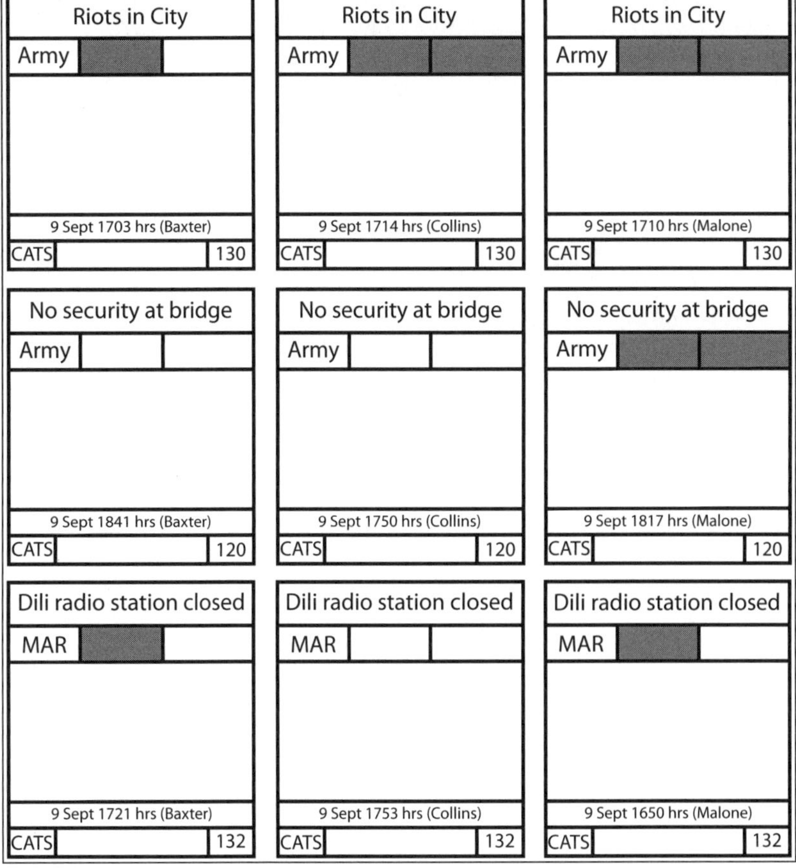

Figure 12.6 Display of IOBs with maximum disparity between individual CATs

Summary

The basic building block of DCODE is the Information Object (IOB) shown in Figure 12.2. The critical difference between IOBs and a typical information analysis item is that IOBs contain both information content as well as the originator's macrocognitions regarding the impact of this item on a particular decision option. A knowledge elicitation template extracts these macrocognitions and displays them as size- color-coded icons (referred to as Cognitive Assessment Tags, CATs) on the IOB.

Figure 12.4 depicts the primary display a team member would use to make their individual option recommendation. The graphical bar chart shows a quantitative summary of all the CATs that this team member has assigned to the retained pool of information items. It has taken qualitative subjective assessments and converted them into numeric baseline values that can compare and contrast all options. Both the individual and team can vary the different weightings assigned to the CATs and examine the resulting effects for each COA on the summary scores graphs.

Figure 12.5 shows the method of displaying the differing individual team member preferences for each COA. With a quick visual scan, all participants can see how the other members ranked each option. Minor ranking differences can then be discussed and resolved.

If there is a lack of consensus due to major differences with regard to individual option recommendation, the team can call up the display shown in Figure 12.6. DCODE does an automated contrasting of the impact/importance CAT scores each individual has assigned to an IOB and then displays those IOBs that have the highest total discrepancy scores between members. This enables the team to immediately focus in on the select set of IOBs that are most contributing to the lack of consensus. The cognitive processes, displayed in Figures 12.3 to 12.6, shows the decision-making sequence from initial assessment of individual information items, pooling the assessments for each individual's overall option recommendation, determining level of group agreement on option selection, and in the event of conflict and lack of consensus, a display to assist the team in reaching a shared understanding of which differing macrocognitions are most contributing to the lack of consensus.

References

Cowen, M. and Fleming R. (2004). 'A Knowledge Management Tool for Collaboration: Subjective Tagging of Information'. Paper presented at the 2004 TTCP HUM-TP9 Human Systems Integration Workshop, Ottawa, Canada.

Fleming, R. (2003). 'Information Exchange and Display in Asynchronous C2 Group Decision Making'. Paper presented at the 8th International Command and Control Research and Technology Symposium, held at the National Defense University, Washington DC.

Fleming, R. and Cowen, M.B. (2004). 'Improving Individual and Team Decisions Using Iconic Abstractions of Subjective Knowledge'. Paper presented at the 9th International Command and Control Research and Technology Symposium, San Diego, CA.

Galitz, W.O. (1996). *The Essential Guide to User Interface Design*. (New York: John Wiley & Sons).

Hofstede, G. (2001) *Culture's Consequences, Comparing Values, Behaviors, Institutions, and Organizations Across Nations.* (Thousand Oaks, CA: Sage Publications).

Jensen, J.A. (2002). 'Joint Tactics, Techniques, and Procedures for Virtual Teams', *Proceedings of the Office of Naval Research, Collaboration and Knowledge Management Workshop, San Diego, CA*. (Arlington, VA: Human Systems Department).

Keel, P. (2005). 'Electronic Card Wall (Ewall)', *Proceedings of the Office of Naval Research Collaboration and Knowledge Management Workshop.* (Arlington, VA: Human Systems Department).

Klein, G.A., Orasanu, J. Calderwood, R. and Zsambok, C. (eds) (1993). *Decision Making in Action: Models and Methods*, pp. 362–388. (Norwood, NJ: Ablex).

Simon, H.A. (1957). *Models of Man – Social and Rational.* (New York: John Wiley and Sons).

Tufte, Edward. (1989). *Visual Design of the User Interface*. (Armonk, NY: IBM Corporation).

Warner, N.W., Wroblewski, E.M. and Shuck, K. (2003). 'Noncombatant Evacuation Operation: Red Cross Rescue Scenario', in *Proceedings of the Office of Naval Research, Collaboration and Knowledge Management Workshop, San Diego, CA*. (Arlington, VA: Human Systems Department).

Chapter 13

Modeling Cultural and Organizational Factors of Multinational Teams

Holly A.H. Handley and Nancy J. Heacox

Introduction

The Collaboration and Knowledge Interoperability program has described a model of team collaboration in order to facilitate the understanding of the cognitive mechanisms underlying collaborative team behavior. The model depicts four collaboration stages that a team traverses during a problem-solving task: knowledge construction, collaborative team problem-solving, team consensus, outcome evaluation and revision. The research described in this chapter supports all of these collaboration stages in the context of teams who are planning multinational missions. A tool was developed that provides decision support for these planners based on a model of cultural and organizational factors that are relevant to multinational forces. Although the domain used for development was military coalitions, the tool is also applicable to a multinational corporate environment. Corporations have long struggled with the ability to understand work styles encountered in different nations, and planning and coordinating multinational workforces is a frequent challenge. As organizations continue to expand internationally, modeling and simulation tools that can help anticipate the effects of cultural differences will become increasingly important. In this chapter, relevant culture and work processes will be summarized, and the development of the tool will be described.

Support for team collaboration

In *knowledge construction*, the problem task and required relevant domain information are identified. As individual members of a planning team begin developing their own mental models of the problem, cultural and organizational information can be provided to facilitate understanding of the 'ways of doing business' that are applicable to the problem task. This supports the metacognitive process of the individual team member's understanding of the elements, relations and conditions that compose the initial state of the problem. In *collaborative team problem-solving*, team members iteratively collect and analyze information as revised and new information is made available within the problem domain. The members begin to develop a team shared understanding of the situation requirements, constraints and resources. Information can be provided about authorized, pre-developed mission task sets that may be applicable to the situation at hand. Likewise, information can also be provided about characteristics of potential resources that may be assigned as task performers. Alternatives for solutions to reach the mission or organizational goal become viable in the next stage, as team members work from their shared understanding toward *team consensus*. Modeling of projected performance of available personnel units on mission tasks offers criterion-based parameters on which to judge the unit assignments. Finally, in the stage of *outcome evaluation and revision*, team members are supported in the comparison of solutions against the original problem task.

Initially, a model was developed that incorporated empirically observed cultural factors into decision-maker, work process, organization and national culture component models. A tool was then developed to simulate a culturally based decision-making process using an algorithm that draws from these four components. The resulting Integrative Decision Space (I-DecS) allows a work process to be modeled as a series of tasks, along with attributes of the organization that controls the process, the personnel assigned to that task and their nationality and proficiency. It provides a description of the decision environment from the process and organization models and introduces personnel with cultural profiles into the decision nodes. The outcomes from the simulation can be compared for different sets of personnel units that are assigned as task performers. The simulation returns projected outcomes of the work process completed by a baseline of nationally homogeneous personnel units compared to multinational or multi-organizational personnel units, for outcomes of accuracy and speed of performance. In most work processes, changing component model variables will result in the same *output* (that is, the same product will ensue from the process); however these *outcome* values may differ, reflecting different decision styles and work methods. A decision support tool can facilitate the collaboration process by presenting visual representations of the problem space for a mission or organizational goal, tasks, resources and task performers, as well as the attributes that affect their interactions.

Culture and work processes

Research has shown that there are consistent differences in the ways that people from diverse cultures approach and complete tasks; these differences, in turn, affect the performance of work processes. Task performance effectiveness and work process efficiency will be highest when assignment of human resources to the process is based on relevant attributes of the available personnel units. Cultural factors that have been identified from values dimension and world views research literature have been shown to affect behaviors such as decision-making and work methods. Cultural patterns in the way people behave or are motivated have their roots in these values. The influence of cultural values can be observed in organizational settings where dimensional attributes have been identified and associated with decision-making and task completion.

For the cultural component of the I-DecS model, a set of cultural impact variables based on the nationality of task performers were identified. The foundation for these variables is the four dimensions based on the work on Geert Hofstede (Hofstede

1980; Hofstede 2001). Over 50 countries have been profiled with these dimensions, so that comparisons between national cultures are possible. These four dimensions are Power distance (relationship between people at different levels of authority), Uncertainty avoidance (tolerance for ambiguity), Masculinity (expected gender roles) and Individualism (importance of individual vs group accomplishments). Other researchers, notably Shalom Schwartz (Schwartz and Bilsky 1987) and Harry C. Triandis (1972), have also developed schemes of categorizing cultures. However, it is acknowledged by many in the cross-cultural domain that Hofstede's work still represents the most comprehensive investigation into foundational, value-based cultural factors to date. In addition, the work of many other researchers has validated and built upon these factors. The empirical findings of multiple researchers (for instance: Erez 1994; Smith, Peterson and Wang 1996; Offermann and Hellmann 1997; see the Appendix) were consulted for this research.

Related to the four dimensions are eight impact parameters that affect the decision-making process of a task performer at a node. Supervision indicates the level of supervision that is typically required in a decision-maker's culture; Delegation is the level of task designation that is characteristic in a decision-maker's culture; Information flow is the communication pattern the decision-maker will expect; Decision-making is the preference for individual or group decision-making; Reliance on guidance is the type of authority source the decision-maker uses at the node; Adherence to rules is the extent to which a decision-maker adheres to rules of the process; Risk-taking is the level of uncertainty that a decision-maker will tolerate and still perform; Decisiveness is the resolution with which the decision-maker reaches a decision. An assessment that shows empirically established relationships of cultural dimensions to organizational impacts is included in the Appendix.

Along with cultural differences based on national culture, task performers also bring to multinational coalitions or corporations the business methods from their home organizations. While national culture is generally defined as a 'collective mental programming' of the people of any particular nationality (Hofstede 1980, 1991), organizational or corporate culture covers many facets of organizational life, such as management styles, appraisals, rewards and communication styles used by employees. An organization's culture is conceptually tied to the predominant national culture of the organization and it manifests itself in multiple ways; for example, different role structures and levels of hierarchy, different communication patterns at nodes where information is exchanged and varying performance metrics that are expected to be met. The culture of an organization can be represented by parameters in the work process in which the task performers are participating. By identifying organizational cultural characteristics, especially in situations where there has not been an opportunity for the assigned personnel units to train together and learn each other's processes, the potential effects of mismatches in organizational characteristics can be highlighted. For example, Merritt and Helmreich (1996) found striking differences in the structure and functioning of communications within the cockpits of Asian and US airliners. The ability to communicate, especially in an emergency, might be negatively impacted if a US crewmember were to be assigned to fill a billet in an Asian cockpit or vice versa.

For the organizational component of the I-DecS framework, parameters are determined by the relationship between each assigned personnel unit and the organizational leader.

- Interface culture represents the degree of congruity between the nationality of the unit and the nationality of organizational leader.
- Command authority indicates whether the unit is under the leader's command and control structure. In work processes that involve staff from multiple organizations, multiple relationships concerning command and control affect how the process will flow – for example units who need to report back to their own headquarters for authority slow the flow of activities.
- Training indicates whether the unit has trained with the multinational organization.
- Doctrine refers to the degree of agreement between the organizational guiding principles of the unit with those of the organizational leader. Different doctrines may contain fundamental differences regarding the degree of information that should be reported, the detail that should be contained in directives, or the degree to which subordinates are allowed to take initiative.
- Service represents the impact of different command arrangements for different military services and non-governmental organizations.

The cultural and organizational components make up two of the four component models that are the foundation of I-DecS. The third component, the process model, contains the sequence of tasks in a work process; the tasks are characterized as types of activities that must be performed. The US Department of Labor coding scheme is used to specify characteristics of these activities. Subsets of the Universal Joint Task List and the Navy Unified Task List have been characterized by this scheme and included in I-DecS. The fourth component, the decision-maker model, represents either individual or team task performers, depending on the scale of the specified decision process. Example attributes of task performers are their training and experience as applicable to the nodes of the process.

The 'decision space' is where the four models are integrated and interact. The decision space predicts process outcomes from combinations of component model parameters – cultural, organizational, process and decision-maker. In this way, potential conflicts between decision nodes can be highlighted and variations in the output due to manipulations of assignment of task performers can be observed.

I-DecS decision support tool

The catalyst for development of the I-DecS decision support tool was a gap in the ability of the US military to model aspects of work processes that are important in the multinational coalition environment. At present, there is little assistance available to operational planners about expected work behaviors or organizational factors affecting personnel from other nations, beyond personal experience and anecdotal evidence. Given this lack of information about human resources, political considerations often become the sole criteria for staffing mission tasks. Many examples of performance problems in coalitions have been documented that are related to unanticipated differences in work styles or organizational procedures (such as Siegel 1998). The US military decision-maker needs to have the ability to incorporate organizational and cultural factors into the building of effective multinational mission forces.

Coalitions are often designed and staffed quickly in order to meet a pressing situational need, such as a natural disaster or an act of war. Nations step up to the plate and make resources available to staff the mission. The resources are assigned by operations planners of the coalition's lead nation – often the US – in consultation with the other involved nations. In time-critical situations, this is accomplished during the doctrinally specified Crisis Action Planning (CAP) process (The Joint Staff 1995). The planning process is conducted at the headquarters of a military combatant command. Based on knowledge of this planning process, I-DecS was developed for CAP staff to allow organizational and cultural factors to be incorporated into mission staffing decisions. Planners can build alternate configurations of a mission plan in I-DecS as sets of tasks and staffing assignments. As additional nations offer resources, planners can assess where these resources fit best in the process.

A user-centered design process was employed to develop the I-DecS interface, to ensure that the goal of congruence between the CAP process and the I-DecS tool was met. Feedback was solicited on early design plans from members of the target user community – multinational mission planners. Then subject matter experts (SMEs) assisted in structuring the user's tasks within I-DecS so that the information required by the tool matched the information available during the planning process. It was important to fully understand the structure and the information requirements of the work process in which the user would be engaged in when using I-DecS. These are reflected in the steps the user is asked to perform when using the tool, in the type of information requested of the user and in the format and content of the tool's databases. SMEs were also instrumental in designing the format of the output so that it would have maximum relevance to the planners. The result was that resource allocation decisions ensuing from the planning process are reflected in the tool output, depicted as simulations of alternate task sets and staffing assignments.

The I-DecS model was implemented as a web tool, based on the requirements of the coalition-wide area network (COWAN). Many development decisions were driven by the anticipated multinational user community. The technology was developed conservatively – that is, to the constraints of the most realistic technology environment available to the multinational planners. For example, while a client–server application would result in more user-friendly features, the web-based application eliminates the need for local software. Also, since the planning process will occur on a secure network but implementation on SIPRNET would limit participation by non-US users, COWAN was chosen as the host network. The I-DecS user interface was designed as a web template to allow easy entry of the variables needed by the underlying models and algorithm, with only simple customizations required. Users are provided with preset options to input as model parameters, and the output formats are designed to allow the user to play 'what if' by varying some values of one parameter in a case. The simulation output consists of two sections: the top is a grid-type output that allows the user to track the outcome impact, and the bottom is a detailed description of the ensuing mission plan (see Figure 13.1).

In addition, aids to planning a multinational mission force are available for user exploration. These include cultural databases that provide guidance for working with personnel from different cultures (see Figure 13.2), national level parameters for an overview of cultural characteristics, operational level parameters for the impact of working closely with different nations, and cultural traditions and norms to help avoid *faux pas*.

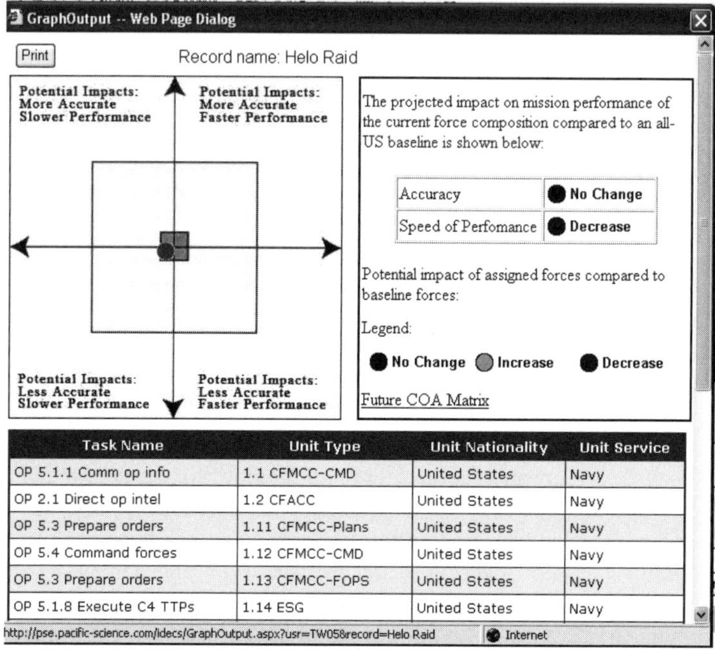

Task Name	Unit Type	Unit Nationality	Unit Service
OP 5.1.1 Comm op info	1.1 CFMCC-CMD	United States	Navy
OP 2.1 Direct op intel	1.2 CFACC	United States	Navy
OP 5.3 Prepare orders	1.11 CFMCC-Plans	United States	Navy
OP 5.4 Command forces	1.12 CFMCC-CMD	United States	Navy
OP 5.3 Prepare orders	1.13 CFMCC-FOPS	United States	Navy
OP 5.1.8 Execute C4 TTPs	1.14 ESG	United States	Navy

Figure 13.1 I-DecS simulation output

Impacts -- Web Page Dialog

Print

Compatibility of Ways of Doing Business -
Based on Nationality of Austria vs. USA Baseline

Description of Characteristic	Similar	Different
Level of *supervision* customarily provide to workers.	√	
Amount of *delegation* customarily employed.	√	
Type of *communication patterns* customarily used.	√	
Rigidity with which *business rules* are typically followed. * Please note: Non-US workers may be accustomed to working more closely 'by the book.'		√
Level of tolerance for *uncertainty* that may be encountered. * Please note: Non-US workers may be more likely to seek additional information in uncertain situations, especially from experts or authority figures.		√
Amount of *decisiveness* in decision-making that is typically found.	√	

http://pse.pacific-science.com/idecs/Impacts.aspx Internet

Figure 13.2 I-DecS 'compatibility of ways of doing business'

While the current emphasis is on cultural aspects, the I-DecS framework can be adjusted to focus more on task complexity, unit skills, or organizational climate. I-DecS has been customized to incorporate different mission characteristics and priorities. Both Unified Joint Tasks (UJTL) and Navy Tactical Tasks (NTL) have been used as the building blocks for the mission. Performance projections have included synergy and interoperability, as well as accuracy and speed of performance. Additional mission attributes have been included, such as transportation, weapons, and networks. Each component model supplies unique data to the system to enable a planner to predict the performance of a work process. The categories of data in each component model originated from the empirical literature, and thus are valid and traceable. Variables have been chosen based on their relationships with performance outcome parameters. The architecture of the system allows for ease of insertion, modification, or deletion of variables as development progresses. Two methods were used to ensure the validity of the constructs underlying the component model framework. The first method used experimental data and case studies to validate the component tables of the model. The second method employed a regression analysis and correlation between the parameters in the framework to verify the algorithmic relationships.

Usability and utility assessments

Usability and utility assessments determine how well a system meets its design goals and provides value within a particular environment. Systems are subjected to structured evaluations of (a) effectiveness (how accurately and completely users can complete tasks), (b) efficiency (the relationship between accuracy and completion of goals, including the time to complete a task), and (c) user satisfaction (whether the system is easy to understand and use). Two venues for subject testing were conducted in order to address assessment issues: (1) a structured experiment at the NAVAIR Cognitive and Automation Research Laboratory (CARL) in order to assess the system usability; and (2) a focus group at the Naval Postgraduate School (NPS), in order to assess its military utility.

The structured experiment was held at CARL over a two-week period in order to identify any usability concerns, that is, to assess whether the system was easy to learn and use. Additionally, this was the first assessment of I-DecS with non-developer, naive users; it was important to observe that the I-DecS prototype program worked consistently and without programmatic errors. Thirty undergraduate students from a community college in southern Maryland served as participants in this study. Subjects were randomly assigned to a face-to-face, unaided control group (five teams) or a face-to-face group using the I-DecS tool (five teams). Each team consisted of three participants who were required to develop together a course of action for a collaborative problem-solving task. An existing Non-combatant Evacuation Operation (NEO) scenario was modified to include multinational military resources. Students were instructed, as a team, to develop an operational plan to rescue three Red Cross workers trapped in a church basement on a remote island in the South Pacific. Required elements of the final plan included a list of military forces to be used in the rescue, transportation and weapons to be used and a specific timeline of events. All teams were given a brief introduction to the NEO scenario and the experimental task, which included a detailed explanation of each component of the required final plan (personnel, transportation, weapons, critical times and the detailed plan).

In order to assess the participants' perceptions of the tool and its effect on their completion of the exercise, two surveys were completed by the participants. All participants completed the NASA Task Load Index (TLX), a subjective, *post hoc* workload assessment tool. Additionally, the I-DecS users completed the System Usability Scale (SUS)[1] while the control teams completed a post experiment survey. The usability evaluation consisted of four explicit hypotheses in the following areas: workload (H_1), efficiency (H_2), confidence (H_3), and satisfaction (H_4). The hypotheses were tested by comparing the survey results of the teams using I-DecS with the teams not using I-DecS. Analyses of the results from participant surveys on their perceptions of the use of I-DecS indicated that I-DecS lowers perceived workload (H_1) and increases efficiency (H_2) by lowering temporal demand. However, no significant difference was found either on the confidence level (H_3) or the satisfaction level (H_4) between teams using and not using I-DecS to compose the mission plan. Overall, the user comments and survey results expressed that the tool was easy to learn and use.

A follow-on focus group was held at the Naval Postgraduate School, (NPS) in Monterey, California, as part of the Large Scale Systems and Experiment Design Class. The objective was to elicit specific comments from military personnel on the tool's utility. Six military officers and one civilian, all students currently assigned to the class and responsible for planning an upcoming exercise, participated in the focus group. The participants were given a scenario entitled, 'Connectivity and Collaboration for Radiation Awareness, Biometrics Fusion and Maritime Interdiction Operations'. This was a class experiment scenario that had been modified to include multinational military personnel. Students were instructed to use I-DecS as a planning tool to develop an operational plan to conduct the Maritime Interdiction Operation for their experiment. Required elements of the final plan included a list of military forces to be used, aligned with specific tasks, and interoperability requirements.

The participants were allowed to work uninterrupted for a short period of time (half an hour). The facilitators and students then participated in a question and answer focus group forum for an additional two and a half hours. Finally, the participants completed two surveys: the NASA TLX and the SUS. The focus group also validated the use of I-DecS in a web server mode; I-DecS was accessed through the Internet from a host server with multiple users simultaneously. It confirmed that I-DecS was appropriate for military personnel, some of which were non-US, and allowed extensive discussion among researchers and participants. Personnel assessments of current and potential future output features of I-DecS were also obtained. The forum discussion provided direct feedback on the use and utility of the tool.

The NASA TLX survey responses of the military subjects were compared to the responses of NAVAIR non-military subject responses. These indicated that the military personnel were comfortable working with the I-DecS tool: mental demand, temporal

1 ©Digital Equipment Corporation, 1986.

demand and effort were all reported significantly lower by the military group. These were the same three categories that were reported lower by the team with I-DecS (compared to without I-DecS) at NAVAIR. This verified that I-DecS lowers perceived workload (H_1) and increases efficiency (H_2) by lowering temporal demand.

Integration with decision support system for coalition operations

The Decision Support System for Coalition Operations (DSSCO) is a suite of software tools for planning multinational collation operations (Heacox, Quinn, Kelly and Gwynne 2002). Coalition Operations require detailed, accurate planning before the start of a mission as well as the ability to revise those plans in response to changing conditions after a mission has started. Large amounts of information must be collected, analyzed, categorized, integrated and applied when developing a mission plan. During these information processing activities, planners need to maintain situation awareness of developing circumstances that could dictate changes to the developing plan. DSSCO provides planners with information they need to maintain awareness of the state of the planning process.

The DSSCO Planning Module shows the current status of the planning process and links the steps in that process to information that is used to create and visualize planning products. Within this planning environment, I-DecS facilitates the task of multinational resource mapping by providing planners with the ability to evaluate culture and organizational attributes and their impact on the overall plan; I-DecS is being integrated into DSSCO to provide this functionality to coalition planners. As the planners begin to describe COAs, the I-DecS module will be available to choose tasks, link coalition resources and identify cultural or organizational constraints. The addition of I-DecS to the DSSCO planning suite adds robustness to planners' ability to identify cultural parameters that affect the developing coalition mission plan.

Conclusion

Multinational coalitions and corporations bring together task performers with different cultural backgrounds, operating procedures and decision-making processes. Cultural models may capture the effect of national differences but lack the contribution of the individual's training and experience. Likewise a model of an individual task performer may capture expertise and cultural biases, but it does not capture the role in the organization or the decision process. Different organizational structures also influence the decision process. Moreover, the task performer's rank status within the structure affects his performance and the performance of staff with whom he interacts. While these distinct factors that influence the decision process are reasonably well understood in separation, in order to model cultural differences in human decision-making, all components must interact together. In order to understand how these organizational, process, cultural and individual attributes affect coalition operations and collaborative decision-making, a model was developed to incorporate these predispositions into the decision-making process and predict their effect on the outcomes of work processes. I-DecS is the first decision support tool of its kind.

It quantifies and integrates aspects of culture – an important foundation of human behavior but one that has remained evasively subjective.

I-DecS supports the four collaboration stages of the team collaboration model. Using I-DecS, task, resources, and required relevant domain information can be captured, assisting the knowledge construction stage. Information can be iteratively collected and analyzed facilitating collaborative problem-solving. Alternatives for solutions to reach the mission or organizational goal can be evaluated using the I-DecS outcome parameters, guiding team members toward team consensus. Finally, in outcome evaluation and revision, I-DecS supports the comparison of solutions against the original problem task.

I-DecS is a tool for planners of multinational operations in military and corporate environments. It assists in the design of coalition missions or outsourced processes by providing templates to choose tasks and assign units. I-DecS provides several cultural databases that can assist planners in determining the best fit of resources for a multinational mission force. It provides planners with comparative performance projections of the alternate task-staffing configurations that may be constructed with selected task and resource sets. Alternate task-staffing configurations can be simulated iteratively until the desired combination of staff diversity and process outcomes is achieved. I-DecS helps planners to include cultural considerations in the design of work plans for military coalitions and multinational corporations.

Acknowledgment

This work was supported by the Office of Naval Research under grant no. N00014-04-C-0392. The views, opinions, and findings contained in this article are the authors and should not be construed as official or as reflecting the views of the Department of Defense.

References

Heacox, N.J., Quinn, M.L., Kelly, R.T., Gwynne, J.W. and Sander, S.I. (2002). *Decision Support Systems for Coalition Operations: Final Report*. Technical Report 1886. (San Diego, CA: SSC San Diego).

Hofstede, G. (1980). *Culture's Consequences: International Difference in Work-related Values*. (Beverly Hills, CA: Sage Publications).

Hofstede, G. (1991). *Cultures and Organizations*. (New York: McGraw-Hill).

Hofstede, G. (2001). *Culture's Consequences: Comparing Values, Behaviors, Institutions and Organizations Across Nations*, 2nd edn. (Thousand Oaks, CA: Sage Publications).

The Joint Staff (1995). *Joint Pub 5-0 Doctrine for Planning Joint Operations*. (Washington, DC: Joint Doctrine Branch).

Merritt, A.C. and Helmreich, R.L. (1996). 'Human Factors on the Flight Deck: The Influence of National Culture', *Journal of Cross-Cultural Psychology* 27(1), 5–24.

Schwartz, S.H. and Bilsky, W. (1987). 'Toward a Universal Psychological Structure of Human Values', *Journal of Personality and Social Psychology* 53, 550–562.

Siegel, P.C. (1998). *Target Bosnia: Integrating Information Activities in Peace Operations*. (Washington, DC: Institute for National Strategic Studies).

Triandis, H.C. (1972). *The Analysis of Subjective Culture.* (New York: Wiley).

Appendix: Cultural Dimensions and Organization Impacts

		Associated Organization Structures:														Associated Cognition:
		Hierarchy:		Centralization:		Standardization:		Formalization:	Specialization							Style of Thinking
		Organization Impact														
Cultural Dimension	Level	Supervision	Delegation	Information Flow	Decision Making	Reliance on Guidance	Adherence to Rules (& intolerance to deviation)	Directives Preference		Accomplishment	Risk-Taking		Decisiveness	Preference	Regard for Expertise	Over-Confidence
Power Distance	High	Close	Low		> emerg	rules						No	High			
	Low	General	High		>emerg	Exper.						Yes	Low			
		[2] [3]	in 6]	[8]		in 6]						[22]	[3]			
Uncertainty Avoidance	High		Low				High				Low			Detail	High	
	Low		High	Ad hoc			Low				High			Gen.	Low	
			[3]	[8]	[6]		[3]	[3]			in 6]			[8]	[3]	
Masculinity (aka Mastery)	High												High			
	Low												Low			
					[3]											
Individualism	High				Indiv d-m >emerg	Exper/ trng				Pragmatic						
	Low				Group d-m >emerg	Formal rules				Relation-ship orient						
Long-Term Orientation	High															More likely
	Low															Less likely
																[17] [18]
Activity Orientation	Inde-pendent			Direct comms						Pragmatic						
	Interde-pendent			Indirect comms						Relation-ship orient						
				[8]						[8] [19] [see Indiv]						

References for Appendix

[1] Erez, M. (1994). 'Toward a Model of Cross-cultural Industrial and Organizational Psychology', in H.C. Triandis, M.D. Dunnette, and L.M. Hough, (eds), *Handbook of Industrial-Organizational Psychology: Volume 4*, 2nd edn, pp. 559–607. (Palo Alto, CA: Consulting Psychologists Press, Inc).

[2] Cummings, T.G. and Worley, C.G. (1994). *Organization Development and Change*, 5th edn. (Minneapolis/St. Paul, MN: West Publishing Company).

[3] Hofstede, G. (1980). *Culture's Consequences: International Difference in Work-related Values.* (Beverly Hills, CA: Sage Publications).

[6] Hofstede, G. (2001). *Culture's Consequences: Comparing Values, Behaviors, Institutions, and Organizations Across Nations*, 2nd edn. (Thousand Oaks, CA: Sage Publications).

[8] Sutton, J.L. and Pierce, L.G. (2003). 'A Framework for Understanding Cultural Diversity in Cognition and Teamwork'. Presented to the 2003 International Command and Control Research and Technology Symposium, June 17–19, 2003, National Defense University, Washington, DC.

[9] Smith, P.B., Peterson, M.F., Akande, D., Callan, V., Cho, N.G., Jesuino, J., d'Amorim, M.A., Koopman, P., Leung, K., Mortazawi, S., Munene, J., Radford, M., Ropo, A., Savage, G. and Viedge, C. (1994). 'Organizational Event Management in Fourteen Countries: A Comparison with Hofstede's Dimensions', in A.M. Bouvy, F.J.R. Van de Vijver, P. Boski and P. Schmitz, (eds), *Journeys into Cross-cultural Psychology*, pp. 364–373. (Lisse, Netherlands: Swets and Zeitlinger).

[10] Pavett, C. and Morris, T. (1995). 'Management Styles within a Multinational Corporation: A Five Country Comparative Study', *Human Relations* 48, 1171–1191.

[11] Merritt, A.C. and Helmreich, R.L. (1996). 'Human Factors on the Flight Deck: The Influence of National Culture', *Journal of Cross-Cultural Psychology* 27(1), 5–24.

[12] Smith, P., Peterson, M.F. and Wang, Z.M. (1996). 'The Manager as Mediator of Alternative Meanings: A Pilot Study from China, the U.S.A. and U.K.', *Journal of International Business Studies* 27(1), 115–137.

[13] Offermann, L.R. and Hellmann, P.S. (1997). 'Culture's Consequences for Leadership Behavior: National Values in Action', *Journal of Cross-Cultural Psychology* 28, 342–351.

[14] Sagie, A. Elizur, D. and Yamuchi, H. (1996). 'The Structure and Strength of Achievement Motivation: A Cross-cultural Comparison', *Journal of Organizational Behavior* 17, 431–444.

[17] Azar, B. (1999). 'How do Different Cultures Weigh the Same Decision?', *APA Monitor* 30(5), 12.

[18] Yates, J.F., Lee, J., Shinotsuka, H., Patalano, A.L. and Sieck, W.R. (1998). 'Cross-cultural Variations in Probability Judgment Accuracy: Beyond General Knowledge Overconfidence?', *Organizational Behavior and Human Decision Processes* 74(2), 89–117.

[19] Klein, H.A., Pongonis, A. and Klein, G. (2000). 'Cultural barriers to multinational C2 decision making'. Presented to the 2000 Command and Control Research and Technology Symposium, 11–13 June 2000, Naval Postgraduate School, Monterey, CA.

[22] Heacox, N.J., Gwynne, J.W., Kelly, R.T. and Sander, S.I. (2000). *Cultural Variation in Situation Assessment: Influence of Source Credibility and Rank Status*. Technical report 1829. (San Diego, CA: SSC).

[23] Hofstede, G. (1997). 'Organization culture', in A. Sorge and M. Warner, (eds), *The IEBM Handbook of Organizational Behavior*, pp. 193–210. (London: International Thomson Business Press).

Chapter 14

CENTER: Critical Thinking in Team Decision-making

Kathleen P. Hess, Jared Freeman and Michael D. Coovert

Introduction

The ease with which a problem can be solved is influenced by many aspects of both the environment and the decision-maker(s) (Jensen 2002). As problems, technologies, and organizations become more complex, teamwork skills are increasingly important for effective problem-solving. Working in distributed teams can pose unique challenges – requiring team members to make contributions to assessments and plans from different locations and at different times. Time and space hinder timely, critical dialogue over assessments and plans (Paulus 2000) and, as such, are barriers to effective decision-making. In particular, such distributed teams may be particularly vulnerable to social loafing (Latane, Williams and Harkins 1979) because the collectivist aspect of teams is diminished. Further, it becomes more difficult to assess individual contributions to the team in a timely and relevant manner due to such group phenomena as communications bottlenecks and failures as well as various forms of process loss (Steiner 1972). Collaboration among asynchronous, distributed, culturally diverse teams is further complicated by the heterogeneous knowledge of specialists, hierarchical organizational structure and (especially in the military) rotating team members.

Their vulnerabilities notwithstanding, teams are necessary. No individual has enough knowledge or cognitive capacity to fully address complex mission problems in real time. In these cases, a team effort is necessary to ensure that key information is gathered and considered, assumptions are revealed and tested and plausible interpretations and plans are considered. Accordingly, the military relies on asynchronous, spatially distributed teams to analyze intelligence, develop courses of action and perform many other tasks.

One strategy to better understand team collaboration, particularly collaboration to critique and refine intellectual products such as assessments and plans, is to begin by taking *individual cognition* as an analogue for intellectual collaboration. The empirical research literature shows that individuals succeed in solving complex problems in part through experience, which develops the capacity to recognize problems and retrieve solutions from memory (Klein 1993), and in part by thinking critically about their understanding of the problem at hand and their solutions to it (Cohen and Freeman 1997; Cohen *et al.* 1996, 1998).

The current research refers to the team analog of individual critical thinking as *Collaborative Critical Thinking* (CCT). Collaboratively critical teams encourage productive, timely, critical dialogue between team members – a process which should result in higher-quality decisions that achieve greater rates of mission success.

This chapter discusses the initial development of the CCT construct and an attempt to develop training and a software tool to facilitate the use of CCT. We begin with a theoretical background that provides a framework for understanding CCT within an existing structural model of collaboration (Warner, Letsky and Cowen 2005; Warner and Wroblewski 2004). Next we discuss the measures, feedback, tools and training that make it possible for decision makers to develop CCT. We further discuss attempts at validating a polling application developed to enhance CCT: the Collaboration for ENhanced TEam Reasoning (CENTER) tool. Finally we discuss the fielding process for CENTER and conclude with a discussion of future plans for the tool and the training.

Collaborative critical thinking theory

A fundamental challenge of military organizations is to operate decisively and synchronously in highly uncertain and dynamic settings. There is strong evidence that *individuals* succeed in these settings in part by thinking critically about the tasks before them; that is by critiquing their understanding of the situation, improving it and adapting their decisions accordingly. CCT is the process by which individuals *work as a team* to apply this cognitive skill to the group level.

Prior to the current research, there has been little, if any, systematic development of theory to guide training, measuring, monitoring, and managing CCT. There are useful foundations for such theory, however, in recent research concerning individual critical thinking and team performance on information-intensive tasks.

A theory of how individual warfighters make decisions under uncertainty has been validated in research concerning *individual critical thinking*. The Recognition–Metacognition framework (Cohen and Freeman 1997; Cohen *et al.* 1996, 1998) asserts that expert warfighters *monitor* for opportunities to critique their assessments and plans, *identify* sources of uncertainty (i.e., gaps, untested assumptions and conflicting interpretations), and *reduce or shift* that uncertainty by gathering information, testing assumptions, formulating contingency plans and other cognitive and material actions. The current work is extending this theory to team settings.

Research conducted under ONR's program in Adaptive Architectures for Command and Control (A2C2) has formally modeled teams in decision-making settings in order to optimize collaboration (e.g., MacMillan, Entin and Serfaty 2002). This is achieved largely by concentrating – in as few individuals as possible – the information required to perform interdependent tasks (Vroom and Jago 1988), while balancing this with the requirement to distribute task execution skills and responsibility widely enough to process large numbers of tasks in parallel. A2C2 has developed measures of collaboration: however, these measures are fairly gross in scale – describing the frequency and periodicity of coordination, but not the cause, content, or methods. The present work develops measures of finer granularity, while focusing on team member interactions that critique and refine assessments, plans and team processes.

The context of collaborative critical thinking

Collaboration builds on the conceptualization of groups as information processors (Hinsz, Tindale and Vollrath 1997) where collaboration can be defined as a social process by which people exchange information and perspectives, create knowledge (Rawlings 2000) and discuss and integrate the implications of these processes. More specifically, during collaboration, people with complementary skills interact to create assessments (shared understanding of events) and plans of action that are often more robust than would have been achieved by any individual alone.

The current work shares its focus on collaboration processes with recent research by Warner *et al.* (2005) and Warner and Wroblewski (2004). Warner and his colleagues propose that collaboration occurs in four stages:

1. Knowledge-base construction – identifies the human, information and technological resources required for successful collaboration and mission execution;
2. Collaborative problem-solving – develops solutions to the problem at hand;
3. Consensus – to achieves team agreement on the common work products;
4. Outcome evaluation and revision – analyzes, tests and validates the team solution against requirements and exit criteria.

Each stage of this model engages processes of metacognition, information processing, knowledge-building and communication.

Within this model of collaboration, the current research posits that CCT is relevant to all stages of the model (Figure 14.1), but is particularly relevant to three: Problem-solving, Evaluation and revision, and Knowledge-base construction. In problem-solving and evaluation, team members must voice their reservations about their joint products, investigate the most serious of those deficiencies and define actions to handle the risks they uncover. In knowledge-base construction, team members engage in similar critical thinking interactions to define the team's structure and processes. Thus, CCT can be seen as a global process that is available in all of the collaboration stages posited by Warner and colleagues.

A theory of collaborative critical thinking

Collborative critical thinking (CCT) is the interaction between team members that manages uncertainty by revealing it, identifying its sources and devising ways to test its depths or diminish it. The effect of managing uncertainty is to improve estimates of risk so that plans can be verified, made more robust to failure, or discarded. CCT consists of four unique activities that each team member engages in:

1. *monitoring* interactions that alert other team members to the existence of uncertainty,

2. *assessment* interactions in which team members evaluate the opportunity (e.g., available time) and need (e.g., priority or stakes) to resolve the uncertainty,
3. *critiquing* interactions in which team members identify the source of uncertainty, i.e., *gaps* in knowledge (missing information), *conflicting interpretations* of the evidence at hand (e.g., alternative assessments or plans), and *untested assumptions* that shape the inferences from explicit knowledge (Cohen and Freeman 1997; Cohen *et al.* 1996, 1998), and
4. devising *actions* that reduce uncertainty, at best, or that compensate for irresolvable uncertainty.

CCT can be used to handle any uncertainty that the team encounters; however, it is most frequently applied to two objects: the specific *mission* or problem at hand, and the team *processes* in achieving that mission or solving the problem. A focus on the mission involves critiquing *assessments* (e.g., of enemy intent, or the state of 'friendly' forces) and critiquing *plans* (as is done in Course of Action development and assessment) that are instrumental in mission success. A focus on the team's process involves a critique of *goals*, the *plans* (or strategies) for achieving those goals, and the state of *tasks* that constitute the plan.[1] The behaviors that constitute CCT and their objects are represented in Table 14.1.

CCT is a success when it decisively tests theories or fills critical gaps in knowledge; that is, when it devises *actions* that produce useful results that *reduce uncertainty*. However, CCT is also a success when monitoring and assessment interactions *increase awareness of uncertainty* and its importance. For example, monitoring may uncover gaps in knowledge about the enemy's true strength (order of battle) so that intelligence assets can be tasked to focus on this issue. A process that unveils uncertainty may counterbalance decision biases – such as representativeness, availability and anchoring and adjustment (Tversky and Kahneman 1974). These biases are adaptive by design, in that they enable people to decide and act in the face of uncertainty; however, in some circumstances they can be maladaptive.

Organizations often formalize CCT processes and the structures that support it. Military organizations define roles such as scout, air surveillance operator (e.g., the Air Control Officer [ACO] in an E2-C), and tactical intelligence analyst (e.g., imagery analysts). The warfighters who fill these roles serve a *monitoring* function. The procedures (implicit collaboration) and communication protocols they use are designed in part to handle new but unevaluated (uncertain) entities and events and to supply relevant and timely information to their teammates.

Assessment and *critiques* of mission information are to some extent built into the jobs of individuals (e.g., the ACO has procedures for evaluating unidentified entities), and to some extent addressed by interactions at the command level. For example,

1 While it is reasonable to think that teams might also focus on the structure of their organization (who communicates with whom, who controls which assets, etc.), in practice teams rarely adapt their structures on the fly to fit the problem at hand, and they rarely do so successfully (Hollenbeck *et al.* 1999; Entin *et al.* 2004). CCT might encourage rapid and successful adaptation of team structure; however, we have not focused on this application of CCT in the present work.

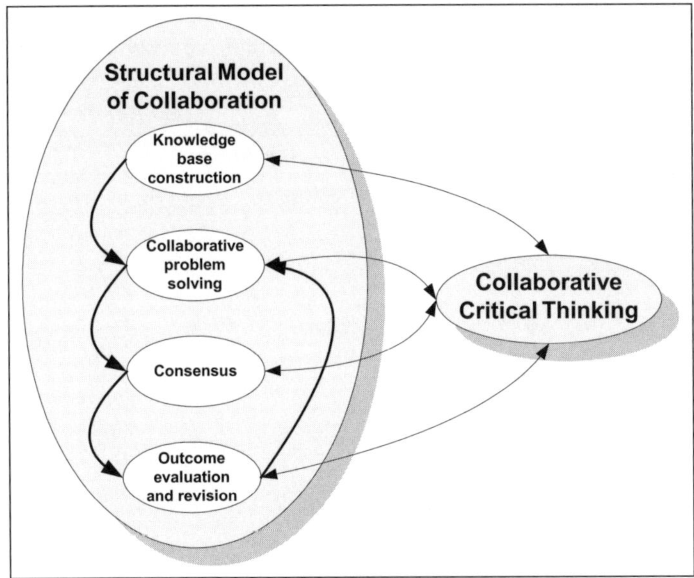

Figure 14.1 A structural model of collaboration (Warner, Letsky, and Cowen, in press) to which collaborative critical thinking serves as a global resource.

critical incident interviews in the TADMUS program provide some evidence that the Commanding Officer and Executive Officer in the Combat Information Center (CIC) revealed or negotiated conflicts in their assessments, and critiques of the intent of unidentified aircraft that were engaged in nominally threatening behaviors (Cohen *et al.* 1998). Assessment and critique are more common and observable in team activities that unfold at a slower pace than CIC operations. Mission planning, for example, includes phases of Intelligence Preparation of the Battlefield, Course of Action (COA) generation, and COA evaluation that are designed to identify what is known, unknown, inferred and (to a lesser extent) assumed.

The selection and execution of *actions* to resolve uncertainty are often formulated as standard operating procedure in tactical operations. For example, watchstanders in the CIC have a few, well understood actions with which to resolve uncertainty concerning track identity and intent. These include such things as radio queries, visual identification, and inference from kinematic information (e.g., the acceleration or turn rate of an aircraft). In less well-defined domains, such as military operations on urban terrain (MOUT), intelligence analysis and mission planning, decision-makers have a vast range of techniques for gathering needed information, testing for information that discriminates between conflicting assessments and testing assumptions. MOUT soldiers, for example, may test for the presence of enemies within a room by looking through the door, listening through walls, applying video cameras through windows and other holes and observing the behavior of nearby civilians who may have observed the enemy's entry to the room. The primary challenge in selecting actions

Table 14.1 Collaborative critical thinking behaviors defined and applied to two objects of analysis: team products and team process

<table>
<tr><th colspan="2" rowspan="2">Collaborative Critical Thinking (CCT)</th><th colspan="5">Object of CCT</th></tr>
<tr><th colspan="2">Team Products</th><th colspan="3">Team Processes</th></tr>
<tr><th>CCT Behaviour</th><th>Definition</th><th>Assessment</th><th>COA/Plan</th><th>Goals</th><th>Plans</th><th>Tasks</th></tr>
<tr><td>Monitoring</td><td>Detect instances in which critical thinking is needed due to high uncertainty and high stakes</td><td colspan="2" rowspan="6"></td><td colspan="3" rowspan="6"></td></tr>
<tr><td>Assessment</td><td>Detect opportunities to handle uncertainty because time is available to think critically & the problem is of sufficient high priority (relative to other problems)</td></tr>
<tr><td>Critiquing for gaps</td><td>Find gaps in information that lead to uncertainty</td></tr>
<tr><td>Critiquing for assumptions</td><td>Find untested assumptions that lead to uncertainty</td></tr>
<tr><td>Critiquing for ambiguity</td><td>Find ambiguous information that supports conflicting interpretations, a source of uncertainty</td></tr>
<tr><td>Action</td><td>Determine what info or tests will resolve uncertainty, or what contingency plans will manage it, then implement them</td></tr>
</table>

is, of course, determining which actions will efficiently and reliably produce useful information, information that bears on the identified source of uncertainty.

Thus, the ability to exercise critical thinking credibly and productively is in part a function of domain expertise; CCT is not a process for novices. CCT leverages expert knowledge of what to monitor, how to assess opportunities and need for critical thinking, how to critique a solution (e.g., what information has *not* been considered, what assumptions have *not* been tested), and which *actions* resolve deficiencies in assessments and plans. Furthermore, although highly effective expert teams can make collaboration look easy, collaboration is not an innate skill. Individuals have been trained to exercise better critical thinking processes and thus make better decisions – the effects of which were increased accuracy of their judgments about complex problems (Cohen *et al.* 1998) and more appropriate language used to express them (Freeman, Thompson and Cohen 2000; Freeman, Cohen and Thompson 1998). There are strong indications that CCT can be trained; empirical research by Entin and Serfaty (1999) demonstrated that training to help teams reduce uncertainty about (or focus on) future goals and tasks produced better team performance and mission outcomes. The current research posits that even novice teams can learn interaction processes that facilitate CCT.

Finally, it is important to note that CCT is necessarily a cyclical process (Cohen and Freeman 1997; Cohen *et al.* 1996, 1998) (see Figure 14.2). Actions produce changes in the state of the perceived environment or knowledge about it. This informs monitoring – ideally reducing alerts concerning the presence of uncertainty. The process is also variable in its structure (as indicated by the shortcut around critiquing, in the figure). The function of assessment is to determine when to invest team effort in critiques, and when to act immediately. Assessment is, thus, a branching function that accommodates rapid, recognitional decision-making (Klein 1993) on the short path, and more deliberate analytic decision-making on the other.

CENTER

CENTER is a software system designed to improve team members' collective knowledge and decisions by enhancing CCT. The CENTER tool (1) enables a leader to query members of the organization concerning the state of mission knowledge and decisions, (2) elicits brief responses and summarizes them statistically, and (3) presents these measures to leaders with guidance concerning the issues on which leaders should focus their attention and that of members. In short, CENTER helps leaders to measure, monitor and manage CCT about team knowledge and decisions.

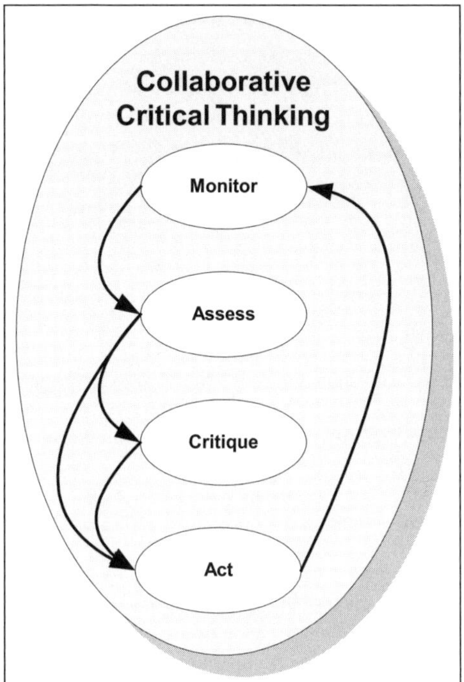

Figure 14.2 Collaborative critical thinking consists of four types of team member interactions in a cyclical and adaptive process

Measuring knowledge

One management adage states that 'You can't manage what you can't measure.' Neither can you understand it, train it, or support it with job aids. The object of the current research is to understand CCT, to develop training that supports CCT and to create tools that enable teams to manage CCT. To do this requires measures of the phenomenon.

CCT is a direct derivative of the literature on individual critical thinking, specifically the work of Cohen and colleagues (Cohen and Freeman 1997; Cohen *et al.* 1998), which empirically validated that individuals who engage in several specific critical thinking behaviors outperform others on tactical assessment, planning and decision-making tasks. CCT involves interaction between team members to reveal uncertainty concerning knowledge or decisions, identify the source(s) of this uncertainty, and devise ways to diminish or accommodate the uncertainty. These collaborative activities may help team members improve estimates of risk and refine plans to accomplish missions in the face of risk. Thus, to create measures of the state of CCT within a team, the current research focuses on measures of its impact on team products (e.g., mission plans) and processes (e.g., team synchronization).

Specifically, the definition of the space of behavior that must be measured can be described as the intersection of two dimensions of CCT (see Table 14.2) – the collaborative activities that constitute critical thinking (i.e., monitoring, assessment, critiquing – for gaps, assumptions and ambiguity – and action) and the objects to which it is applied (i.e., team products and processes). Taken together these measures provide a comprehensive assessment of collaboration effects.

Measures within this space have been defined as ratings of agreement (strongly disagree to strongly agree) with assertions or probes (below) that can be customized [by adjusting the terminology in the brackets] for a given mission. For example, probes relevant to the marked cells in Table 14.2 are:

A1: The team's assessment [of _____] is correct.

A6: The team is taking actions to resolve problems with the assessment [concerning _____].

C2: The team has time to critique and refine the plan [regarding _____].

P4: The team has identified key assumptions that have yet to be tested concerning its strategy [for _____].

G6: The actions of team members are consistent with the mission goals [concerning _____].

T1: Team members seek feedback on their tasks [concerning _____].

T3: The team is completing all tasks [concerning _____].

T5: Team members seek to resolve ambiguity in task assignments.

Table 14.2 CENTER measures critical thinking activities applied to team products and processes

	Objects of CCT				
	Team Products		Team Processes		
CCT Behaviour	**Assessment**	**COA/Plan**	**Goals**	**Plans**	**Tasks**
Monitoring	A1				T1
Assessment		C2			
Critiquing for gaps					T3
Critiquing for assumptions				P4	
Critiquing for ambiguity					T5
Action	A6		G6		

CENTER allows a leader or facilitator to select a subset of probes appropriate to the mission at hand, and customize them or use them in their generic form. The facilitator can then trigger the delivery of each probe to networked members of the team. Each probe appears in a small window in a member's workspace (see Figure 14.3) with a rating scale and two buttons: one to add textual comments and one to send the response back to the facilitator. Each window disappears after a specified period, and a countdown reminds the team member of this.

In experimental research conducted at the University of South Florida, participants found these probes to be useful in a teamwork exercise and non-disruptive of taskwork if they occurred at least three minutes apart.

Monitoring knowledge

CENTER converts responses to each probe into numeric values, and summarizes them as a mean and range (see Figure 14.4). Leaders or facilitators can view the numeric responses to all probes, or drill down to inspect the responses – both ratings and comments – to any one probe. In this way, leaders can monitor the organization's state of collaborative critical thinking with respect to mission-specific issues.

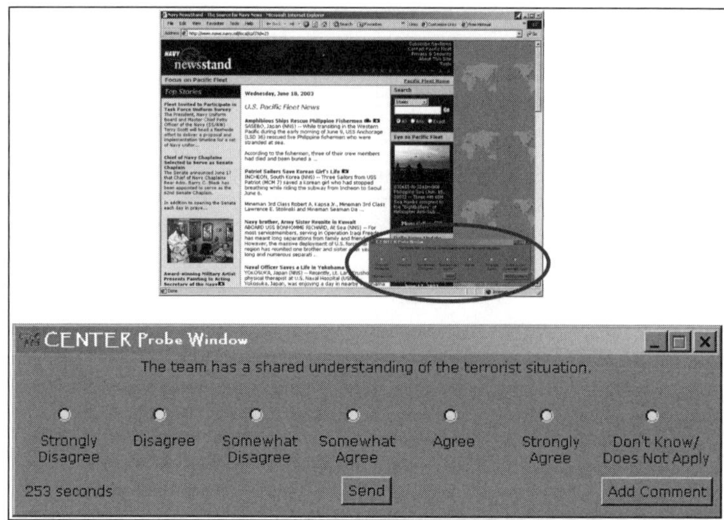

Figure 14.3 CENTER probes are simple, unobtrusive and rapidly addressed

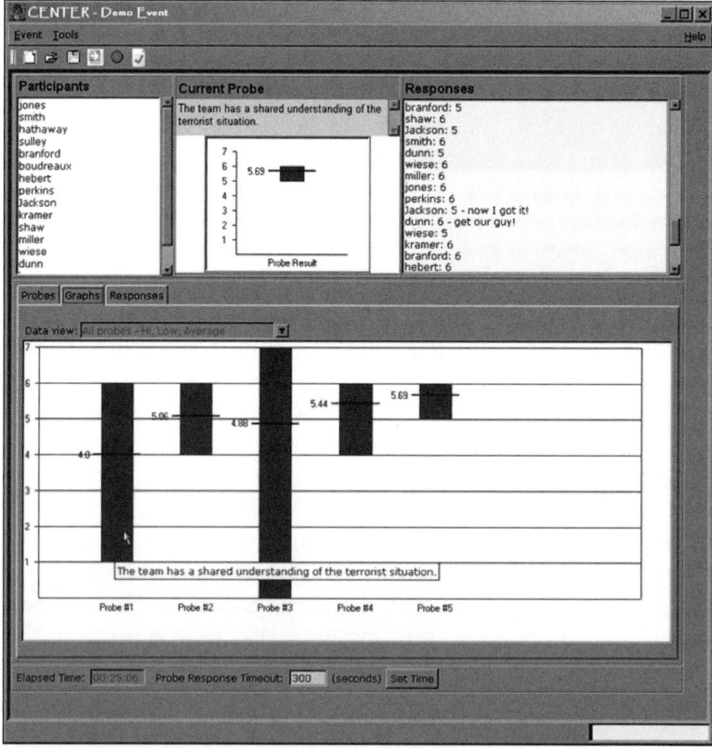

Figure 14.4 Responses to CENTER probes are summarized as means and range

Managing knowledge

CENTER helps leaders interpret measures of collaborative critical thinking and take action to improve it. It does so by analyzing distribution patterns in each response and presenting guidance to the leader. For example, assume that this probe – 'The team has the information it needs to plan' – is administered to the team well into a long mission planning task. The distribution of responses across the team may have a high (positive) mean and low variance, indicating that there is near unanimous agreement with the probe statement. In this case, CENTER advises the leader as follows:

The team members believe that they have the information they need to plan. Suggest that they move on. If there is a large team then probe for lone dissenters, if found, engage them, for example: 'Would you like to add anything?'

If the responses across the team exhibit an average mean and high variance (indicating that some people agree with the probe while others disagree), CENTER returns the following guidance:

> The team members do not agree whether the team has the information it needs to plan. Seek to understand why there is so little consensus. Tell those who disagreed with the probe: 'Share your concerns regarding insufficient information needed for planning with the other team members. Tell them what information seems to be missing. See if they have that information.' Tell the team members who agreed with the probe 'Not all team members believe that there is enough information to plan. Find out what information is missing.' Help the team quickly get the information it needs.

The facilitator can then send messages to the team members, or the facilitator can engage all or a subset of team members in an instant message session (see Figure 14.5) to facilitate more complete knowledge management.

CENTER validation study

Data were collected to better understand CCT and the CENTER tool. Specifically, a study was designed and implemented to gather validity evidence for the CENTER tool. This study was fully described in Hess *et al.* 2006. What follows is a high-level description of the study, analyses and the results and implications for future research, development and use of the CENTER tool.

Study

A study was conducted at a large south-eastern university with 160 college students (30 men, 130 women) to assess the utility of (1) training developed to increase a team's ability to practice CCT, and (2) the use of probes to facilitate team CCT during a team-based task. The seven experimental conditions are provided in Table 14.3.

The probes, developed to facilitate critical thinking, are listed below. They were administered with the CENTER tool either during a preplanning session (prior to actual task performance) or during the running of the task. It was hypothesized

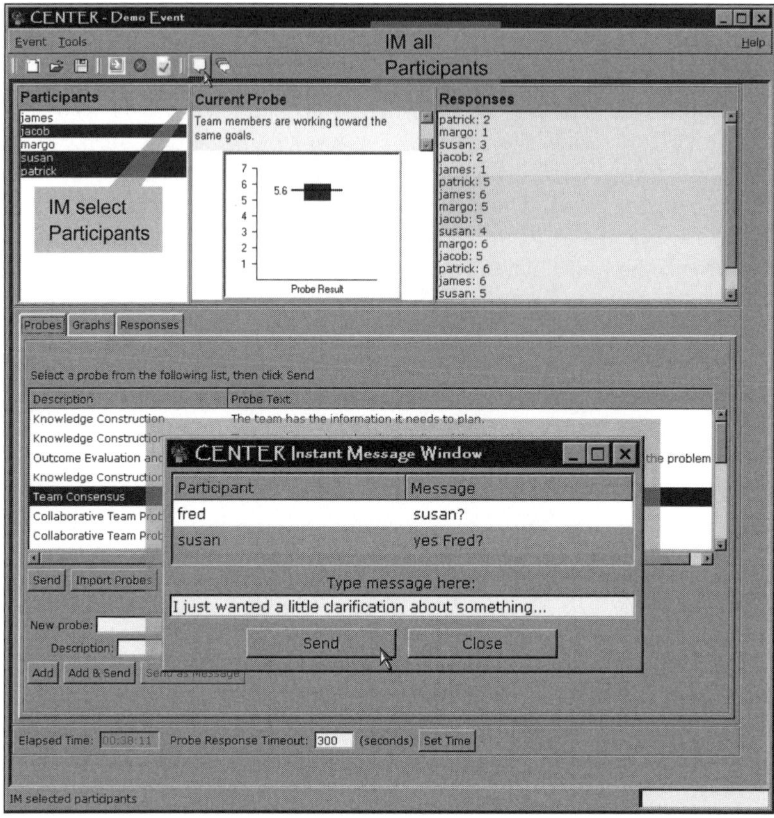

Figure 14.5 CENTER's instant message capability

Table 14.3 Seven experimental conditions

Condition	Description
Baseline	No CCT training, no 'control' training
'Control'/no probes	'Control' training only (no probes administered)
'Control' – preplan probes	'Control' training and probes were administered during the preplanning phase
'Control' – task probes	'Control' training and probes were administered during the performance of the task
CCT/no probes	CCT training (no probes administered)
CCT – preplan probes	CCT training and probes were administered during the preplanning phase
CCT – task probes	CCT training and probes were administered during the performance of the task

that administering the probes during the planning session would further stimulate critical thinking and thereby facilitate performance, whereas administering the probes during the performance of the task may overwhelm participants and lead to decreased performance. A probe was sent every 3 minutes until the 42-minute mark of the task. Participants responded to each probe on a 5-point Likert-type scale.

The 10 probes used in the experiment were:

1. The team has the information it needs to plan.
2. The team has a shared understanding of the situation.
3. The team is taking actions to resolve problems with the plan.
4. The team has clearly communicated its goals.
5. The team members are working towards the same goals.
6. The team has considered possible, alternative plans.
7. Alternative plans have been evaluated, according to team goals.
8. The team's strategy for solving the problem is correct.
9. There is an opportunity to reassign tasks within the team.
10. The team has identified critical assumptions that have yet to be identified.

It was further hypothesized that both CCT training and the use of the CENTER tool would result in better team performance than no training, 'control' training, or no tool use, with the combination of CCT Training with the use of the CENTER tool resulting in the highest performance.

CCT training CCT training, fully described in Hess *et al.* (2006), was presented to the participants through a power point presentation. The presentation gave some background on what CCT is and why it is important. The concept of schemas was introduced to the participants; then two short exercises on recognizing schemas were performed by the participants. At the end of each exercise the experimenter provided an explanation.

'Control' training For the 'Control' training, a different PowerPoint presentation was presented to the participants. In this scenario, the participants were to imagine themselves on vacation in the jungle in a foreign country (such as Mexico's Yucatan Peninsula). The bus the participants were traveling on was hijacked, but they managed to escape. They were provided a list of materials that were with them and instructions on how to use the natural surroundings to their advantage (i.e., a rock as a hammer or a vine as a tourniquet). The participants were instructed to have a seat at the lab's central round table and spend the next 10 minutes trying to determine the best course of action to survive in the jungle and eventually make it to safety. This training was used because the setting gave it face validity (i.e., made it appear to be related to the simulation used for the experiment), without overtly training any behaviors related to CCT.

Once the appropriate training session had ended, all of the participants were informed that they would begin training for the 'Artic Survival task.'

The team's task

To assess team performance, teams participated in a computer simulation, or game, called Arctic Survival. The participants were told that the goal of the task was to use simulated 'snow cats' and other resources to locate an unmanned aerial vehicle (UAV) and a lost team that is stranded somewhere in the Arctic. In this scenario there are three separate color-coded team members, red, blue and purple. The administrator was blue. The participants were instructed to direct any questions during the running of the simulation to the blue snow cat.

The red and purple snow cats and resources were controlled by participants; these were the only snow cats able to complete the tasks of the scenario. Communication between the red and purple participants was accomplished through a messenger system built into the simulation. This was the participants' only means of communication. Each red and purple team consisted of a snow cat, a medic, a technician, a scout and a mechanic. All four could be put onto a snow cat and transported to various locations. In addition, each color-coded team member (medic, technician, scout and mechanic) started out with a certain amount of usable units. For example, the medic started out with 15 medic units. If a task required three medical units to complete, and the red medic was assigned to a task that indicated it needed three medical units, then after completing the task the red medic would have 12 medic points. This is similar for each of the other four color-coded team members who each started with a certain amount of points. Each team, red and purple, had the same personnel and each of these personnel had the same amount of points as their counterparts on the other team (i.e., red technician had 15 units, so purple technician had 15 units).

Scoring Scoring was divided into three different areas. Objective scoring was accomplished by simply looking at the scores each team received in accordance with the simulation. Each team had an opportunity to score points on their tasks. The point system was follows:

- 300 points: Find the unmanned aerial vehicle (UAV) or the lost team.
- 100 points: Render emergency assistance (-100 from both team members if emergency assistance is not rendered in the allotted time period).
- 50 points: Assist with repair or medical requests.
- 10–80 points: Process seismic monitors.

These tasks popped up at predetermined intervals and were relayed to the red and purple snow cats via the blue snow cat. The participants were told by satellite messages relayed from the blue snow cat how much time they had to complete each task. Some tasks popped up even before the time had expired for other task. If the participants successfully completed a task, they were awarded the above amount of points, depending on the task. If the participants were unable to complete the task in the given time, they were deducted that amount of points. An emergency could be neutralized by processing it with only part of the needed resources. In this case, the team is neither penalized nor rewarded and therefore received 0 points for attending to that particular emergency. The running time for the DDD Arctic Survival task was 75 minutes.

Once the simulation task was completed, all the participants were instructed to sit at the round table in the center of the room for questions and debriefing. Participants were instructed that they could talk to others about the experiment, but to not mention specific details of their specific conditions. After all questions were answered we thanked the participants for lending us their time and informed them the experiment was over and they were free to go.

Analyses and results

A full and thorough discussion of analysis and results can be found in Hess at al. (2006). For the purposes of this chapter, the discussion remains abstract. Analyses were performed at both the team and the individual level. Team level analyses were performed using analyses of variance (ANOVAs) to examine the performance of the teams in the different experimental conditions. Table 14.4 presents the performance scores and the sample size for each experimental condition.

In brief, the team-level analyses showed support for a positive effect of training (both the CCT and 'Control') on the team's performance. Solely at the team level, the benefit of using of the CENTER tool was not apparent. Several possible explanations for these findings have been speculated. It may be that the population chosen for the study was inappropriate since CCT requires a level of expertise that college students have yet to achieve; or perhaps the disproportionate number of female participants influenced the outcome – it may be that these participants were exceptionally social in nature and this neutralized the effects of the probes on the team performance, whereas male participants, or other participants less social, may benefit more from the use of these probes. Further research is warranted to more fully understand these findings.

Because the data collected for this study were organized hierarchically, (e.g., individuals nested within teams) multilevel analyses were also performed. The team score variable remains the dependent variable. There were two level 1 predictor-variables: IQ (measured by SAT or ACT), and the personality variable agreeableness.

Table 14.4 Summary of means and sample size by condition

Condition	Mean Score	STD	Subject N
Baseline	13.75	291.16	32
'Control' no probes	248.89	166.66	18
'Control'/preplanning probes	103.33	575.66	6
'Control'/task work probes	137.78	261.66	18
CCT no probes	407.50	352.16	32
CCT preplanning probes	220.67	331.96	30
CCT task work probes	102.86	166.48	14

Agreeableness is a dimension of the Big 5, and generally reflects such attributes as: trust, straightforwardness, altruism, compliance, modesty and tender-mindedness.

The multilevel analyses began with a baseline or null model that served as a comparison against which to compare other substantive model solutions. Models were compared using the deviance statistic. Deviance reflects the overall lack of fit of the model to the data. Models with lower overall deviance terms are preferred, all other things being equal. Of course it is important not to add complexity to a model without ensuring the additional complexity significantly reduces overall deviance. This is easily examined as the deviance statistic for nested models is distributed as a chi-square. So the importance of the addition of a term to an equation (or other modification) can easily be checked by examining the decrease in overall deviance. The result of these multilevel analyses indicated that both individual-level variables (IQ and agreeableness) contributed significantly to team performance.

A further strength of multilevel modeling is the ability to examine cross-level interactions. This allows the researcher to ask questions about the impact of variables at one level on variables at another level. To test for these effects, variables that reflect the product of level 2 by level 1 variables were computed. We were interested in whether cross-level interactions between level 2 variables might exist (CCT training coded 0 = no training, 1 = training; probes coded 0 = no probes, 1 = received probes) and level 1 variables (IQ, agreeableness). The results of these analyses suggested both level 1 variables interact with both level 2 variables. This was seen by a significant reduction in model deviance for the models that include the respective interaction terms: IQ by CCT, IQ by probes, agreeableness by CCT and agreeableness by probes. One interpretation is that the use of probes strengthens the relationship between individual-level variables (IQ and agreeableness) and team performance.

Fielding CENTER

The CENTER tool was developed to benefit any team involved in a planning-intensive situation. Some examples of the types of teams that should benefit include the following:

- *Intelligence analysis teams* gathering and interpreting data to infer adversary location, identity, capabilities, plans and intent.
- *Operational planning teams* developing and evaluating courses of action.
- *Management consultants* or *board members* engaged in developing organizational strategy and policy.
- Any other *collaboration team* creating an intellectual product.

However, operational personnel (i.e., Marines and the Expeditionary Strike Group [ESG] 1 aboard the USS Tarawa) approached to discuss the tool's possible use in their mission planning, and the benefits that the tool's use would have on their mission performance agreed that the use of probes during mission planning would be too distracting. The correct place to use the CENTER tool is during *training* sessions to train teams how to use CCT in their jobs and missions. They all felt that CCT was an important skill for a team to have, and they saw the utility of the

CENTER tool in training CCT, but they felt that training with the CENTER tool had to happen long before the team actually needs to use CCT. With this insight, and the supporting evidence from the earlier study, efforts are now focused more on finding opportunities to insert the CENTER tool into training situations, where teams have more time, and are more motivated, to develop their CCT skills and behaviors. For example, Aptima Inc. has successfully incorporated CENTER into After Action Review (AAR) and debrief processes that are being developed for the Army and the Air Force Research Lab.

Conclusion

By using macrocognitive processes as a solid theoretical base to develop CENTER, the CENTER technology measures the state of knowledge within teams as well as team judgments about knowledge and decision-making, and enables leaders to monitor and manage such knowledge. These capabilities are important to help leaders leverage the capabilities of modern information systems, by giving them insights into the use of information from these systems and judgments about it. Thus, CENTER may help to integrate data systems with social systems. Technologies such as this may be particularly important in distributed organizations, and in virtual organizations,[2] in which leaders cannot easily observe interactions – such as 'buzz' about specific information or arguments over decisions – that convey the state of team knowledge. In these environments, technology is needed to make team knowledge state accessible.

Acknowledgments

The authors thank Dr. Michael P. Letsky of the Office of Naval Research for his support of CENTER through ONR's Collaborative Knowledge Management program. Although work described in this chapter was funded by the ONR grant N00014-02-C-0343, the opinions expressed here are the authors' and do not necessarily reflect the views of the Navy or Department of Defense.

References

Cohen, M.S. and Freeman, J.T. (1997). 'Improving Critical Thinking', in R Flin *et al.* (eds), *Decision Making Under Stress: Emerging Themes and Applications*, pp. 161–169. (Brookfield, VT: Ashgate Publishing Co).

Cohen, M.S., Freeman, J.T. and Thompson, B.T. (1998). 'Critical Thinking Skills in Tactical Decision Making: A Model and A Training Method', in J Canon-Bowers and E. Salas, (eds), *Decision-Making Under Stress: Implications for Training and Simulation*, pp. 155–189. (Washington, DC: American Psychological Association Publications).

2 Virtual organizations are *ad hoc* alliances or opportunistic, temporary alliances of individuals from different formal organizations.

Cohen, M.S., Freeman, J.T. and Wolf, S. (1996). 'Meta-recognition in Time-stressed Decision Making: Recognizing, Critiquing, and Correcting', *Journal of the Human Factors and Ergonomics Society* 38(2), 206–219.

Entin, E.E., Weil, S.A., Kleinman, D.L., Hutchins, S.G., Hocevar, S.P., Kemple, W.G. and Serfaty, D. (2004). 'Inducing Adaptation in Organizations: Concept and Experiment Design', in *Proceedings of the Command and Control Research and Technology Symposium, San Diego, CA*, pp. 1–11. (Arlington, VA: Office of the Secretary of Defense).

Entin, E.E and Serfaty, D. (1999). 'Adaptive Team Coordination', *Human Factors* 41, 312–325.

Freeman, J.T., Thompson, B.T. and Cohen, M.S. (2000). 'Modeling and Assessing Domain Knowledge Using Latent Semantic Indexing', *Special Issue of Interactive Learning Environments* 8(3), 187–209.

Freeman, J., Cohen, M.S. and Thompson, B.T. (1998). 'Effects of Decision Support Technology and Training on Tactical Decision Making', *Proceedings of the 1998 Command and Control Research and Technology Symposium, Monterey, CA*, pp. 1–13. (Arlington, VA: Office of the Secretary of Defense).

Hess, K.P., Freeman, J., Olivares, O., Jefferson, T., Coovert, M., Willis, T., Stillson, J.F. and Grey, A. (2006). *Team Collaboration in Critical Thinking: A Model, Measures, and Tools; Final Technical Report*. Unpublished Manuscript.

Hinsz, V.B., Tindale, R.S. and Vollrath, D.A. (1997). 'The Emerging Conceptualization of Groups as Information Processors', *Psychological Bulletin*, 121, 43–64.

Hollenbeck, J.R., Ilgen, D.R., Moon, H., Shepard, L., Ellis, A., West, B. and Porter, C. (1999). 'Structural Contingency Theory and Individual Differences: Examination of External and Internal Person–team Fit'. Paper presented at the 31st SIOP convention, Atlanta, GA.

Jensen, J. (2002). 'Operational Requirements for Collaboration'. Paper presented at the TC3 Workshop: Cognitive Elements of Effective Collaboration, San Diego, CA.

Klein, G.A. (1993). 'A Recognition-Primed Decision (RPD) Model of Rapid Decision Making', in G.A. Klein, J. Orasanu, R. Calderwood and C.E. Zsambok, (eds), *Decision Making in Action: Models and Methods*, pp. 138–147. (Norwood, NJ: Ablex Publishing Corporation).

Latane, B., Williams, K. and Harkins, S. (1979). 'Many Hands Make Light the Work: The Causes and Consequences of Social Loafing', *Journal of Personality and Social Psychology* 37, 822–832.

MacMillan, J., Entin, E.E. and Serfaty, D. (2002). 'From Team Structure to Team Performance: A Framework', *Proceedings of the Human Factors and Ergonomics Society 46th Annual Meeting, Baltimore, MD*, pp. 408–412. (San Francisco, CA: Human Factors and Ergonomics Society).

Paulus, P.B. (2000). 'Groups, Teams, and Creativity: The Creative Potential of Idea Generating Groups', *Applied Psychology: An International Review* 49, 237–262.

Rawlings, D. (2000). 'Collaborative Leadership Teams: Oxymoron or New Paradigm?', *Consulting Psychology Journal* 52, 151–162.

Steiner, I.D. (1972). *Group Processes and Productivity*. (New York: Academic Press).

Tversky, A. and Kahneman, D. (1974). 'Judgment Under Uncertainty: Heuristics and Biases', *Science* 185, 1124–1131.

Vroom, V.H. and Jago, A.G. (1988). *The New Leadership: Managing Participation in Organizations*. (Englewood Cliffs, NJ: Prentice Hall).

Warner, N., Letsky, M. and Cowen, M. (2005). 'Cognitive Model of Team Collaboration: Macro-Cognitive Focus', *Proceedings of the 49th Annual Meeting of the Human Factors and Ergonomic Society*, pp. 2–19. (Santa Monica, CA: Human Factors and Ergonomics Society)

Warner, N.W. and Wroblewski, E.M. (2004). 'The Cognitive Processes used in Team Collaboration During Asynchronous, Distributed Decision Making', *Proceedings of the 2004 Command and Control Research and Technology Symposium, San Diego, CA*, pp. 1–24. (Arlington, VA: Office of the Secretary of Defense).

Chapter 15

Measuring Situation Awareness through Automated Communication Analysis

Peter W. Foltz, Cheryl A. Bolstad, Haydee M. Cuevas, Marita Franzke, Mark Rosenstein and Anthony M. Costello

Introduction

As the military undergoes significant changes including moving toward smaller, more deployable, dispersed forces, the need to find new methods to analyze and assess team performance has increased significantly. In this new modernized military, warfighters operate in a distributed fashion, sharing ever-increasing volumes of information, making it more difficult to establish the high degree of shared understanding and situation awareness (SA) that is needed to function effectively. Moreover, prior research has established SA as a promising method to predict team performance, and as an insightful theory to analyze when and why team performance might break down (Endsley 1995). Consequently, valid and reliable assessment of the team's SA is critical for evaluating team performance and diagnosing performance deficits.

To address this need, in this chapter, we describe our approach to assessing team performance. This primarily involves first observing team actions and communications and then linking these observations to team performance directly or to surrogates and/or components of team performance, such as decision-making and SA. While this limits our explanatory power to those aspects of performance being modeled (such as SA or certain communication events), in many cases of training or operational settings, this level of detail is sufficient. For instance, in a training situation, this approach can be used to notify an instructor that one team is using communication that is off-topic or that another team has lost SA, or in a cockpit situation, notify the flight crew that the team has not discussed an important checklist item.

We begin this chapter with a theoretical overview of the SA construct, distinguishing among the different levels and types of SA. We then review the theoretical and empirical research underlying our measures of SA, focusing on the application of the TeamPrints methodology to analyze team communications. The emphasis of this chapter is to present recent promising research that suggests how team communication analytical techniques can be combined to develop an automated online analysis of team performance. We conclude with implications for the design of tools for automated monitoring and training.

Situation awareness and team performance

Numerous studies have highlighted the vital role of SA to ensure successful performance in complex operational environments (e.g., Artman 2000; Endsley 1993; Furniss and Blandford 2006). SA can be defined as 'the perception of the elements in the environment within a volume of time and space, the comprehension of their meaning and the projection of their status in the near future' (Endsley 1995, p. 36). Building SA, therefore, involves *perceiving* critical factors in the environment (Level 1 SA), *comprehending* what those factors mean, particularly when integrated together in relation to the individual's goals (Level 2 SA), and, at the highest level, *projecting* what will happen in the near future (Level 3 SA). Although alone it cannot guarantee successful decision-making, SA does support the necessary input processes (e.g., cue recognition, situation assessment, prediction) upon which good decisions are based (Artman 2000). The higher levels of SA are especially critical for timely, effective decision-making.

At the group level, SA has been investigated in terms of *team SA* and *shared SA*. Team SA can be defined as 'the degree to which every team member possesses the SA required for his or her responsibilities' (Endsley 1995). Implicit in this statement is the idea of consistency and synchronicity between individual team member's SA. The success or failure of a team depends on the success or failure of each of its team members. In contrast, shared SA refers to the 'degree to which team members possess the same SA on shared SA requirements' (Endsley and Jones 1997). Ideally, each team member shares a mutual understanding of what is happening on those SA elements that are in common. Unlike team SA, shared SA requirements exist as a function of the essential interdependency of the team members.

Considerable research has documented the importance of SA for effective teamwork across a broad range of domains, from aviation (e.g., Hartel, Smith and Prince 1991; Nullmeyer, Stella, Montijo and Harden 2005) to emergency medical dispatch centers (e.g., Blandford and Wong 2004; Rhodenizer, Pharmer, Bowers and Cuevas 2000) to military command and control (e.g., Gorman, Cooke and Winner 2006; Kaempf, Klein, Thordsen and Wolf 1996). For example, in their study of CRM behaviors in military MC-130P crews, Nullmeyer and Spiker (2003) found that SA stood out as the crew resource management behavior most strongly associated with mission planning and that crews receiving high SA ratings often also demonstrated good mission performance.

Our approach to linking SA with team performance is consistent with the structural model of team collaboration described earlier in this volume (see Chapter 2). The model was created to aid in understanding the relationships between cognitive mechanisms and team collaborative problem-solving abilities. It provides a framework from which further research can be performed to assess the underlying processes of team performance, such as SA or communication. The model delineates four main collaboration stages:

1. team knowledge-base construction,
2. collaborative team problem-solving,
3. team consensus, and
4. outcome evaluation and revision.

SA is used throughout each stage to construct a shared understanding of the situation at hand and to decide on the appropriate course of action. During each of these stages, individual, team, or shared SA will play a more prominent role depending upon the task being performed. For example, during the first stage, team knowledge-base construction, teams strive to gather the necessary information to form an assessment of the situation (Level 1 SA). Their assessments are then shared with one another to form a more complete common operating picture (shared SA). This shared SA may then be used by the team to solve the problem at hand. Thus, their performance may be based on the accuracy of their shared and team SA.

Team communication and situation awareness

Communication is an essential mechanism for achieving the critical cognitive processes at each of the four main stages of the team collaboration model (see Chapter 2, this volume). Independent of operational context, team communication, either in the form of verbal transmissions or electronic text exchange (as in e-mail or chat) is central to the ongoing collaboration of any team. Communication is necessary to share individual cognitive representations, including developing SA, building a joint knowledge base, solving problems, forming consensus, evaluating outcomes and revising plans and actions. As such, communication is indicative of the individual and joint cognitive processes that underlie the stages of team decision-making.

Notably, research has shown that communication increases team effectiveness by helping teams form shared SA and shared mental models (e.g., Brannick, Roach and Salas 1993; MacMillan, Entin and Serfaty 2004). Team communication (particularly verbal communication) supports the knowledge building and information-processing that leads to SA construction (Endsley and Jones 1997). Thus, since SA may be distributed via communication, utilizing machine learning, models can be created that draw on the verbal expressions of the team as an input and provide performance estimates as an output. The logical next step is to attempt to develop technology that would make real-time automatic analysis of team communication with respect to SA and team performance possible. To achieve this goal, two promising methods of predicting team performance were joined to enhance each other's predictive and inferential power. Specifically, we combined the explanatory capacity of the SA construct with the predictive and computational power of the *TeamPrints* methodology. Next, we discuss the theoretical background and empirical support for TeamPrints.

TeamPrints: LSA-based analysis of real-time communication

TeamPrints is a system that uses computational linguistics and machine learning techniques coupled tightly with Latent Semantic Analysis (LSA) to analyze team communications. LSA is a fully automatic technology for modeling and matching discourse content. This technique's special capabilities for communications analysis include its ability to represent the entire conceptual content of verbal communication rather than surrogates such as keyword, titles, abstracts, or overlap counts of literal words. In addition, LSA's representation of the similarity of two words, sentences, utterances, passages, or documents closely simulates human judgments of the overall similarity of their meanings. LSA also represents two passages on the same topic, but phrased in different vocabulary as similar.

LSA presumes that the overall semantic content of a passage, such as a paragraph, abstract or coherent document, can be closely approximated as a sum of the meaning of its words:

$$\text{Meaning of paragraph} \approx \text{meaning of word}_1 + \ldots \text{meaning of word}_n$$

Mutually consistent meaning representations for words and passages can thus be derived from a large text corpus by treating each passage as a linear equation and the corpus as a system of simultaneous equations. In standard LSA, the solution of such a system is accomplished by the matrix decomposition Singular Value Decomposition (for details on the theoretical and mathematical underpinnings of LSA, see Landauer, Foltz and Laham 1998). LSA's effectiveness in simulating similarity of meaning for humans has been empirically demonstrated in many ways. For example, by matching documents with similar meanings but different words, LSA improves recall in information retrieval, usually achieving 10–30 percent better performance *cetera paribus* by standard metrics (Dumais 1994). After training on corpora from which humans learned or might have, LSA-based simulations have passed multiple choice vocabulary tests and textbook-based final exams at student levels (Landauer *et al.* 1998). LSA has been found to measure coherence of text in such a way as to predict human comprehension as well as sophisticated psycholinguistic analysis, while measures of surface word overlap fail badly (Foltz, Kintsch and Landauer 1998). By comparing contents, LSA predicted human ratings of the adequacy of content in expository test essays nearly as well as the scores of two human experts predicted each other, as measured by ~90 percent as high mutual information between LSA and human scores as between two sets of human scores.

Typically, LSA ignores word order within documents. However, in much of the present work, additional statistical Natural Language Processing (NLP) techniques are used in conjunction with LSA to account for the syntactic and grammatical differences found in sentences (Foltz and Martin 2004). This combination of LSA and NLP techniques to model team communication is what we have termed TeamPrints.

Predicting team performance

TeamPrints has been evaluated favorably in terms of its ability to predict team performance. For instance, as a proof of concept, TeamPrints was able to successfully predict team performance in a simulated unmanned aerial vehicle (UAV) task environment based only on communications transcripts (Foltz and Martin 2004; Gorman, Foltz, Kiekel, Martin and Cooke 2003). Using human transcriptions of 67 team missions in a UAV environment, TeamPrints reliably predicted objective team performance scores (LSA alone, $r = 0.74$; LSA combined with additional computational linguistic analysis measures, $r = 0.79$).

In a similar study, Foltz, Martin, Abdelali, Rostenstein and Oberbreckling (2006) modeled Naval air warfare team performance in simulated missions consisting of eight six-person teams. Sixty-four transcribed missions were obtained from the Navy Tactical Decision-Making Under Stress (TADMUS) dataset collected at the Surface Warfare Officer's School (see Johnston, Poirier and Smith-Jentsch 1998). In this scenario, a ship's air defense warfare team performed a detect-to-engage sequence on aircraft in the vicinity of the battle group, and reported it to the tactical action officer and bridge. Associated with the missions were 16 subject-matter experts-generated performance measures, including elements associated with SA, such as the passing of information, seeking of information and updating the team as to the situation. All predictive models for each of these 16 measures were highly significant with correlations to performance measures ranging from $r = 0.45$ to 0.78, demonstrating that the models can provide accurate predictions for a range of different independently rated team performance scores, including communication quality, SA, coordination and leadership.

Predicting situation awareness

Having established the efficiency and effectiveness of the TeamPrints methodology, we next focused on applying this approach to predict SA. As discussed earlier in this chapter, communication is a critical component of the process of team decision-making as a whole, and of the component processes of developing SA, building a joint knowledge base, problem-solving, consensus-building, and outcome evaluation. These components should be extractable and quantifiable from recorded streams of team communication. In particular, we propose that individual and team SA is latent in team communications and, thus, should be measurable through an analysis of the communication stream. The research previously described in this chapter provides support for the utility of TeamPrints for reliably predicting objectively measured task performance as well as measures of SA. The objectives of the current research effort, therefore, were to

1. replicate these findings using new tasks,
2. focus on using TeamPrints to predict SA, and
3. attempt to build a TeamPrints-based automatic tagging method that would parse communication streams into larger groupings and automatically assign SA levels to these communication units.

In order to achieve these objectives, we followed a two-pronged approach. First, we identified an existing data set, the Non-combatant Evacuation Operation (NEO) Mission Scenario collected as part of the Collaborative Knowledge in Asynchronous Collaboration (CASC) Phase II Project (Warner, Wroblewski and Shuck 2003), to help us define our analyses methods. With the knowledge gained through these exploratory analyses, we then created an experimental plan using the C^3Fire simulation package that allowed us to collect team communications and SA and task performance data in a more optimal way to assure that the necessary analyses could be conducted. Each of these efforts will be described next.

Exploratory analyses: The NEO mission scenario data set

Our exploratory analysis of the NEO data set has been published elsewhere (Bolstad, Foltz, Franzke, Cuevas, Rosenstein and Costello 2007), and, thus, will only be summarized here to highlight the application of the TeamPrints methodology. The NEO mission scenario asks participants, as part of a three-person team, to develop a plan to rescue stranded personnel from a church basement on a remote island in the middle of guerilla warfare (for a detailed description of the original study, see Warner et al. 2003). Our analyses of this data set focused on the written and recorded communications of the teams while creating a plan, and a set of independently derived performance scores. We evaluated how well our analysis of the communication data using TeamPrints would successfully predict objective measures of team performance and SA level.

Predicting performance using TeamPrints

Each team's performance on the NEO mission scenario task was scored by independent raters on a 100-point scale. The average score across all 32 teams was 83.8 (SD = 7.2). Using variables based on LSA nearest neighbors (transcripts that were near in semantic space), we were able to predict team performance of 16 teams with a jack-knife correlation of 0.77 (for a description of the jack-knife technique, see Yu 2003). The model with the best fit to the data used four variables, all measuring different aspects of the similarity of the team communications of the predicted team to successful teams. As in earlier work (e.g., Foltz and Martin 2004), the closer (more similar) a transcript was to its nearest neighbor the better the team's performance, how well a transcript's nearest neighbors performed was a positive indicator of how well this team did, and the more variability there was in the performance of the nearest neighbors, the worse a team was likely to perform.

We also evaluated whether the communication data alone would predict the team's experimental condition (static vs dynamic). In dynamic conditions, teams received new information halfway into their planning process. Intuitively, the communication of teams in this condition should be different, because they had to adjust their joint knowledge and decision-making based on this information. As hypothesized, TeamPrints predicted assignment to static or dynamic condition with 75 percent accuracy using a discriminant analysis. Indeed, our TeamPrints analysis of the communication data made more accurate predictions of condition than the team performance scores, which had been assigned by human raters (56 percent accuracy).

In summary, we achieved the first goal of our plan, that is, we replicated our earlier findings that TeamPrints analyses of the communication data stream allows for reliable prediction of team performance. We also extended these initial findings to show that our analysis was sensitive enough to detect systematic differences between conditions.

Predicting SA level using TeamPrints

Because the NEO data set was collected for related but different purposes, no traditional online or post-experimental measures indicating actual or perceived SA were administered. Accordingly, an alternative method of assessing SA *post-hoc* was utilized (for further details, see Bolstad *et al.* 2007). Seven of the team communication transcripts were analyzed manually by two human raters. The raters first grouped communication exchanges into meaningful, coherent units and then tagged each grouping by SA level (SA Level 1, 2, or 3). SA level was then predicted in a series of hold-one-out experiments, where a model was created by training TeamPrints using six of the seven transcripts, and the SA levels of the held out transcript were predicted.

The average exact agreement obtained across this set of experiments was 51.1 percent, which is less reliable than TeamPrints has been in other applications (see, for example, the analysis of the TADMUS data set, described earlier in this chapter). Still, this estimate is better than a pure random assignment, which would have produced an exact agreement of only 33.33 percent. Further, the robustness of the TeamPrints model may have been constrained by: first, the relatively small training set of only seven transcripts; and second, having to infer SA based on *post-hoc* analysis of team communications, without real-time objective SA measures from the actual experiment for comparison. Given the preliminary nature of this exploratory analysis, these findings are promising nonetheless.

Predicting team performance using SA levels

The last step in this exploratory series of analyses was to attempt to predict team performance using the human- and machine-produced SA-level tags. Teams with higher levels of SA would be expected to show better performance. However, analyses did not reveal any significant correlations between the different levels of SA and team performance, neither for the human- nor the machine-derived tags. Nevertheless, the correlations between the SA levels and performance were in the predicted directions, where higher proportions of Level 1 SA tended to be correlated with lower performance, and higher proportions of Level 2 and 3 SA (indicative of higher levels of data processing) tended to be correlated with higher levels of performance.

Summary of NEO data exploratory analyses

Through our exploratory analyses of the NEO data set, we were able to provide evidence to support that analyses of communication streams using TeamPrints could reliably predict team performance as well as detect systematic differences between conditions. We also found some evidence to suggest that analysis of team communications using TeamPrints can be used to identify SA levels among teams. Finally, although the correlations between the different SA levels and team performance were not significant, our analysis demonstrated that it is possible to use TeamPrints to tag SA level automatically, and that these automatically derived tags should, in principle, predict performance similar to human tags. Clearly, we would expect more convincing results in an experimental situation that allowed measuring SA level with more proven methods. The experiment described next was designed to create this type of data set.

C3Fire simulation experiment

The computer game C3Fire (http://www.c3fire.org) was selected as the testbed for investigating team collaboration, communication and SA. C3Fire was developed as a dynamic team simulation used for team training and research of command, control and communication – C3 – in a collaborative environment. The C3Fire computer-based system generates a task environment possessing complex, dynamic and opaque characteristics, similar to cognitive tasks normally encountered in real-life systems.

Experimental design

The study employed a one-factor within-groups design, with scenario complexity serving as the independent variable. Dependent variables include assessment of participants' communication exchanges, SA and task performance derived from the simulation data.

Participants

Twenty-four students (male = 23; female = 1; mean age = 32.42) from the Naval Postgraduate School, Monterey, CA, voluntarily participated in this experiment for course credit. Participants were randomly assigned to three-person teams, creating a total of eight teams. Treatment of all participants was in accordance with the ethical standards of the American Psychological Association.

C3Fire scenarios

The C3Fire scenarios created for this study involved a three-person team task played using standard desktop PC-based computers. C3Fire simulates the outbreak of a fire and the team's primary goal was to put the fire out, saving as many houses and as much terrain as possible. This required participants to coordinate their efforts to control the fire through the use of fire, fuel and water trucks. Two participants were each assigned the role of Fire Chief and were responsible for controlling a varying number of fire trucks. A third participant was assigned the role of Water/Fuel Chief and was responsible for controlling a varying number of water and fuel trucks. The objective of the C3Fire game was to keep the map area clear of fire by closing out any existing cells (squares) on fire and preventing any further spread to other cells. The shading of the cells on the map indicated the status of the fire, which included one of four states:

- Clear (no shading) – area is clear of fire
- Fire (red) – area is burning (on fire)
- Burned Out (black) – area is no longer burning; fire burned itself out
- Closed Out (brown) – area is no longer burning; fire successfully put out.

Participants were geographically distributed in different rooms and could only communicate with each other using USB headsets and Voice-Over Internet Protocol (VOIP) software. Participants viewed the fire via networked interfaces and controlled their trucks via the GUI. All participants could see the map showing the developing fire and the assets that needed to be protected. However, each participant could only see their own trucks and their status indicators (fill level). Their team members' trucks were only visible when these were in the nine-square radius on the map surrounding the participant's own trucks. This set-up made it necessary for participants to verbally communicate their positions, fill-levels and their (intended) actions to coordinate their plans for fighting the fire.

Six scenarios were created varying in complexity to elicit targeted communication and coordination team behaviors and create opportunities to assess SA. Scenario complexity was manipulated by varying the number of trucks assigned to each participant and other C3Fire parameters, such as wind speed, number and placement of assets (e.g., houses to be protected), initial fill level of the trucks and burn rate of the fires. Increasing scenario complexity inherently heightened the necessity for more frequent and succinct communications as the scenarios became more difficult.

Objective SA measure: SAGAT queries

Using the Situation Awareness Global Assessment Technique (SAGAT) (see Endsley 2000), 13 queries were created to assess participants' SA during each scenario. These queries were derived from a Goal-Directed Task Analysis (GDTA) (for a description on GDTA creation, see Endsley, Bolte and Jones 2003) of the C3Fire domain. Seven queries assessed participants' Level 1 SA (perception), such as, for example, 'Indicate the province to which Fire Truck 3 (F3) is most closely located.' Six queries assessed participants' Level 2 and 3 SA (comprehension and projection), such as, for example, 'Indicate which province is most threatened by the fire.' The SAGAT queries were presented to each participant on their computer screen once during each scenario, with presentation order and time varied randomly across trials. Participants' responses were compared to 'ground truth', as recorded by the computer, to provide an objective evaluation of their SA. Team-level SA scores were calculated by averaging across individual team members.

Subjective SA measure: PSAQ

Participants' subjective SA was assessed by administering the Post-Trial Participant Subjective Situation Awareness Questionnaire (PSAQ) (Strater, Endsley, Pleban and Matthews 2001) at the end of each scenario. The PSAQ consists of three items designed to solicit self-report of participants' perceived *workload* ('Rate how hard you were working during this scenario'), *performance* ('Rate how well you performed during this scenario'), and *awareness* of the evolving situation (i.e., their SA) ('Rate how aware of the evolving situation you were during the scenario'). Ratings were recorded on a 5-point scale, with responses ranging from 1 (low value) to 5 (high value). Team-level scores were calculated by averaging across individual team members.

Team communication

Audio recordings of the verbal team communications occurring during the scenarios were transcribed and were annotated to indicate scenario start and end times as well as to note when the SAGAT queries were administered. The transcripts were then analyzed using the TeamPrints methodology (as described later in the results).

C3Fire task performance

Task performance data was extracted from the event logs automatically generated by the C3Fire system. Performance was measured in terms of how well the team was able to close out the fire, with minimal loss to critical objects. Specific metrics included determining the number of cells on the map that were *burning* (red), *burned out* (black), and *closed out* (brown). These values were calculated twice for each team: at each SAGAT stop and at the end of each scenario.

Results

The C3Fire experiment data was analyzed using the same rationale as used with the NEO data. Specifically, team performance was predicted directly using TeamPrints alone and with the two SA measures, PSAQ and SAGAT. TeamPrints was also used to predict SA level (as measured using SAGAT). Finally, the TeamPrints model of SA was used to predict performance.

Communication protocols Each of the eight teams ran through the six scenarios. Data from one scenario was lost due to technical problems, leaving communication data from 47 transcripts. Table 15.1 shows a brief exchange illustrating the richness of the verbal data collected from these exercises.

These transcribed utterances from the scenarios were used in the modeling efforts. Scenario transcripts contained an average of 1217 words (SD = 429). The

Table 15.1 Illustrative example of team communications during a C3Fire scenario

Water/fuel	Do you need fuel?
Fire Chief 1	No. Ah, looks like we got our fire contained
Water/fuel	Yeah. Do you need fuel?
Fire Chief 1	No
Water/fuel	1?
Fire Chief 1	Red 1 is all good for trucks 1, 2 and 3
Fire Chief 2	Yeah, my fuel is good, for red 2
Water/fuel	What about water?
Fire Chief 2	I'm at above 50% for all of them

complete communication transcript for each scenario was used to model overall team performance and team-level PSAQ responses. To predict SAGAT scores, only the utterances up until the SAGAT stop were used. Average scenario word count at the SAGAT stop was 668 (SD = 264).

C3Fire team performance As indicated earlier, the team's primary goal was to control the fire as well as protect houses from the fire. Performance measures, therefore, focused on the total number of cells burned or burning (later termed *lost* cells) and the number of houses that were burning or burned out (*lost* houses). Greater values indicated worse performance. Figure 15.1 represents the total number of lost cells and lost houses at the time of the SAGAT stop. One striking aspect of this graph is the lack of variability in the number of houses lost. At the SAGAT stop, 42 out of 47 scenarios showed only 1, 2 or 3 lost houses. Given the variability in the language of the teams, this finding makes predicting performance very difficult. Thus, we focused on modeling the total number of lost cells as our main dependent variable.

This performance measure was derived both at the time of the SAGAT stop as well as at the end of the scenario to allow meaningful modeling of the SAGAT measure. Results showed that performance at the SAGAT stop predicted end-of-scenario performance well ($r = 0.88, p < 0.01$). If a team had a higher number of lost cells at the time of the SAGAT stop, they also tended to have higher numbers of lost cells at the end of the scenario.

As noted earlier, scenarios became progressively more difficult during the experiment. The significant Spearman rank correlation between scenario number and total lost cells ($r = 0.54; p < 0.01$) suggests that the manipulations to increase scenario complexity were effective. In general, participants appeared to find controlling the fire in later scenarios more difficult, leading to higher numbers of lost cells at the end of these scenarios (see Figure 15.2).

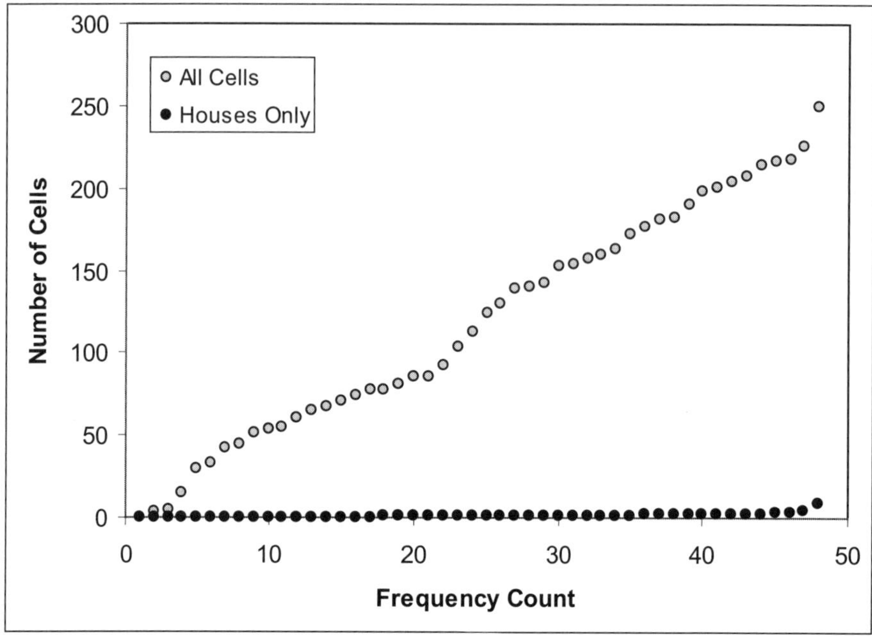

Figure 15.1 Total lost cells and lost houses by team and scenario at SAGAT stop

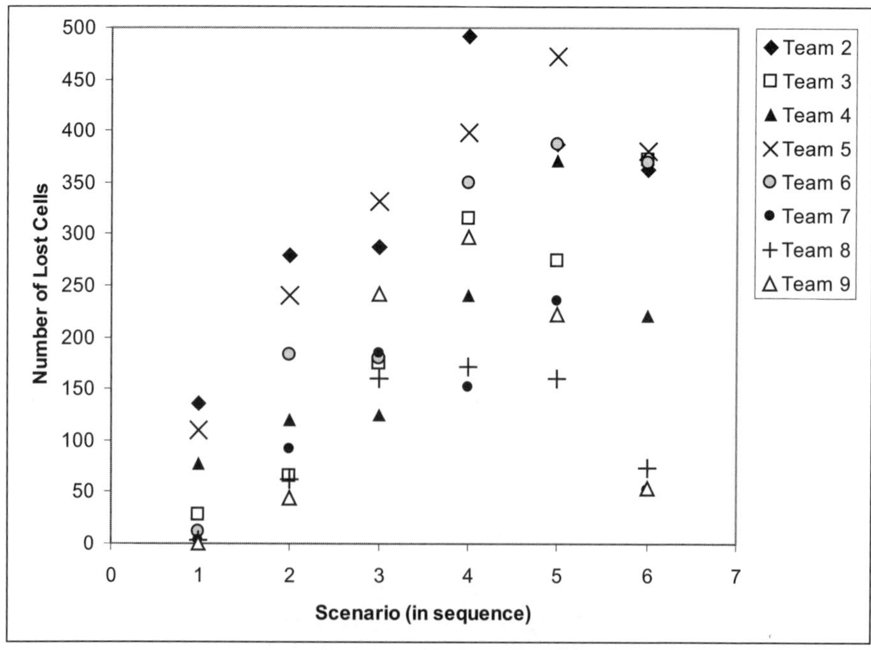

Figure 15.2 Total lost cells by scenario and team number at end of scenarios

PSAQ Results on the analysis of team-level average for the PSAQ responses revealed an interesting pattern of results. Subjective judgments of SA and C3Fire performance level were highly correlated ($r = 0.72$ $p < 0.01$), but correlations between perceived workload and perceived SA ($r = -0.09$, $p > 0.10$), and perceived workload and perceived performance ($r = 0.33$, $p < 0.05$) were not as strong. One explanation for this pattern may be that SA and performance are attributed to self and, thus, are found to be more similar, whereas workload is attributed to the situation, and might, therefore, be judged on a different dimension. Interestingly, only perceived workload was a good predictor of actual C3Fire performance (as measured by number of lost cells) ($r = 0.68$, $p < 0.01$). Correlations between perceived SA and actual performance ($r = -0.15$, $p > 0.10$) and perceived and actual SA (as measured by the SAGAT scores) ($r = -0.38$, $p < 0.05$) were lower.

SAGAT One would expect higher SAGAT scores to be associated with a fewer number of lost cells, indicating better performance. However, although in the expected direction, the correlation between the team-level SAGAT score and C3Fire performance (total lost cells at the SAGAT stop) was not significant ($r = -0.13$, $p > 0.10$). When correlations were evaluated by query type (SA Level 1 vs SA Level 2/3), slightly better predictions were obtained using the SA Level 2/3 queries ($r = 0.21$, $p > 0.10$), but not enough for statistical significance. Thus, the modeling results were obtained at the aggregate level of average team performance across all SAGAT queries.

TeamPrints Having established our measures, we used TeamPrints to create a number of models. All models were variants of the hold-one-out paradigm, where a subset of the data was used in training the model, and remainder were used to test the model's performance. Specifically, we used all but one of the scenarios (46) to predict the held-out scenario. This type of experiment was performed 47 times, so that each scenario was predicted once, using the other scenarios as the training set. The values reported represent the correlations between the 47 predictions and 47 actual values.

Training TeamPrints on the C3Fire performance score, the held-out performance was significantly correlated with the predicted performance ($r = 0.73$, $p < 0.01$), suggesting that the TeamPrints analysis of the communication data was able to predict performance quite well. However, for the TeamPrints model of the SAGAT data, the correlation between the predicted and actual SAGAT scores was not significant ($r = 0.13$, $p > 0.10$). With regard to the correlations between predicted and actual responses on the three PSAQ items, only the model for perceived workload was significant ($r = 0.62$, $p < 0.01$). No significant correlations were found for either perceived performance ($r = 0.22$, $p > 0.10$) or perceived SA ($r = 0.22$, $p > 0.10$).

Summary of C3Fire experiment

Overall, the analysis of team communications using TeamPrints revealed several promising results. The number of lost cells was shown to be a valid and reliable objective measure of team performance. Using this as the dependent variable in a number of analyses, we found that subjective judgments of workload, as well as team communications, as analyzed through our TeamPrints models, were reliable predictors of team performance. We were also able to predict the workload measure using the TeamPrints models, suggesting that these three constructs measure closely related team performance components.

Unfortunately, the objective measure of the team's SA during the scenarios (SAGAT queries) did not correlate significantly with the performance measure, nor were we able to use TeamPrints to make reliable predictions about the team's SA. It should be noted that the SAGAT is typically administered multiple times during task performance to truly capture operators' SA in a given situation. However, in this study, SAGAT was administered only once in a single trial, and thus does not represent participants' SA for the scenario, but instead their SA at the time of the stop. Thus, matching a single SA score to a communication measure that is continuous may have caused these low correlations. Nonetheless, although not significant, the positive correlation for the Level 2 and 3 SA queries suggests that queries probing higher levels of SA may yield greater predictive value of team performance. Specifically, teams scoring better on queries assessing situation comprehension and projection were also performing at a higher level. This finding is consistent with the results from our NEO data exploratory analysis, which showed that teams expressing proportionally more Level 1 SA performed worse.

Conclusions and future directions

While the SA construct is widely known, the quantification and measurement of SA is a relatively new field. A critical aspect of SA that also needs to be understood is that it is dynamic in nature and thus continually changing. Many metrics of SA do not take this into consideration, and instead focus on a single SA measurement taken per team task or event. Communication measures not only have the potential to provide a real-time assessment of SA, they can also be used to provide diagnostic information on why a team is performing a certain way.

Overall, the results of our exploratory analyses and empirical investigations support the utility of the TeamPrints methodology for assessing and predicting team performance and only minimal support for assessing SA. We found evidence to suggest that analyses of team communications using TeamPrints can reliably be used to detect systematic differences between conditions as well as potentially identify SA level among teams. Finally, the models derived using TeamPrints reliably predicted perceived workload.

While automated communication analysis helps our understanding of macrocognitive processes such as SA, this methodology also has implications for the development of products to take communication input, analyze and predict cognitive measures in near real time. TeamPrints can serve as a useful diagnostic tool to organizations trying to understand team cognition and performance. It would be particularly beneficial to organizations in which the team members are continually changing or its characteristics are changing, such as the experience of the team members or the number and types of task they perform together. By providing relevant diagnostic information on team cognition and SA, the system can provide an indication of the impact of such changes on the resultant team cognitive performance. Additional validation on novel data sets and different team contexts will still need to be performed in order to assess the generalizability of the methods developed.

The integration of the technology into performance-monitoring systems can further be used to develop systems that can adapt interfaces to provide optimal team performance. By performing real-time monitoring, assessing, diagnosing and adjusting feedback for teams, a host of products can be developed for improving training in simulators as well as monitoring live real-time communication. Much of the expense in training is due to having knowledgeable human trainers to monitor the teams and be able to provide feedback at appropriate times. The technology described in this chapter can alleviate some of this expense while still providing accurate and effective feedback to teams and their trainers.

Acknowledgments

Work on this chapter was partially supported by a Phase II Small Business Innovative Research Contract (N00014-05-C-0438) awarded to Cheryl Bolstad and Peter Foltz from the Office of Naval Research (ONR). The views and conclusions contained herein, however, are those of the authors and should not be interpreted as representing the official policies, either expressed or implied, of the ONR, US Navy, Department of Defense, US Government, or the organizations with which the authors are affiliated. We thank an anonymous reviewer for helpful suggestions on improving this chapter.

References

Artman, H. (2000). 'Team Situation Assessment and Information Distribution', *Ergonomics* 43(8), 1111–1128.

Blandford, A. and Wong, W. (2004). 'Situation Awareness in Emergency Medical Dispatch', *International Journal of Human–Computer Studies* 61, 421–452.

Bolstad, C.A., Foltz, P., Franzke, M., Cuevas, H.M., Rosenstein, M. and Costello, A. (2007). 'Predicting SA from Team Communication', *Proceedings of the 51st Annual Meeting of the Human Factors and Ergonomics Society*, pp. 789–793. (Santa Monica, CA: Human Factors and Ergonomics Society).

Brannick, M.T., Roach, R.M. and Salas, E. (1993). 'Understanding Team Performance: A Multimethod Study', *Human Performance* 6(4), 287–308.

Dumais, S.T. (1994). 'Latent Semantic Indexing (LSI) and TREC-2', in D. Harman, (ed.), *The Second Text REtrieval Conference (TREC2)*, pp. 105–116. (National Institute of Standards and Technology Special Publication 500–215).

Endsley, M.R. (1993). 'A Survey of Situation Awareness Requirements in Air-To-Air Combat Fighters', *International Journal of Aviation Psychology* 3(2), 157–168.

Endsley, M.R. (1995). 'Toward a Theory of Situation Awareness in Dynamic Systems', *Human Factors* 37(1), 32–64.

Endsley, M.R. (2000). 'Direct Measurement of Situation Awareness: Validity and Use of SAGAT', in M.R. Endsley and D.J. Garland, (eds), *Situation Awareness Analysis and Measurement*, pp. 147–173. (Mahwah, NJ: Lawrence Erlbaum Associates).

Endsley, M.R., Bolte, B. and Jones, D.G. (2003). *Designing for Situation Awareness: An Approach to Human-Centered Design.* (London: Taylor and Francis).

Endsley, M.R. and Jones, W.M. (1997). *Situation Awareness, Information Dominance, and Information Warfare.* No. AL/CF-TR-1997-0156. (Wright-Patterson AFB, OH: United States Air Force Armstrong Laboratory).

Foltz, P.W. and Martin, M.A. (2004). 'Automated Team Discourse Annotation and Performance Prediction Using LSA', *Proceedings of HLT/NAACL 2004*, pp. 97–100. (Boston, MA: Human Language Technology Conference/North American Chapter of the Association for Computational Linguistics).

Foltz, P.W., Kintsch, W. and Landauer, T.K. (1998). 'The Measurement of Textual Coherence with Latent Semantic Analysis', *Discourse Processes* 25(2–3), 285–307.

Foltz, P.W., Martin, M.J., Abdelali A., Rosenstein, M.B. and, Oberbreckling, R.J. (2006). 'Automated Team Discourse Modeling: Test of Performance and Generalization', *Proceedings of the Cognitive Science Conference,* Vancouver, CA.

Furniss, D. and Blandford, A. (2006). 'Understanding Emergency Medical Dispatch in Terms of Distributed Cognition: A Case Study', *Ergonomics* 49(12–13), 1174–1203.

Gorman, J.C. Cooke, N.J. and Winner, J.L. (2006). 'Measuring Team Situation Awareness in Decentralized Command and Control Environments', *Ergonomics* 49(12–13), 1312–1325.

Gorman, J.C., Foltz, P.W., Kiekel, P.A., Martin, M.J. and Cooke, N.J. (2003). Evaluation of Latent-Semantic Analysis-based Measures of Team Communications', *Proceedings of the Human Factors and Ergonomics Society 47th Annual Meeting*, pp. 424–428. (Santa Monica, CA: Human Factors and Ergonomics Society).

Hartel, C.E.J., Smith, K. and Prince, C. (1991). 'Defining Aircrew Coordination: Searching Mishaps for Meaning'. Paper presented at the 6th International Symposium on Aviation Psychology, Columbus, OH.

Johnston, J.H., Poirier, J. and Smith-Jentsch, K.A. (1998). 'Decision Making Under Stress: Creating a Research Methodology', in J.A. Cannon-Bowers and E. Salas, (eds), *Making Decisions under Stress: Implications for Individual and Team Training*, pp. 39–59. (Washington, DC: American Psychological Association).

Kaempf, G.L., Klein, G., Thordsen, M.L. and Wolf, S. (1996). 'Decision Making in Complex Naval Command-and-Control Environments', *Human Factors* 38(2), 220–231.

Landauer, T.K., Foltz, P.W. and Laham, D. (1998). 'Introduction to Latent Semantic Analysis', *Discourse Processes* 25, 259–284.

MacMillan, J., Entin, E.E. and Serfaty, D. (2004). 'Communication Overhead: The Hidden Cost of Team Cognition', in E. Salas and S.M. Fiore, (eds), *Team Cognition: Understanding the Factors that Drive Process and Performance*, pp. 61–82. (Washington, DC: American Psychological Association).

Nullmeyer, R.T. and Spiker V.A. (2003). 'The Importance of Crew Resource Management Behaviors in Mission Performance: Implications for Training Evaluation', *Military Psychology* 15(1), 77–96.

Nullmeyer, R.T., Stella, D., Montijo, G.A. and Harden, S.W. (2005). 'Human Factors in Air Force Flight Mishaps: Implications for Change', *Proceedings of the 27th Annual Interservice/Industry Training, Simulation, and Education Conference*, paper no. 2260. (Arlington, VA: National Training Systems Association).

Rhodenizer, L., Pharmer, J., Bowers, C.A. and Cuevas, H.M. (2000). 'An Analysis of Critical Incident Reports from a 911 Center' [Abstract], *Proceedings of the XIVth Triennial Congress of the International Ergonomics Association and 44th Annual Meeting of the Human Factors and Ergonomics Society* 4, 441. (Santa Monica, CA: Human Factors and Ergonomics Society).

Strater, L.D., Endsley, M.R., Pleban, R.J. and Matthews, M.D. (2001). *Measures of Platoon Leader Situation Awareness in Virtual Decision Making Exercises*. No. Research Report 1770. (Alexandria, VA: Army Research Institute).

Warner, N., Wroblewski, E. and Shuck, K. (2003). *Noncombatant Evacuation Operation Scenario.* (Naval Air Systems Command, Human Systems Department (4.6), Patuxent River, MD).

Yu, C.H. (2003). 'Resampling Methods: Concepts, Applications, and Justification', *Practical Assessment, Research and Evaluation* 8, 19.

Chapter 16

Converging Approaches to Automated Communications-based Assessment of Team Situation Awareness

Shawn A. Weil, Pacey Foster, Jared Freeman, Kathleen Carley, Jana Diesner, Terrill Franz, Nancy J. Cooke, Steve Shope and Jamie C. Gorman

Introduction

Collaboration enables people to execute tasks that are beyond the capabilities of any one of them. Each member of an organization or team has a set of skills, roles and responsibilities that, when executed accurately and in synchronization with other members of that organization or team, enables them to accomplish the work of the organization. In complex, dynamic environments, team collaboration is more than the simple aggregation of the work products of individuals; collaboration also requires complex exchanges of information, largely through spoken or written language. Linguistic communication is the choreography of team performance.

Modern networked information systems support synchronous and asynchronous communication among globally distributed team members via telephone, e-mail, instant messaging and text chat rooms, making possible the coordination of activities that would have been impossible or impractical in the past. The distributed project coordination common in commercial organizations would not be tenable without these technologies.

However, these technologies do not ensure highly effective organization. Cognitive collaboration quality varies significantly between teams, whether they are collocated (McComb 2005; Isaacs and Clark 1987; Warner and Wroblewski 2004), or distributed over networks or in virtual environments (Cooke 2005; Entin and Serfaty 1999; Cooke, Gorman and Kiekel, Chapter 4, this volume). When collaboration tools do make a difference, it is not always positive. These tools increase the opportunity for information overload; errors of commission (miscommunication) can increase relative to errors of omission (non-communication); and decisions and actions sometimes – and sometimes tragically – are mis-coordinated (Woods, Patterson and Roth 2002; Weil *et al.* 2004).

The Department of Defense (DoD) uses networked collaboration technologies to coordinate distributed, heterogeneous forces for both wartime and peacetime activities. Such coordination is a key element of Network-Centric Warfare or Network-Centric Operations (NCO), a theory of warfare in the information age. Among the primary tenets of NCO is the belief that networked information and collaboration increases 'shared situation awareness', a common understanding of the state of the mission environment. This, in turn, is predicted to enhance the effectiveness of forces (Alberts 2002). Some critics foresee a concurrent increase in the errors cited above. The debate is important to the nation's economic health, as well as its defense. However, NCO concepts are not the sole province of military organizations; they are also being applied in large and economically important commercial organizations. Effective work product requires that the individual contributors align their understanding of the state of the commercial environment, the corporate mission within it and their responsibilities. Tools such as data dashboards, knowledge portals, distributed conferencing applications and chat are often installed to facilitate this. Corporations, like DoD, are putting NCO concepts and technology to use to build situation awareness.

DoD has increased funding to develop common operating picture displays, collaborative environments and communication tools, explicitly to improve situation awareness. However, aspects of situation awareness are not often considered when these tools are developed. Nor have there been many attempts to assess whether, how, or how much these tools improve situation awareness. Thus, there is little evidence from which to determine whether and how to (re)design the tools, the organizations that use them, or the procedures for doing so.

To address this gap in knowledge, the authors are developing IMAGES – the Instrument of the Measurement and Advancement of Group Environmental Situation awareness. IMAGES is a software tool that gives mission personnel and researchers access to multiple, complementary communications assessment techniques that provide insight into both the content and process of collaboration in an organization. The capability to analyze content should enable leaders of organizations and designers of systems to measure shared situation awareness (SA) unobtrusively (Weil 2006). The capability to assess collaborative processes should enable them to manage the development of SA. In this chapter, we describe the approaches used to assess team performance via analysis of communications rather than describing the software.

Situation awareness and macrocognition

Individual, team and shared situation awareness

Effective collaboration and the utility of the NCO doctrine are predicated on the concept of situation awareness. Endsley (1988) defines three levels of individual situation awareness:

1. perception of elements in the environment,
2. comprehension of meaning in those elements, and
3. use of that understanding to project future states.

High situation awareness is associated with good performance, as the individual can anticipate the actions of elements in their environment. Low situation awareness is not desirable, as actions taken by the individual will likely be inappropriate given the true state of the environment.

When assessing an organization rather than an individual, situation awareness requires further definition. Organizational situation awareness is not simply the aggregate situation awareness of all constituent members. Endsley has distinguished between *shared situation awareness* (shared SA) – 'the degree to which team members possess a shared understanding of the situation with regard to their shared SA requirements' (Endsley and Jones 1997) – and *team situation awareness* (team SA) – 'the degree to which every team member possesses the SA required for his or her responsibilities' (Endsley 1995). In complex situations, it is often counterproductive for every member of an organization to have total knowledge of the state of the environment (i.e., completely shared SA) – nor is it advantageous for information and knowledge to be perfectly partitioned among members of the organization without overlap (i.e., perfect team SA). For teams, a balance of shared and team SA is required (e.g., Cooke, Salas and Cannon-Bowers 2000). Discovering the right balance for a given team – and measuring SA to determine whether that balance is being met – is an unsolved problem. It is not simply a matter of 'more is better'.

It is equally challenging to identify the communication processes that produce a given level of SA balance of SA types (shared vs team). This diagnosis of process is required to prescribe a remedy, a change to communication processes that improves SA in the long run.

Individual situation awareness is a largely internalized construct. The outward behaviors of individuals may reflect some aspects of situation awareness, but a true empirical evaluation typically requires interviews, surveys, or highly instrumented and artificial experiments. However, the elements of *collaboration* that produce team and shared SA are external; they are evident in communications content and process (Cooke 2005). Thus, they are observable and measurable in operational settings, without the intrusive methods cited above.

Macrocognition

The approach to assessing situation awareness described in this chapter relates to several of the macrocognitive processes described elsewhere in this volume (Warner, Letsky and Wroblewski, Chapter 2). The behaviors of teams and organizations – both externalized behaviors associated with joint cognitive systems and internalized events related to the perception and cognition of individuals – have been associated with the construct of macrocognition (Warner *et al.* 2005; Klein *et al.* 2003; Cacciabue and Hollnagel 1995). Macrocognition describes the cognitive functions that are performed in natural decision-making settings, typically by multiple, interacting individuals. Warner *et al.* (2005) have attempted to formalize this construct, positing four collaboration stages – knowledge construction, collaborative team problem-solving, team consensus and outcome evaluation and revision. Within these stages are between 10 and 20 distinguishable macrocognitive processes that relate to the process of problem-solving, mental model creation and information exchange.

The externalized communication behaviors posited to be related to organizational SA relate to several of the macrocognitive processes described in Warner *et al.* (2005). The process of consensus-building (the macrocognitive process of *knowledge interoperability development*) is supported by linguistic interchange among members of a collaborating team – the same communications content we will use to assess SA. The *team shared understanding development* macrocognitive process reflects some of the distinctions between shared and team SA described in the preceding section. The implied, cyclical form of the model is consistent with the notion that communications process (not just static content) predicts overall trends in SA.

However, the level of aggregation implicit in the large organizations on which we are focused – dozens to hundreds of individuals – seems to be incompatible with some of the macrocognitive processes. For example, the macrocognitive process '*convergence of individual mental models to team mental model*' posits a single mental model for the group. We argue that this is neither possible nor advantageous in large groups. Instead, we advocate for individual mental models or situational assessments that are aligned with the tasks required.

Whether developing a theory of macrocognition or designing technology to facilitate macrocognition, assessment is central. To illustrate the point, consider the concept of human intelligence. Without assessment in the form of various intelligence quotient (IQ) tests, we would have very little to say about a theory of intelligence and no evidence as to whether interventions improve intelligence or not. Likewise, assessment of macrocognition is necessary for theory development and for testing the success of interventions designed to facilitate it.

Two forms of assessment are required to understand and manage SA. Assessments of the state of SA are descriptive; they are valuable for comparing and evaluating new technologies, techniques and organizational forms prior to acquisition or implementation. Assessments of the process of SA are diagnostic; they enable us to understand the root causes of success and failure of macrocognitive processes (communication processes, in particular) and to manage those processes. The methods and tools described in this report focus on communication assessment as a means of assessing macrocognition.

Multiple communications assessment approaches

In the research literature, communication analysis methods focus either on communication content (i.e., what is said; Carley 1997; Kiekel, Cooke, Foltz, Gorman and Martin 2002; Kiekel, Cooke, Foltz and Shope 2001), communication flow (who talks to whom; Kiekel, Gorman and Cooke 2004), or communication manner (style, intonation, gestures, face-to-face or not, etc.; Queck *et al.* 2002). Communication content and communication flow have been the primary focus of team researchers.

Content methods attempt to summarize the content of discourse and then associate the content with team performance or macrocogntive processes. Although there have been successes in applying content-based communication analysis techniques to the understanding of teams, there are some drawbacks including the need to translate the spoken record into words (i.e., transcription) either manually (a tedious and resource-intense process) or by speech recognition software (an inexact process). In some cases the terminology used is domain-specific such that an ontology for that domain must be developed prior to interpreting the transcribed terms. Once the transcription and initial domain analysis is complete, the content-based methods pay off by providing a good source of information about the content (the state) of macrocognition.

Another source of information that is perhaps more tied to macrocognitive processes is the analysis of communication flow. Specific interaction or flow patterns in communications have been associated with effective team performance or team SA (Kiekel, Gorman and Cooke 2004). Similarly some patterns may signal a loss of team situation awareness (Gorman, Cooke, Pedersen *et al.* 2005; Gorman, Cooke and Winner 2006). Algorithms and software have been developed to extract patterns in the specific timing and sequences of interactions. Although the data used in this analysis is very basic, it is easy and inexpensive to collect and analyze and is conducive to automation. Thus there is a trade-off between ease of analysis and potential automation and richness of the resulting data.

The authors have taken a multifaceted approach to measurement of situation awareness through the assessment of communications. The first employs a set of networks aimed at representing both content and interactivity in an organization. The second is inherently temporal in nature, relying on regularities in turn-taking behavior in communication. These two methods are described in detail below.

Network representations of content and structure

As speakers of a language, humans can often read a transcript or listen to a recording of an event to understand the meaning of interchanges, follow conversational threads as they propagate through an organization, surmise the level of situation awareness within the team communicating and interpret the relationships between issues being discussed and action in the world. Distilling this content into forms that lend themselves to measurement and analytic interpretation would allow greater insight into situation awareness in large, complex organizations.

Several representations of the content and structures that underlie an organization's behavior can be extracted from standard communications media. Transformed into matrices, metrics can be developed that can aid in interpretation. In the current effort, we use the approach developed by Carley (2002) in which entity extraction techniques, embedded organizational ontologies and semi-automated thesaurus construction are used to extract and link both social networks and semantic networks. In this *meta-matrix approach,* the social structure of an organization is derived from mention of pertinent organizational elements (e.g., agents, resources, locations), within a corpus of texts. These components are classified into an ontology that structures them into a model of a social system. This model allows investigation of the composition of a social system, and identifies the connections among organizational components. In a given domain, this type of inquiry becomes highly automatized; models of organizations can quickly be extracted from large amounts of text, whereas manual extraction of an organizational model from text would be laborious and error prone. The meta-matrix scheme, derived from organizational research, provides such ontology for modeling the social and organizational structure of teams, groups, distributed teams, clans, organizations and so on (Carley 2003,

2002; Krackhardt and Carley 1998). In the meta-matrix approach the entities of a social system are agents, organizations, knowledge, resources, tasks and events and locations. Previously, Diesner and Carley (2005) have described an approach for combining networks that reflect the content of communications (i.e., map analysis) with the meta-matrix model. The resulting integrative technique we refer to as Meta Matrix Text Analysis (Diesner and Carley 2005). This technique enables analysts to extract not only knowledge networks, but also social and organizational structure from texts.

We have identified five network types that may provide insight into organizational situation awareness and other macrocognitive processes.

1. The Knowledge Network The knowledge network is a representation of the externalized knowledge or understanding of those individuals communicating. One way to represent the semantic content of communication is through the use of Network Text Analysis (NTA; Popping, 2000). NTA is based on the notion that language and knowledge can be represented as a network of words and the relations among them (Sowa 1984). Map analysis (Carley 1993, 1997b; Carley and Palmquist 1992), as achieved using AutoMap (Carley, Deisner and Doreno 2006) is one method to create these semantic networks. The major concepts in a text are extracted and become nodes in a network. These nodes are construed as 'concepts'. The arcs among the nodes – or the relationship among concepts – are defined by the proximity of those concepts to each other in the text. Pairs of concepts are construed as 'statements'. Given sufficient text, a complex web or network of concepts and statements is created (Carley 1997b).

This knowledge network reflects some of the complexity of the semantic and syntactic structure of the original texts, but in a form that allows for easier manipulation and perhaps automated interpretation. One way to construe these knowledge networks is as proxies for the aggregate knowledge of the individuals who framed the communication. As communication is the observable engine of team cognition, the semantic networks based on that communication become a representation of team cognition and/or organizational understanding. By selecting and comparing the semantic networks for different groups of people within and organization and specified time periods, you can assess the similarity or divergence of the knowledge of the individuals communicating. It is in this way that team and shared situation awareness can be automatically derived. In the parlance of the meta-matrix, this is known as a 'knowledge × knowledge' network, as all of the nodes are given the 'knowledge' attribute.

2. The Social Network A social network describes the relationships among individuals in an organization. In a corpus of e-mails, a social network can be easily generated from the 'To', 'From' and 'CC' lines. This social network is representation of interaction among members of an organization. Within the meta-matrix ontology, each of the nodes is given the attribute 'agent,' and the resulting network becomes an 'agent × agent' network.

3. The 'Agent × Knowledge' Network Combining the social network and the knowledge network allows inquiry into the relationship between patterns of interaction and the resulting change in situational understanding. As the meta-matrix model allows nodes to be identified by ontological type or class, a new 'agent × knowledge' network can be created in which some nodes represent the individuals communicating (i.e., 'agents'), while other nodes represent the 'knowledge' being communicated by those individuals.

4. The Implicit Meta-matrix While the text of communication can be distilled into a knowledge network, many of the nodes in that network can be further described using the meta-matrix attribute labels (e.g., location, agent, task, event, etc.). A new network representation is thus created, an implicit meta-matrix, which adds additional specificity and affords additional measurement possibilities. We have used the term *implicit* because the relationships among the nodes does necessarily refer to real world connections among the nodes (as is the case in the social network described above) but instead could be construed as the aggregate mental model of those communicating.

5. The 'Agent × "Implicit Meta-matrix" Network' Finally, a network can be created in which the social network derived from message headers is combined with the implicit meta-matrix derived from the content of messages. The resulting network is an 'agent × "implicit meta-matrix" network.' As in 'agent × knowledge' network above (3), the combination of a network based on patterns of interactivity with one representing content allows researchers and operators to gauge how knowledge is affected by different types of interactivity. This 'agent × "implicit meta-matrix" network' allows more nuanced assessment of changes in knowledge, as both the change in meta-matrix nodes and the structural arrangements of those nodes can be assessed.

Use of Meta-matrix Networks The purpose for creating these networks is to enable inquiries into the state of knowledge of an organization at a given time and the relationship of those states to patterns of interactivity among members of that organization. The degree of shared situation awareness between two groups within an organization can be determined by measuring the degree of overlap in the knowledge networks or implicit meta-matrices created for each of those groups. Comparing networks derived from the same individuals at different time periods provides insight into organizational change, and can be used in conjunction with knowledge of the world to correlate patterns in communication/interactivity with real-world consequences.

Flow analysis

The network approaches described in the previous section emphasize content and interactivity. However, they are relatively static in nature – dynamism is implied only when comparing networks based on different time spans. However, communication flow is inherently dynamic. To complement the network approaches, we have also included an explicitly temporal approach to assessing situation awareness. One set of algorithms that has been developed to process flow data is called FAUCET (Flow Analysis of Utterance Communication Events in Teams; Kiekel 2004). FAUCET metrics have been developed and validated in a UAV command-and-control scenario and more recently in the context of Enron e-mails. One of the FAUCET metrics is called ProNet (Cooke, Neville and Rowe 1996). ProNet is a method for reducing sequential event data that relies on the Pathfinder algorithm (Schvaneveldt 1990) as its kernel. ProNet has been recently applied to communication data with some success (Kiekel 2004). The result of this analysis is a graph structure in which communication events that occur together frequently are connected by directed links. ProNet, like Pathfinder, is limited in the sense that the multiple-link paths represented in the network structure are only certain to exist on a pair-by-pair basis. ChainMaster is a software tool that implements the ProNet algorithm, but that extends it by doing tests for the existence of chains at multiple lags. With these tests, the likely multiple-link paths can be highlighted. More information about these approaches can be found in Cooke, Gorman and Kiekel (2007).

Assessing situation awareness in large organizations: the Enron example

The purpose of the current effort was to explore ways of automatically assessing situation awareness of large organizations using converging, complementary communications assessment methods. To illustrate this assessment, we required a suitable corpus of communications. As most military corpora are classified, we chose a corpus of e-mail from a publicly available corporate entity, the Enron corporation. This corpus was a reasonable approximate of the types of corpora we would expect to see in large military organizations: there are several hundred people interacting, there are hundreds of thousands of messages over a multi-month period, and there are observable events in the public record to correlate with patterns in the data. This section describes the Enron corpus, our assessment methodology, the preliminary results and a high-level interpretation.

The Enron accounting scandal

The Enron accounting scandal of 2001 was one of the largest (and most widely known) cases of corporate malfeasance in US. history. Beginning in the summer of 2001, revelations about the scope and extent of Enron's accounting practices started to become public. While the events precipitating the collapse had started many years before with a series of illegal accounting practices, the fall of 2001 represented the beginning of the end for Enron. Between August and December 2001 Enron CEO Jeffrey Skilling resigned, a series of Wall Street Journal articles reported on the the scandal, stock prices fell from over $80 per share in January 2001 to junk bond status by January 2002, and the Security and Exchange Commission launched a formal investigation into Enron's accounting practices.

These calamitous events culminated in Enron's declaration of bankruptcy in December, 2001. In the end, the Enron accounting scandal wiped out $68bn of market value and caused irreparable damage to investor confidence as well as eliminating over $1bn of employee retirement funds held in Enron stock. The aftermath of this crisis sent shock waves through the stock market and has led to sweeping legal and regulatory changes, such as the Sarbanes-Oxley Act, as well as many years of litigation.

An important product of the Enron investigation was a large corpus of e-mail communication among Enron executives and employees prior to its bankruptcy in December 2001. Because it represents a rare look at the real-time communication inside an organization as it managed a life-threatening crisis, the Enron e-mail corpus is ideally suited to the development of automatic tools for real time assessment of group-level situation awareness.

The Enron e-mail corpus was originally released by the Federal Energy Regulatory Commission (FERC) during its independent investigation. It captures data extracted from the e-mail folders of 151 Enron employees obtained during the FERC investigations. However, because of the *ad hoc* nature of the sample and the raw format of the data, the raw corpus presented challenges for subsequent researchers. Some of these challenges were related to the unstructured nature of the data. For example, employees with multiple e-mail addresses appear in the original data as different people while some e-mail addresses (such as automatically generated responses from servers) do not represent people at all. Several different research groups have addressed these problems in various ways, leading to a proliferation of versions of the Enron corpus over time.

Deisner, Frantz and Carley (2005) provide a comprehensive overview of the iterations of this data set which we will not duplicate here. Suffice it to say that the primary differences between Enron corpora have to do with the methods used to clean the data. For the purposes of the current study, we obtained a later generation of the Enron corpus which was cleaned by researchers at the University of Southern California and placed in a SQL server. Each of the research teams used a subset of this corpus for their work, providing triangulation among multiple views of the situation awareness of Enron employees during the final months of the company's life.

The Enron corpus

The Federal Energy Regulatory Commission (FERC) gathered 619,449 e-mails from the computers of Enron employees, mainly senior managers,[1] as part of their investigation. In May of 2002, FERC publicly released the Enron e-mail dataset. For each e-mail, the e-mail address of the sender and receiver(s) were made available, as well as the e-mail's date, time, subject and body, while attachments were not released. The e-mails were written between 1997 to 2004.

SRI International (SRI), Cohen from CMU, Corrada-Emmanuel from the UMASS, Shetty and Adibi from the ISI group at USC, UC Berkeley Enron Email Analysis Project, and Corman and Contractor together with further members of the Organizational Communication Division of the International Communication (ICA) each played a hand in reducing and refining the Enron dataset and providing interfaces to the data. Their resulting database consists of 252,759 e-mails in 3000 user-defined folders from 151 people.

Methods

As described above, research on situation awareness has been largely confined to studies on small groups in command and control settings. Our research seeks to broaden the construct of situation awareness to measure the impact of critical events on knowledge- and information-sharing at the organizational level. Because the measurement and assessment of situation awareness at the organizational level represents new theoretical and methodological terrain, we used inductive, exploratory methods to understand how critical organizational events affected the content and structure of communication during the fall of Enron.

Our basic research question reflects one of the central theoretical assumptions of small group research on situation awareness, namely, that critical events will serve as orienting stimuli that generate increased communication and information sharing as collaborating groups seek to make sense of and coordinate reactions to unexpected events. While this has been a focus of many small group studies on situation awareness, it has not been addressed at the organizational level. To explore this notion at the organization level, we selected a time period containing five critical events in the decline of Enron and compared the structure and content of organizational communication around these events with the communication in times that did not contain critical events. Table 16.1 below depicts five critical events that took place at Enron between August and December of 2001 including the resignation of Jeff Skilling, the announcement of the SEC investigation and the declaration of bankruptcy in December. Each of the five events we selected was large, widely known by members of the organization, and likely served as a focal point of attention and corporate communication. To understand how organizational situation awareness changes around critical events, we also selected five dates between

1 Note that the Enron e-mail corpus contains a plethora of e-mails written by individuals who were not involved in any of the actions that were investigated by FERC.

Table 16.1 Critical events at Enron from August – December 2001

Date	Critical Events (Non Events) at Enron
August 1, 2001	(No event)
August 14, 2001	Jeff Skilling announces his resignation
September 11, 2001	(No event)
September 25, 2001	Ken Lay assures employees that Enron's accounting practices are legal
October 2, 2001	(No event)
October 16, 2001	Enron announces SEC investigation
October 23, 2001	(No event)
November 6, 2001	Enron announces profit overstatement
November 20, 2001	(No event)
December 4, 2001	Enron files for bankruptcy

August and December 2001 when there were no critical events at Enron.[2] Using the three business days before and after these critical and non-critical dates, we are able to compare how the structure and content of organization-wide communication changes in response to critical events.

Results: Semantic and social network analyses of the Enron e-mail corpus

This section describes the results of the network analysis of the Enron e-mail corpus. We begin by describing the data pre-processing that was performed followed with a description of the measures that were calculated to measure changes in group level situation awareness at Enron.

Before texts are analyzed, they can be pre-processed in order to normalize the data and to reduce the data to the terms relevant for a research question. For this particular project a series of pre-processing steps were required to make the e-mail corpus suitable for network analysis. All the texts were automatically cleaned to remove non-content-bearing symbols, such as apostrophes and brackets. Text that was to be used in network text analysis (e.g., all the content of e-mails, but not the header information) was submitted to several additional preprocessing steps.

2 Readers might note that 11 September 2001 is one of these 'non-critical' dates. We selected this date because it represents a useful quasi-control as it was clearly a critical event in the history of the world, but was not directly related to Enron *per se*. This allows us to explore whether organization level situation awareness changes in response to critical internal organizational events as opposed to external events.

First, we performed deletion, which removes non-content bearing conjunctions and articles from texts (Carley 1993). The list of words we built was tailored for this data set and contained 32 entries. Second, we applied a custom thesaurus to resolve common synonym and abbreviation ambiguities (e.g., changed 'U.S.A.' and 'U.S.' to 'United States'). When applying a thesaurus, AutoMap searches the text set for the text-level words denoted in the thesaurus and translates matches into the corresponding words. Because the terminology of a thesaurus depends on the content and the subject of the data set we used a thesaurus developed specifically for analysis of the Enron e-mail corpus.

Our third step was to construct and apply a generalization thesaurus. A generalization thesaurus typically is a two-columned collection that associates text-level concepts with higher-level concepts, e.g., in order to convert multiple noun phrases into single noun phrases (such as Jeff Skilling into Jeff_Skilling). The text-level concepts represent the content of a data set, and the higher-level concepts represent the text-level concepts in a generalized way (Popping and Roberts 1997). The generalization thesaurus we built contained 517 association pairs. To ease the construction of the generalization thesaurus, we also performed named-entity recognition in order to automatically identify names of people, places and organizations, as well as bi-gram detection, which return the most frequent combinations of any two terms in a data set. After this stage of pre-processing, semantic network analyses were run in order to extract Knowledge × Knowledge networks (KK – also referred to as knowledge networks).

Our fourth and final pre-processing step was to construct a meta-matrix thesaurus. This thesaurus is needed in order to perform meta-matrix text analysis, which enables the classification and analysis of concepts in texts according to the meta-matrix model ontology (Diesner and Carley 2005). Example: Jeff_Skilling will be associated with and translated into agent. Since one concept might need to be translated into several meta-matrix categories, a meta-matrix thesaurus can consist of more than two columns. The meta-matrix thesaurus we built associated 482 words with meta-matrix categories. After this stage of pre-processing, meta-matrix text analysis was performed in order to extract implicit meta-matrix (iMM) text networks from the data.

The CASOS Email Parser (CEMAP) enables the extraction of different types of network information from e-mails (e.g., who exchanges information, who provides what information, etc.). The following image shows what types of information can be extracted with CEMAP.

As described above, the social network (SN) represents social network data that can be extracted from e-mail headers. This includes agent–agent networks, where agents are the people who sent and received an e-mail, agent–task networks, where tasks are e-mails, and agent–knowledge networks, where knowledge is the content from the subject line. In SN, nodes represent people, and edges represent exchanged e-mails (frequency count). This network type does not require any text coding in AutoMap.

The knowledge (KK) and implicit meta-matrices (iMM) are extracted by performing semantic text analysis with AutoMap (Diesner and Carley 2004). More specifically, knowledge networks (KK) represent semantic network or mental models that are contained in the bodies of individual e-mails. In KK, nodes represent knowledge items, and edges represent the co-occurrence of terms in text. For iMM, texts are coded in AutoMap according to a taxonomy or ontology (e.g., meta-matrix, while are ontologies can be specified by the user). In iMM, nodes represent instances of categories (e.g., agent, knowledge, resources) of the ontology, and edges represent co-occurrences of terms in texts.

Networks types 4 and 5 result from the combination of SN with KK and iMM, respectively. In type 4, nodes represent people and knowledge, and edges represent e-mails and mental models. In type 5, nodes represent the categories of the taxonomy as specified by the user, and the edges represent the co-occurrence of the terms that represent instances of the taxonomy in the corpus. For the creation of type 4 and 5, the extraction of type 2 and 3, respectively, is mandatory. CEMAP stores all network data as DyNetML files (Tsvetovat, Reminga and Carley 2003) (a derivate of XML). This data can be analyzed with any package that reads DyNetML (e.g., ORA, Carley, Diesner and DeReno 2006).

To explore the impact of critical events on communication network structures at Enron, we generated multiple measures for each of the four matrices described above. A two-sample Welch's *t*-test (variances assumed to be unequal) was conducted on each of the measures for the relevant, corresponding meta-matrices. Despite the use of multiple measures and multiple matrices, in no case could we reject the null hypothesis that the population means of the events and non-events samples were equal. For example, the communications network, which was derived from the e-mail headers information, did not show a statistical difference in the clustering, number of components or cliques between the event and non-event samples. Similarly, the implicit organizational network which is derived from network text analysis of the e-mail message content, does not show a statistical difference in any of the four change-measures inspected (clustering coefficient, weak component count, clique count, the number of groups). The semantic network which is entirely based on the content of the e-mail messages does not show a statistical difference in any of the change-measures inspected. Finally, the results for both the inferred and semantic networks when each is combined with the communications network did not show a statistical difference in any of the four change-measures inspected. Likewise, the

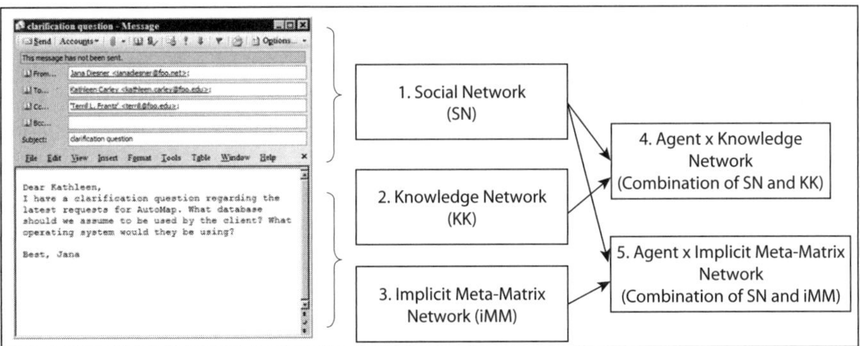

Figure 16.1 Network representations of e-mail data

communication and inferred organization network, which is based on a combination of the content of the e-mail messages and the associated e-mail headers, does not show a statistical difference in any of the seven change-measures inspected. While this analysis was not able to demonstrate that large organizational events shift organizational communication networks, we believe that the lack of statistically significant results may be due to problems with the data which we discuss below.

Communication flow processes in the Enron corpus

Over the last six years CERI has developed a suite of methods to analyze patterns of communication flow data (who is talking to whom and when; Kiekel, Cooke, Foltz and Shope 2001; Kiekel 2004). Patterns in the specific timing and sequences of interactions can be extracted using some custom routines (ProNet, Chums, Dominance). We refer to the set of routines as FAUCET. FAUCET metrics have been developed and validated in a UAV command-and-control scenario and more recently in the context of Enron e-mails. In this project we focus on the application of ProNet (Cooke, Neville and Rowe 1996; Kiekel, Gorman and Cooke 2004).

One advantage of focusing on communication flow is that these data can be collected relatively cheaply and unobtrusively compared to content data that requires either speech recognition routines or laborious human transcription and coding. In addition, specific interaction or flow patterns have been found to be associated with effective team performance or team SA (Cooke, Gorman, Kiekel, Foltz and Martin 2005). Similarly some patterns may signal a loss of team SA. Although the data used in this analysis is very basic, it is also inexpensive to collect and analyze and is conducive to automation. Marrying these data with some content from the AutoMap networks could be done in cases in which one wishes to drill deeper. The combined interaction patterns and content could provide a representation of team coordination of the kind needed to share dynamic information in real time when knowledge and information are distributed across team members.

One of the metrics included in the FAUCET suite is ProNet (Cooke, Neville and Rowe 1996). ProNet is a method for reducing sequential event data that relies on the Pathfinder algorithm (Schvaneveldt 1990) as its kernel. ProNet has been successfully applied to communication data (Kiekel 2004). The result of this analysis is a graph structure in which communication events that occur together frequently are connected by directed links. ProNet, like Pathfinder, is limited in the sense that the multiple-link paths represented in the network structure are only certain to exist on a pair-by-pair basis. ChainMaster is a software tool that implements the ProNet algorithm, but that extends it by doing tests for the existence of chains at multiple lags. With these tests, likely multiple-link paths can be highlighted. The ChainMaster analyses reported here used the following Pathfinder parameter settings: number of nodes = 5, q = 4, and r = infinity. These parameter settings are the default values that the Pathfinder algorithm uses to generate networks from proximity matrices. ChainMaster provides regularly occurring multilink chains as output when provided with XML-formatted flow data as input. In this project, the Enron e-mail database was examined by applying ChainMaster to segments of it to uncover regularly occurring chains. Resulting chains were then examined for changes corresponding to critical corporate events.

Data preprocessing

Using the Enron e-mail corpus described above, we categorized entities into one of nine particular job functions: president, vice president, CEO, director, manager, lawyer, trader, employee, or unknown. The database was then filtered by selecting only entities categorized as 'Executive group' members. The 'Executive group' consisted of those e-mail entities categorized as president, vice president, CEO, director, and manager (Table 16.2). In order to reduce the number of spurious links in the ChainMaster networks due to undirected e-mail traffic such as 'list serves', the data were then filtered by selecting those e-mails only sent within the Executive group.

Segments of Enron e-mail flow data were processed using the ChainMaster software tool using the Pathfinder parameter settings: number of nodes = 5, $q = 4$, and r = infinity. ChainMaster returned regular occurring chains for each segment of e-mails processed.

Results: Communication flow processes in the Enron corpus

The immediate objective was to use ChainMaster to detect shifts in the e-mail flow patterns. In order to accomplish this several analyses were conducted.

First, multiple non-critical controls were identified and compared to five critical Enron events. Of these critical events, one was excluded because stable chains were not identified. Critical event time periods were defined as the critical event (e.g., the day on which the event occurred) plus or minus three business days. Control time periods were also identified as a non-critical day plus or minus three business days. The control time period took place one week before a critical day. The original non-critical events included the dates 1 August, 11 September, 22 October, 23 October and 20 November 2001.

Critical and control chains from the ChainMaster analysis resulted in both common and uncommon chains. An uncommon chain is defined as a chain consisting

Table 16.2. Executive group

ChainMaster Node	Job Category
1	President
2	Vice President
3	CEO
4	Director
5	Manager

of nodes and directional links (e.g., President → CEO) that occurs in either the control or critical time period, but not both. Alternatively, a common chain occurs in both control and critical time periods. In order to measure change between control and critical time periods in e-mail flow, nodes and directed links between nodes for each time period were compared. Because ChainMaster is based on transition matrices for each time period, the difference between transition matrices for baseline versus critical were computed as the C-value, where C-value is computed as C = number of common links/number of unique links. A relatively high C-value indicates small change between baseline and critical chains (small change in communication patterns), while a relatively low C-value indicates large change between baseline and critical chains (large change in communication patterns).

C-values were calculated for all four critical events relative to a non-critical control. The results of this analysis indicate that the smallest C-values (i.e., the biggest difference or change in flow patterns) occurred for Events 1 and 4; respectively when Jeffery Skilling resigned and when Enron filed for bankruptcy (Figure 16.2). Therefore in this analysis the largest change in e-mail flow was detected in the days following and leading up to these two critical events.

A second analysis using a single baseline was undertaken due to overlaps between non-critical control periods and critical event time periods. For this analysis, the non-overlapping 1 August control time period was chosen as the single baseline in order to detect change in the four critical time periods and a single non-critical control (Table 16.3).

C-values were calculated for all four critical events relative to the non-critical baseline. Further, a C-value was calculated for a non-critical control event

Figure 16.2 Similarity of flow patterns between critical Enron events and non-critical controls. High similarities indicate less change from baseline

relative to the non-critical baseline. The C-values for this analysis are presented in Figure 16.3.

The C-value analysis detected the degree of change in *nodes* between time periods. The biggest change occurred (i.e., lowest similarity) when Ken Lay reassured Enron employees (critical event 2). The relatively high C-value for the non-critical control event indicates the presence of consistent e-mail flow patterns between the two non-critical event time periods.

Further analysis was undertaken in order to detect change in the chains (i.e., beyond pair-wise links) between the two time periods. The number of common chains divided by the number of unique chains between time periods was calculated in order

Table 16.3 Dates and descriptions of critical or non-critical events

Date	Event	Description
11-Sep	Non-Critical Control	X
14-Aug	Critical Event 1	Jeffrey Skilling resigns
26-Sep	Critical Event 2	Kenneth Lay reassures Enron employees
22-Oct	Critical Event 3	Enron announces SEC investigation
2-Dec	Critical Event 4	Enron declares bankruptcy

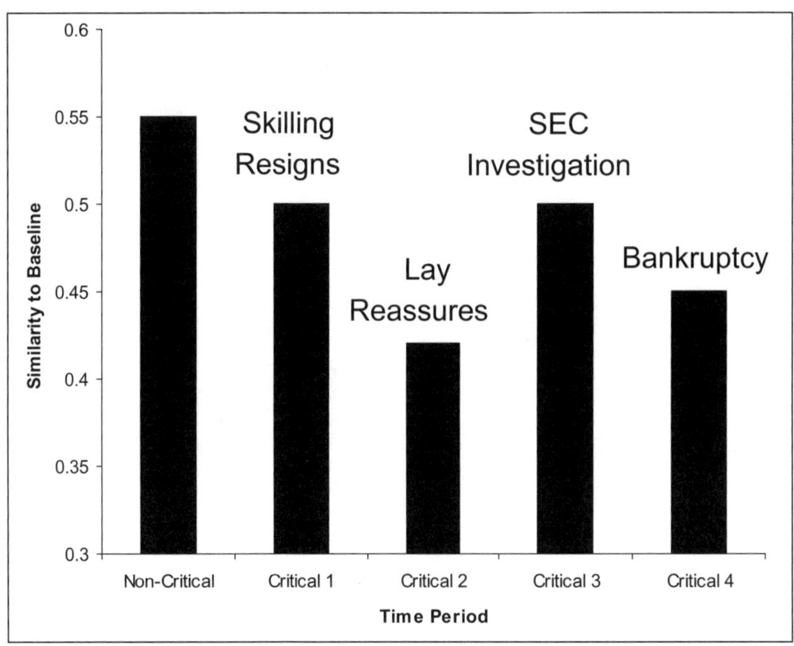

Figure 16.3 Graph of C-values for the critical versus the August 1st baseline and one non-critical event

to estimate change in chains. This analysis was again done using the single baseline time period centered on 1 August. The results are graphically depicted in Figure 16.4.

The results indicate that the chains differ increasingly from the baseline for the critical events as time separation increases from baseline. However, this was not the case for the non-critical control chains. The non-critical chains were actually temporally between critical event period 1 and 2, suggesting the possibility that the non-critical control period and the baseline had similarities that were independent of time. Interestingly, the non-critical (in the context of Enron events) control period also included the date of 11 September 2001. This suggests that changes that are detected here in flow patterns may be specific to the Enron events. More data are needed to support this possibility.

Building on the previous analysis of change in chains, the chains that consistently differed between the critical and baseline time periods were explored. Consistent change was defined as a chain that appeared in more than half of the critical time periods but did not appear in the non-critical 1 August baseline. A chain that was consistently present during the critical events was a president to vice president pattern (4/4), and to a lesser extent, vice president to president (3/4). In the 1 August baseline chains, the president was only linked to the managers. However, in the non-critical control time period, the president to vice president and vice president to president chains were also detected. Overall this pattern suggests more communication among those at the top of the chain of command during critical events.

Discussion of flow analysis

The utility of detecting change in team or organizational dynamics, and the exact nature of this change, is in measuring SA from an interaction-based perspective. Systemic SA (Walker, Stanton, Jenkins, Salmon, Young, Beond, Sherif, Rafferty

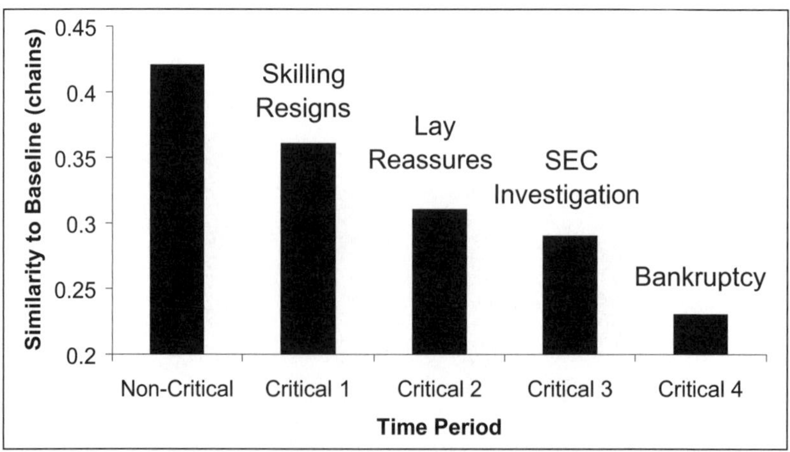

Figure 16.4 Chain difference between non-critical and baseline and chain difference between critical time periods and the August 1st baseline

and Ladva 2006) and Coordinated Awareness of Situation by Teams (CAST; Gorman, Cooke and Winner 2006) are two methodologies for measuring SA that have recently been developed to measure organizational and team SA, respectively. The principle goal of these approaches is not to measure the overlapping or divided situational knowledge of individuals, but rather how the team or organization as a whole changes given an evolving situation.

The ChainMaster tool, within the larger IMAGES framework, shows promise as a method for detecting organizational changes in information flow using low-level, content-free interaction data. Shifts in ChainMaster flow patterns in Enron e-mails seem to correspond with significant corporate events. Based on the process or interaction view, team SA is the coordinated perception and action of team members in the face of change. Thus, coordination shifts as seen in these data would be anticipated in response to critical corporate events. These data indicate that flow shifts may be useful as a signal that an organization is adapting to environmental change. Likewise, lack of response may be indicative of a lack of SA on the part of the organization.

The application of ChainMaster to e-mails stretches the limits of this approach for several reasons. One limitation ChainMaster encountered in analyzing the Enron e-mail corpus involves the detection of spurious links between aliases that were not connected in reality. This limitation is most serious in large, highly unconstrained samples, for cxample when list serves are used. A second limitation ChainMaster encountered was in analyzing relatively small time periods of e-mails (i.e., three days). Specifically, when an unreported analysis was conducted on three-day time periods ChainMaster was unable to detect e-mail patterns in seven out of ten cases for periods that both preceded and followed the critical and non-critical dates.

The results from the ChainMaster analysis of flow patterns demonstrate a means of using communication data to measure team SA within a process-oriented framework. The results also support the validity of a process-oriented view of team SA. That is, critical corporate events are associated with process changes in the organization. Very basic and easy to collect changes in communication flow appear promising as indices to shifts in organizational interaction in response to environmental change. The fact that change can be detected using these very low cost, basic measures, has enormous potential for online monitoring of organizational communications for real-time intervention.

Discussion and conclusions

The goals of the current research effort were simple: explore ways in which the situation awareness of an organization could be derived using only the indicators of behavior resident in captured communications. If organizational SA could be assessed rapidly and largely automatically for trainers, managers and operations, overall organizational performance might improve. These assessments would be especially valuable in complex organizations with critical responsibilities, such as the Expeditionary Strike Group or Air and Space Operations Center.

Situation awareness is not straightforward to define or measure. The three stages of situation awareness described by Endsley (1988) – which roughly correspond to perception, understanding and prediction – are largely internal to the individual. Over the past 20 years, validated measures of situation awareness have been developed for small teams engaged in complex tactical tasks. These measures, involving during-event probes and post-event interviewing techniques, have allowed researchers to better understand the dynamics of these organizations. Unfortunately, the organizations of interest in the current research are significantly different. The timescale of their work is longer; their work is done over the course of hours, days and weeks, not seconds and minutes is the case in aviation or surgical settings. The size of the organization is larger and thus has different knowledge requirements and interactivity patterns than small organizations of only 3–8 members. The sensitivity of tasks to interruption is low, and this makes some current measures of situation awareness (which require that users pause their tasks to respond to polls or probes) inappropriate for real-world use.

Clearly, new measures of situation awareness are needed for large, complex, distributed organizations. In as much as communication enables the work within and between these joint cognitive systems of people and technologies, an analysis of communication can lead to insights into organizational performance, interaction and ultimately situation awareness. The types of analyses possible on communication are many and varied, from time-intensive utterance decomposition based on particular informative taxonomies to complex statistical models of word usage in a multidimensional space. Each technique gives us a glimpse into particular aspects of the inner workings of organizational behavior. When taken together, these perspectives give a more complete picture and understanding of the organization. In the present work, we have developed and integrated tools that facilitate communications analysis at the organizational level. There are other communications analysis tools and metrics that may have value in specific circumstances, and new tools are being developed in academia and industry. IMAGES was designed in a modular fashion, so that it can interoperate with other tools and metrics with relative ease.

We applied IMAGES to the Enron e-mail corpus to evaluate its potential. Although not conclusive, our analysis suggests that we may be able to identify changes in patterns of interactions or network configurations that are indicative of organizations in distress. For example, analyses of networks show critical events seem to isolate different components of an organization – as derived from communications patterns – from each other. These components increasingly focus on different topic areas. These areas may be related to critical events in their areas of expertise. Over time, this segregation abates and normal interactivity resumes.

Similarly, patterns of interactivity among communicators sometimes emerge during critical events that do not arise during more stable periods. In any environment, there is a baseline pattern of interactivity and communication among group members; a level of variability in message flow that is a natural complement to the work required in that organization. Some events disrupt this flow, causing deviations from the baseline that could eventually be identified and used in a predictive manner. Our findings on this topic were made using the FLOW analysis tools, but they are less conclusive than we would like. The medium of communication used in the Enron corpus – e-mail – is an imperfect match with FLOW analysis. Because e-mail is asynchronous and multiple conversational threads interwoven, the ordinality of the messages (the flow) became less clear and thus less informative than other media, such as voice communications and text chat.

The Enron e-mail corpus

The results of this research are promising. However, the reader should consider the limitations concerning our findings. These are due in part to the nature of the Enron corpus itself. The corpus was chosen because of its apparent similarity to our target military organizations. Its availability, size and complexity made it widely used in the machine learning and text-processing areas. However, it is critically lacking in some fundamental respects.

- *The corpus includes only part of the e-mail from part of the organization*. The corpus was constructed by aggregating the e-mail messages found on a small number of machines, not from a central server. It was intended to assist in the court proceedings concerning a few members of Enron, and thus was not intended to comprehensively cover *all* communications among *all* employees of Enron. The intended environment for IMAGES, in contrast, is a domain in which all communications are captured and stored.
- *The corpus systematically omits communication over some channels*. The Enron corpus was collected from e-mails sent in the early 2000s. At that time, e-mail was widely used and accepted as a primary mode of communication. However, much communication at Enron occurred in face-to-face settings, over the telephone and in written form (e.g., point-to-point facsimile, broadcast memoranda). An analysis of an e-mail corpus, even if it were complete, would lead to an imperfect picture of the organization. It is plausible that high-criticality communication occurs in more immediate communication media like face-to-face and telephone communications, leading to spurious results when the messages are analyzed.
- *The critical events lack equivalency*. To demonstrate the ability of the communications assessment techniques to detect changes in organizations, we chose five high-profile events. There is a significant amount of information about these events and their effects on the company, employees and the larger business community. However, we did not make any attempt to ensure that these events were of similar type or scale. Thus, it is likely that different critical events engaged different employees and elicited different reactions from them, leading to different patterns of interactivity and different topic profiles. Future analyses will be enhanced by categorizing critical events and theorizing on the likely effects of those events on communication.

Given these caveats, it is encouraging that we found persuasive trends in these data. Specifically, we found that communications differed between date ranges in which there were critical events, and periods in which there were none (control conditions). This demonstrates the robustness of the approaches used. Further research is needed to explore the full potential of these approaches in conditions that more exactly map to the environments of interest.

The techniques we demonstrated on the Enron corpus can be employed on other large and complex organizations. Given a baseline understanding of the communications characteristics of an organization, users of IMAGES can detect changes in those characteristics and assess deviation from normalcy. When Air Force personnel in an Air and Space Operations Center are being trained using a known scenario, we can compare their communications patterns with previous successful and unsuccessful teams. Both content and flow measures could be used as diagnostic tools, feeding back information to trainers and evaluators. In less predictable environments, such as a deployed expeditionary strike group, the challenge is to discriminate between the normal and abnormal changes in interactivity and communication patterns. This will require additional research. In the interim, gains in operational efficiency may be made simply by presenting changes to operations officers who have an intuitive feeling for congruous and incongruous patterns given the mission changes.

Situation awareness and macrocognition

We made two fundamental arguments at the beginning of this research program. First, we argued that organizational level situation awareness differs fundamentally from what is commonly accepted as situation awareness in tactical environments. As a knowledge construct, organizational SA is the allocation of knowledge and understanding around an organization, with some information shared and some unshared between roles (which have different task demands) and over time. Organizational situation awareness is, thus, a complex mix of shared and team situation awareness. As a process construct, organizational SA consists of patterns (flows) of communications that enable that knowledge to be shared and acted upon in a constantly changing, dynamic environment. It is only when both the knowledge and process aspects of team cognition are considered that an indication of organization situation awareness can be derived.

The second argument fundamental to this research was that organizational situation awareness can be derived from analysis of communications patterns. Teams do not work silently; communication enables information to be passed, confirmations to be made and ambiguities clarified. An analysis of the externalized communications of a joint cognitive system is a window into cognitive collaboration. We have argued that communication *is* team cognition, or at least that it both enables and reflects team cognition to a high degree. We have taken a broad perspective on approaches to communications, including measures of content, context and process. Given the definition of organizational situation awareness, all three of these aspects of communications need to be considered.

For this operationalization of situation awareness, the communications measures we propose are very promising. Even with a suboptimal corpus, trends implied that changes in communications patterns were related to critical changes in the environment. These trends appeared both in context and process measures of communications. While considerable research remains to refine these measures, the potential is clear: analysis of communications offers a window into organizational dynamics and situation awareness.

These same measures of communications have implications for a theory of macrocognition. Communications is an enabler of macrocognitive processes of knowledge interoperability development and shared understanding (Warner *et al.* 2005). Clearly, teams of individuals in problem-solving or decision-making situations must communicate to succeed in their tasks, to gain common ground and to manage ambiguities. Analysis of communications can thus provide insight into critical team processes.

The macrocognition theory of Warner *et al.* (2005) was developed primarily to explain the behaviors of teams. The analyses performed for this project concerned a very large organization, one several orders of magnitude greater in size and perhaps in task complexity that teams. In organizations (unlike teams), consensus is not a goal and multiple macrocognitive processes may occur simultaneously in different parts of the organization. Thus, future research should test the processes and stages of the Warner *et al.* (2005) theory of macrocognition in larger organizations that contain and support the operational teams that accomplish critical missions.

Into the future

The work described in this report represents the first steps on a long journey. It is only in the past decade that communications have been largely capturable and storable, and only in the past few years that researchers have considered analysis of communications a viable approach to performance assessment. In this study, we applied multiple, converging approaches that had never been combined, and used them on a corpus that was orders of magnitude larger and more complex than had been previously attempted for this purpose. The results are promising, with trends in the directions of interest.

As we proceed, those trends need to be investigated further. In an environment in which there is greater control and understanding of the organization and its tasks, we could more accurately correlate changes in communications behaviors with external factors. A large-scale training session or exercise would be an ideal venue in which to apply IMAGES, as the scenario would be carefully scripted and the participants identified. In such an environment, existing measures of performance could be correlated with patterns in communication, allowing a stronger understanding of the antecedents of those patterns.

As discussed above, the Enron corpus – while a useful data set in some respects – was a small sample of the communications and communicators at Enron. This paucity of communications led us to interpret patterns in the data conservatively, as trends. In the future, the same communications measures should be applied to a more complete corpora of communications. An ideal corpus would capture all of the communications, not just those that occur in e-mail form. Unobtrusive methods to capture text chat, recorded/transcribed telephone and face-to-face communication are all being developed. This richer, captured data would be a good proving ground for the communications measures used here. The choice of communications media itself may also be an indicator of team performance, and media measures could be used to complement or confirm measures used here.

Organizations differ in purpose, composition, complexity and a host of other dimensions. As this research proceeds, the individual characteristics of organizations have to be considered as they relate to the analysis of their communications. How do the communications patterns of an expeditionary strike group differ from those of a similarly sized army division? How does it differ from a corporate entity? Understanding these differences will help us refine our interpretation of patterns and their implications.

People have been communicating for millennia; we have only begun to understand how and why. The measures, methods and tools developed in the IMAGES project should improve both our knowledge of communications and our management of teams, whose missions hinge on accurate, effective communication.

Acknowledgments

Work described in this chapter was supported by N00014-05-C-0505 from the Office of Naval Research. The views, opinions, and findings contained in this chapter are the author's and should not be construed as reflecting the official views of the Department of Defense, Aptima, Inc., or Carnegie Mellon University.

References

Alberts, D.S. (2002). *Information Age Transformation: Getting to a 21st Century Military*. (Washington, DC: CCRP Publications). First published 1996.

Cacciabue, P.C. and Hollnagel, E. (1995). 'Simulation of Cognition: Applications', in J.M. Hoc, P.C. Cacciabue and E. Hollnagel, (eds), *Expertise and Technology: Cognition and Human–Computer Cooperation*, pp. 55–73. (Mahwah, NJ: Lawrence Erlbaum Associates).

Carley, K., Diesner, J. and De Reno, M. (2006). *AutoMap User's Guide. Carnegie Mellon University, School of Computer Science, Institute for Software Research, Technical Report, CMU-ISRI-06-114.*

Carley, K.M. (1993). 'Coding Choices for Textual Analysis: A Comparison of Content Analysis and Map Analysis', *Sociological Methodology* 23, 75–126.

Carley, K.M. (1997a). 'Network Text Analysis: The Network Position of Concepts', in C.W. Roberts, (ed.), *Text Analysis for the Social Sciences*, pp. 79–102. (Mahwah, NJ: Lawrence Erlbaum Associates, Inc.).

Carley, K.M. (1997b). 'Extracting Team Mental Models through Textual Analysis', *Journal of Organizational Behavior* 18, 533–558.

Carley, K.M. (2002). 'Smart Agents and Organizations of the Future', in L. Lievrouw and S. Livingstone, (eds), *The Handbook of New Media,* pp. 206–220. (Thousand Oaks, CA: Sage).

Carley, K.M. (2003). 'Dynamic Network Analysis', in R. Brieger and K.M. Carley, (eds), *The Summary of the NRC Workshop on Social Network Modeling and Analysis*, *National Research Council*, pp. 133–145. (Washington DC: National Academies Press).

Carley, K.M. and Palmquist, M. (1992). 'Extracting, Representing, and Analyzing Mental Models', *Social Forces* 70, 601–636.

Cooke, N.J. (2005). 'Communication in Team Cognition', *Proceedings of the Collaboration and Knowledge and Management Workshop*, San Diego, CA. (Arlington, VA: Office of Naval Research Human Systems Department).

Cooke, N.J., Gorman, J.C., Kiekel, P.A., Foltz, P. and Martin, M. (2005). *Using Team Communication to Understand Team Cognition in Distributed vs. Co-Located Mission Environments*. Technical Report for Contract N00014-03-1-0580. (Washington, DC: Office of Naval Research).

Cooke, N.J., Neville, K.J. and Rowe, A.L. (1996). 'Procedural Network Representations of Sequential Data', *Human-Computer Interaction* 11, 29–68.

Cooke, N.J., Salas, E., Cannon-Bowers, J.A. and Stout, R. (2000). 'Measuring Team Knowledge', *Human Factors* 42, 151–173.

Diesner, J. and Carley, K.M. (2004). *AutoMap1.2 - Extract, Analyze, Represent, and Compare Mental Models from Texts*. Technical Report CMU-ISRI-04-100. (*Pittsburgh, PA: Carnegie Mellon University, School of Computer Science, Institute for Software Research).*

Diesner, J. and Carley, K.M. (2005). 'Revealing Social Structure from Texts: Meta-Matrix Text Analysis as a Novel Method for Network Text Analysis', in V.K. Narayanan and D.J. Armstrong, (eds), *Causal Mapping for Information Systems and Technology Research: Approaches, Advances, and Illustrations*, pp. 81–108. (Harrisburg, PA: Idea Group Publishing).

Diesner, J., Frantz, T. and Carley, K. (2005). 'Communication Networks from the Enron Email Corpus "It's Always About the People. Enron is no Different"', *Computational and Mathematical Organization Theory* 11(3), 201–228

Endsley, M.R. (1988). 'Situation Awareness Global Assessment Technique (SAGAT)', *Proceedings of the National Aerospace and Electronics Conference (NAECON)*, pp. 789–795. (New York: IEEE).

Endsley, M.R. (1995). 'Measurement of Situation Awareness in Dynamic Systems', *Human Factors* 37, 65–84.

Endsley, M.R. and Jones, W.M. (1997). *Situation Awareness, Information Dominance, and Information Warfare*. United States Air Force Armstrong Laboratory Technical Report No. AL/CF-TR-1997-0156. (Wright-Patterson AFB, OH: United States Airforce Armstrong Laboratory).

Entin, E.E. and Serfaty, D. (1999). 'Adaptive Team Coordination', *Human Factors* 41, 312–325.

Gorman, J.C., Cooke, N.J. and Winner, J.L. (2006). 'Measuring Team Situation Awareness in Decentralized Command and Control Systems', *Ergonomics* 49, 1312–1325.

Gorman, J.C., Cooke, N.J., Pedersen, H.P., Connor, O.O. and Dejoode, J.A. (2005). 'Coordinated Awareness of Situation by Teams (CAST): Measuring Team Situation Awareness of a Communication Glitch', *Proceedings of the Human Factors and Ergonomics Society 49th Annual Meeting*, pp. 274–277. (Santa Monica, CA: HFES).

Isaacs, E.A. and Clark, H.H. (1987). 'Referencew in Conversation Between Experts and novices', *Journal of Experimental Psychology: General* 116, 26–27.

Kiekel, P.A. (2004). *Developing Automatic Measures of Team Cognition Using Communication Data*. Ph.D. Thesis, New Mexico State University.

Kiekel, P.A., Cooke, N.J., Foltz, P.W. and Shope, S.M. (2001). 'Automating Measurement of Team Cognition through Analysis of Communication Data', in M.J. Smith, G. Salvendy, D. Harris and R.J. Koubek, (eds), *Usability Evaluation and Interface Design*, pp. 1382–1386. (Mahwah, NJ: Lawrence Erlbaum Associates).

Kiekel, P.A., Cooke, N.J., Foltz, P.W., Gorman, J.C. and Martin, M.J. (2002). 'Some Promising Results of Communication-based Automatic Measures of TEAM Cognition', *Proceedings of the Human Factors and Ergonomics Society 46th Annual Meeting, Baltimore, MD*, pp. 298–302. (Santa Monica, CA: Human Factors and Ergonomics Society).

Kiekel, P.A., Gorman, J.C. and Cooke, N.J. (2004). 'Measuring Speech Flow of Co-located and Distributed Command and Control Teams During a Communication Channel Glitch', *Proceedings of the Human Factors and Ergonomics Society 48th Annual Meeting*, pp. 683–687. (Santa Monica, CA: Human Factors and Ergonomics Society).

Klein, G., Ross, K.G., Moon, B.M., Klein, D.E., Hoffman, R.R. and Hollnagel, E. (2003). 'Macrocognition', *IEEE Intelligent Systems* 18, 81–85.

Krackhardt, D. and Carley, K.M. (1998). 'A PCANS Model of Structure in Organization', *Proceedings of the 1998 International Symposium on Command and Control Research and Technology*, pp. 113–119. (Vienna, VA: Evidence Based Research).

McComb, S. (2005). 'Mental Model Convergence (MMC)', *Proceedings of the Office of Naval Research, Collaboration and Knowledge Management Workshop.* (Arlington, VA: Office of Naval Research Human Systems Department).

Popping, R. and Roberts, C.W. (1997). 'Network Approaches in Text Analysis', in R. Klar and O. Opitz, (eds), *Classification and Knowledge Organization: Proceedings of the 20th annual conference of the Gesellschaft für Klassifikation, University of Freiburg*, pp. 381–898. (Berlin, New York: Springer).

Quek, F., McNeill, D., Bryll, R. Duncan, S. Ma, X., Kirbas, C., McCullough, K.E. and, Ansari, R. (2002). 'Multimodal Human Discourse: Gesture and Speech', *ACM Transactions on Computer–Human Interaction (TOCHI)*, 9, 171–193.

Schvaneveldt, R.W. (1990). *Pathfinder Associative Networks: Studies in Knowledge Organization.* (Norwood, NJ: Ablex).

Sowa, J.F. (1984). *Conceptual Structures: Information Processing in Mind and Machine*. (Reading, MA: Addison-Wesley).

Tsvetovat, M., Reminga, J. and Carley, K. (2003). DyNetML: Interchange Format for Rich Social Network Data. NAACSOS Conference 2003, Day 2, Electronic Publication, Pittsburgh, PA.

Walker, G.H., Stanton, N.A., Jenkins, D., Salmon, P., Young, M.S., Beond, A., Sherif, O., Rafferty, L. and Ladva, D. (2006). 'How Network-enabled Capability Changes the Emergent Properties of Military Command and Control', *Proceedings of the Human Factors and Ergonomics Society 50th Annual Meeting*, pp. 2492–2496. (Santa Monica, CA: HFES).

Warner, N., Letsky, M. and Cowen, M. (2004). 'Cognitive Model of Team Collaboration: Macro-Cognitive Focus', *Proceedings of the 49th Human Factors and Ergonomics Society Annual Meeting, September 26-30, 2005. Orlando, FL*, pp. 269–273. (Santa Monica, CA: Human Factors and Ergonomics Society).

Warner, N.W. and Wroblewski, E.M. (2004). 'The Cognitive Processes Used in Team Collaboration During Asynchronous, Distributed Decision Making', *Proceedings of the Command and Control Research and Technology Symposium, San Diego, CA*, pp. 1–24. (Arlington, VA: Office of the Secretary of Defense).

Weil, S.A., Carley, K.M., Diesner, J., Freeman, J. and Cooke, N.J. (2006). Measuring Situational Awareness through Analysis of Communications: A Preliminary Exercise', *Proceedings of the Command and Control Research and Technology Symposium, San Diego.* (Arlington, VA: Office of the Secretary of Defense).

Weil, S.A., Tinapple, D. and Woods, D.D. (2004). 'New Approaches to Overcoming E-mail Overload', *Proceedings of the Human Factors and Ergonomics Society Conference, New Orleans, LA*, pp. 547–551. (Santa Monica, CA: Human Factors and Ergonomics Society).

Woods, D.D., Patterson, E.S. and Roth, E.M. (2002). 'Can we Ever Escape from Data Overload? A Cognitive Systems Diagnosis', *Cognition, Technology, and Work* 4, 22–36.

Chapter 17

Shared Lightweight Annotation TEchnology (SLATE) for Special Operations Forces

Mark St. John and Harvey S. Smallman

Introduction

As the US military transforms itself in response to the changing array of global threats and the fast pace of technological advances, two clear trends have emerged. First, networked technology has grown increasingly capable and accessible to personnel in the field. This capability is sometimes referred to as bringing 'power to the edge'. Second, the role of Special Operations Forces (SOF) – always an important component of modern military operations – has increased. The SOF's specific focus on conducting agile and covert operations makes them a potent force for combating asymmetric threats. The combination of these trends holds tremendous promise for providing every member of the SOF with vistas of networked, mission-relevant, constantly updated, battlefield knowledge to help them out-maneuver and out-coordinate their opponents. However, innovative collaboration technologies are needed to harness the potential of networked technology for the agile uses required for SOF operations.

We are developing a mobile spatial messaging and collaboration tool for the SOF that operates on both palmtop and laptop devices. It offers several innovative capabilities for enhancing team collaboration and situation awareness beyond contemporary communication tools. The tool is called SLATE (Shared Lightweight Annotation TEchnology) for keeping teams 'on the same slate'. SLATE represents the application and integration of cutting edge collaborative technology concepts from cognitive science and knowledge management, including several from the Collaborative Knowledge Interoperability program sponsored by the Office of Naval Research in the US Department of Defense.

In this chapter, we first introduce the innovative capabilities of SLATE. Then we outline the collaboration and communication requirements of SOF operations and show how current networked tools and technologies do not meet these requirements. Finally, we describe SLATE in terms of the cognitive science and knowledge management concepts that underlie it, and we describe how these concpets were implemented and integrated into the SLATE design to more effectively support SOF operations.

SLATE's two key innovations

SLATE possesses two innovative capabilities to enhance shared situation awareness and tactical decision-making: (1) methods for creating multimedia annotations, and (2) attention management tools to recover situation awareness following interruptions. The first SLATE innovation is the capability to create spatial, voice and text annotations and embed them within maps and other mission documents. Users can refer to locations simply by pointing rather than finding and spelling out coordinates, and users can communicate a path with landmarks by naturally drawing it out rather than verbally describing it. The capability to combine graphical annotation with text and voice while embedding the annotations into the evolving mission context makes communication more natural to create and understand, which in turn facilitates macrocognition and collaboration among team members. The intended result is faster and more accurate tactical SOF decision-making (see Figure 17.1).

The second SLATE innovation is a set of attention management tools to support asynchronous collaboration in spite of interruptions. SLATE announces the arrival of new messages by displaying graphically encapsulated information objects ('infobs'). The infobs allow users to quickly scan and prioritize messages for more detailed review. When team members must collaborate asynchronously due to communication or task interruptions, this overview and prioritization capability helps users recover situation awareness quickly. The infobs also control access to the message content, including multimedia annotations, and this capability allows users to manage the presentation of messages, the removal of obsolete messages and to focus on new and relevant information (see Figure 17.2).

The combination of these two innovations represents a new role for annotations in collaboration tools. Annotations are typically treated as passive and static comments that simply reside within mission documents. In SLATE, they are connected to infobs that announce their arrival and control their presentation. In this way, annotations become active, dynamic components that have a central role within the spatial messaging and collaboration tool. The integration of these innovations with other collaboration concepts results in a messaging tool for SOF teams that is easy to use in order to communicate efficiently and that is effective for supporting distributed, networked missions.

SOF characteristics and requirements

What are the communication and collaboration requirements of SOF operations that need to be satisfied and enhanced? And what constraints does the SOF environment specifically impose on the design of a collaborative tool?

SOF operations encompass a wide variety of actions from reconnaissance to direct action to local government liaisons. A fictitious yet quasi-realistic scenario that illustrates how the SOF operates and collaborates can be found in Wroblewski and Warner (Chapter 22, this volume). In general, missions are composed of several phases, from initial planning, through execution, to after-action review. During mission execution, SOF teams exchange messages to relay information among

Figure 17.1 The laptop/desktop version allows users to visually track and interact with multiple mission threads for command centers and reach-back assets

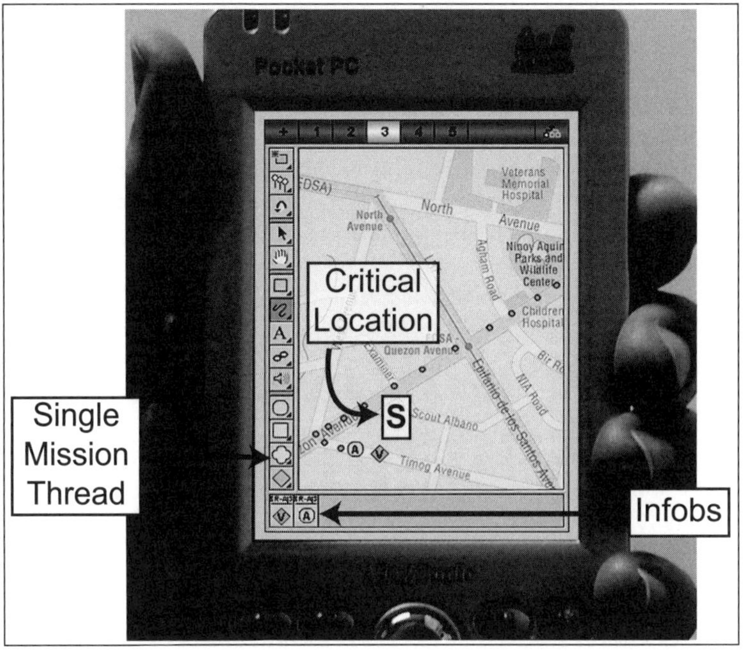

Figure 17.2 The palmtop version maximizes usability for the reduced screen real estate available on a mobile palmtop display and shows a single mission thread for dismounted, fielded users

themselves and their command center in order to stay coordinated across time and space, to respond to rapidly changing events and intelligence, and to re-plan on the fly. As missions unfold, SOF teams refer to specific parts of maps, sketches, photographs and other mission planning documents, and communicate new information and updated plans related to them.

Team members are often distributed, with some members in the field while others reside in local or remote command centers or reach-back facilities that provide mission support. Fielded usage of collaboration tools imposes a number of constraints, including size, mobility, interaction efficiency and limited network bandwidth. Palmtop and even head-mounted displays are better suited to these conditions than larger laptop or desktop devices. In turn, these small displays impose limits on what information and how much information can be displayed. Command centers, on the other hand, relax these constraints and support laptop/desktop devices. The greater screen size and bandwidth of these devices can then be exploited to support additional collaboration features.

Communications may also be distributed in time as well as space. While many communications occur in real time, there are many occasions when SOF teams are engaged in activities that limit communication. Further, SOF teams often prefer to communicate in short, intermittent bursts to avoid detection. This burst mode can make long synchronous discussions difficult and ill-advised. To address these conditions, the ability to intermittently receive and review team communications and regain situation awareness following interruptions is crucial.

Thus, some of the key capabilities that a collaborative tool for SOF operations must possess are:

1. support for distributed collaboration,
2. support for asynchronous collaboration,
3. support for both mission execution and re-planning in the field,
4. support for annotating and referring to a variety of mission-related representations such as maps, photographs, sketches, timelines and other documents,
5. be physically lightweight (fielded version), and
6. be lightweight in terms of communication bandwidth (fielded version).

Limitations of current networked tools and technologies

Today's networked technologies offer impressive capabilities for connecting distributed team members with each other, command centers, and reach-back assets. Future advances in networked communications will only increase these capabilities. Yet contemporary collaboration technologies, several of which are used by the SOF, are limited in several regards. First, radio, chat and even shared whiteboards are still primarily geared towards real-time collaboration. Second, radio, chat and e-mail are limited to verbal or textual communication, which can make discussing and accurately co-referencing spatial information difficult, time-consuming and error-prone. While many work-around strategies have been developed, such as verbal labels and standardized codes, good spatial representations are hard to match. Third, although whiteboards do allow for annotating and commenting on spatial representations, such as annotating maps with updated situation information and plans, they suffer from other problems. Whiteboards offer little or no control over the presentation or hiding of annotations, and they offer no information about the sequencing of annotations. The result can be a 'visual spaghetti' of annotations that clutter and obscure more than they illuminate. Fourth and finally, none of these technologies facilitate an overview of the collaboration session, nor do they provide information that would allow users to prioritize messages for detailed review. Overview and prioritization, however, are important capabilities for aiding situation awareness recovery following frequent interruptions and intermittent communications during dynamic tasks (Smallman and St. John 2003; St. John, Smallman and Manes 2005; St. John 2008).

SLATE: Integration of innovative collaboration concepts

SLATE was developed as an attempt to solve this mismatch between SOF operation characteristics and contemporary collaboration technologies. A user-centered design process was followed to design the tool. First, as described above, SOF operation characteristics and requirements were elucidated with a highly experienced subject matter expert. Next, two use-cases focusing on collaboration during SOF missions were drafted, both to illustrate the collaboration requirements of SOF operations and to drive the design of SLATE. SLATE's design elements were then storyboarded and progressively refined, driven by these use-cases and drawing on a range of collaboration and knowledge management concepts. These cutting edge concepts range broadly across collaboration conceptualizations, from macrocognitive formulations of team collaboration to the cognitive science of linguistic discourse. Below, these contributing concepts are discussed from broader macrocognitive concepts to finer level descriptions of message exchanges.

1. *Stages of collaborative problem-solving* – how collaboration is involved throughout the problem-solving process: knowledge building, problem-solving, consensus-building and evaluation–revision.
2. *Transactive memory* – how knowledge is distributed, found and shared across teams and organizations.
3. *Spatial annotations* – how spatial information can be conveyed using annotations and embedded within the mission context.
4. *Discourse theory* – how communications are grounded in context, and methods for facilitating contextualization.
5. *Coordinating representations* – how joint activity can be facilitated by using shared representations to structure and guide collaboration activities.
6. *Information encapsulation* – how information can be graphically chunked and visually organized as shared 'information objects' to intuitively and visually promote shared situation awareness and knowledge management.
7. *Grounded transactions* – a framework that we developed specifically to facilitate designing collaboration tools for command and control at the level of individual messages exchanges.

Collaboration stages

The concept of collaboration stages breaks the collaborative problem solving process into four unique but interdependent macrocognitive stages (Warner, Letsky and Cowen 2005; Warner, Letsky and Wroblewski, Chapter 2, this volume). Team members first build their own distinctive information regarding a problem. Then they make their knowledge available to the group in an interoperable, mutually reuseable format, and work collaboratively on the problem until a consensus emerges. The consensus solution may then be evaluated or revised as the team deems appropriate.

SLATE facilitates the information sharing and consensus building processes by making it easy for team members to compose messages and annotate maps and other mission documents directly within the evolving mission context. Similarly, SLATE makes it easy for team members to understand their collaborators' messages because the messages arrive embedded within their mission context. Problem-solving should be more effective and consensus achieved more quickly because the situation is understood better and more quickly by each team member.

Further, the dynamic and agile nature of SOF operations requires capabilities to allow team members to evaluate and revise plans on the fly. The rich spatial messaging of SLATE should allow this new information to be conveyed and understood quickly within the relevant context. Revisions can then be suggested quickly and a new consensus can be achieved. Contemporary collaboration tools that rely primary on text that is disjoint from mission maps and other documents make revision and agile re-planning more difficult.

SLATE is designed to be used across all phases of a mission – planning, execution, and review – and each phase can involve the four collaboration stages. SLATE can be used to problem-solve mission parameters such as ingress and egress routes and reconnaissance positions collaboratively during mission planning. It provides sophisticated capabilities for maintaining team situation awareness over the course of mission execution and for supporting agile, collaborative adaptation of plans as new issues arise. Finally, following a mission, the record of SLATE messages can be replayed to provide a framework for after-action review, reporting, and documentation. This multiphase capability minimizes that need to maintain and transition among different systems for each mission phase.

Transactive memory

Transactive memory further characterizes the nature of collaborative teams and how they exchange information. A central concept is that many teams, from families to corporations to military units, are composed of individuals possessing different expertise. Effective problem-solving requires transacting, or sharing, information among the members of the team; that is, getting the right information from the right people to the right people. Much research has explored how to characterize transactive memory systems, how expertise becomes distributed among members of the system and under what conditions transactions occur or breakdown (e.g., Moreland 1999; Wegner 1986). A key finding is that in unstructured situations, team members tend to discuss information that they have in common rather than sharing information that each holds exclusively (Stasser, Stewart and Wittenbaum 1995). SLATE's annotation capability addresses this problem by allowing each team member to see how each new message relates to the evolving mission context. If team members have a clearer understanding of the current mission situation and any new issues that arise, then they should better understand how their exclusive information can contribute to the group problem-solving process.

Supporting and improving transactive memory should improve knowledge interoperability. This capability should primarily benefit the macrocognitive processes involved in the knowledge construction and team problem solving stages of collaboration.

Spatial annotations

The collaborative benefits of spatial annotations are well known (Heiser, Tversky and Silverman 2004; Kirsh, Chapter 10, this volume). SLATE integrates concepts from this research that allow many types of spatial information, such as planned routes and the locations of forces and objects, to be communicated in their natural medium as spatial annotations, rather than having to laboriously translate them into verbal descriptions. Team members can point to locations rather than finding and spelling out coordinates, and can draw rather than verbally describe a path or landmarks. This natural communication improves the speed of composition, through low effort drawing instead of laborious text entry, and accurate comprehension, through integrated, direct co-referencing instead of disjoint, indirect labeling. For example, Heiser, Tversky and Silverman (2004) found that when participants were able to interact over a common map to plan an emergency rescue route, annotations such as pointing and tracing improved collaboration. Kirsh (Chapter 10, this volume) found that annotations improved participants' ability to convey the current situation and issues involved in a video war game to a collaborator taking over the task. These improvements should enhance a SOF team's situation awareness and ability to respond agilely to changing events.

Spatial annotation, of course, is a feature often found in contemporary collaboration tools. Annotations are used in a variety of software for making and editing documents. One well known example is the track changes and commenting features of Microsoft Word. In these situations, however, annotations are construed as static, primarily textual comments that reside within a document (see for example, Cadiz, Gupta and Grudin 2000). The concept of operation for their use and sharing is that when one user is finished commenting, the annotated document is shared with other users, who then view the set of annotations all at once and revise on their own, at their own pace. SLATE construes annotations as messages that are shared with team members as they are created. This alternative concept of operations turns annotations from a spatial commenting tool into a dynamic spatial messaging tool.

Shared whiteboards also treat annotations as dynamic messages. Used in real time, whiteboard annotations allow team members to exchange spatial messages about problems or missions. However, SLATE offers several advantages over shared whiteboards.

First SLATE, like other advanced annotation systems, allows many types of documents to be marked up using text and drawing annotations. There is a concept of 'digital ink' for creating and sharing drawings and markings within presentations and other documents (Anderson, Hoyer, Wolfman and Anderson 2004). Similarly, SLATE's spatial annotations are not limited to maps. They can be applied to images, hand-drawn sketches, text documents and other representations, with similar benefits. This capability extends the contextualized grounding of messages so that each message can be placed in its most appropriate and useful context.

Second, SLATE integrates spatial annotations with additional annotation modalities to create a more sophisticated and a more effective collaboration tool. Users can speak and simultaneously draw on a shared map in a natural communication style that is easy to compose and easy to understand. Anderson *et al.* (2004) describe a system, for example, for combining text and voice to annotate lecture presentations. They also present a taxonomy of annotation types that are useful for an annotation system to capture, such as attentional marks, diagrams and text.

Third, SLATE makes annotations available on-demand, enabling a task-centric, user-tailorable operational picture (see the section below on information encapsulation). This capability allows users to intelligently declutter the operational picture to focus on the most important annotations and the current problem-solving issues of the team. This capability is not available with conventional whiteboard technologies, which quickly become cluttered and unusable. With whiteboards, users have the option to erase the board, but that blunt operation does not differentiate among annotations, and it cannot be used to tailor the picture for more recent or relevant information. SLATE, in contrast, allows control over the presentation of individual annotations.

Fourth, SLATE offers the ability to extend collaboration using spatial annotations from real-time interaction to asynchronous interaction. SLATE's annotations are separated into individual messages (see the section below on information encapsulation). This separation allows users to reconstruct the sequence of annotations that accrue over time as well as to control which are displayed at any given time. With a whiteboard, if a user is distracted for a time, annotations pile up and can result in a confusing 'visual spaghetti' of annotations that is difficult to parse and understand. The separation and control over annotations that SLATE provides allows users to replay a sequence of annotations in order to understand each additional annotation in turn.

SLATE's annotation capabilities should primarily benefit the problem-solving and consensus-building stages of collaboration. It can improve shared understanding and team pattern recognition by highlighting and emphasizing important information within shared maps and documents.

Discourse theory

SLATE is a communication tool and, as such, it must support the discourse of its users as they communicate. Although the wide literature on psycholinguistics deals almost exclusively with verbal communication, either written or spoken, there are lessons that can be applied from this literature to the design of the primarily visual

SLATE tool. A fundamental concept in the psycholinguistics of discourse is the distinction between topic and comment. The topic provides a context, or grounding, for understanding the comment, which is the new information that a message provides. Much of the initial work performed by participants in a conversation is to provide the context for later comments. Once the conversation is grounded, the topic can be assumed to be understood, and participants can focus on comments.

There are a number of linguistic discourse structures that provide efficient shorthand methods for referring back to mutually understood context (Clark 1985; Gernsbacher 1994; Searle 1976). For example, pronouns hark back to previously introduced topics, and they provide a shorthand for reintroducing those topics. They also signal that the topic was in fact previously introduced and is not entirely new. Effective communication, and likewise, effective collaboration, includes providing efficient means for team members to point to mutually understood context so that message senders can focus on new information and message receivers can efficiently ground it in its context.

SLATE applies these concepts of topic-comment and grounding by embedding annotations within the evolving mission context. This grounding allows comments to build on one another in a natural way. Embedding the annotations within the evolving mission context provides a significant advantage over contemporary email and chat tools in which messages arrive stripped of their context (or the context is embedded in the message body), and where a good portion of the communication involves explicitly re-establishing that context. For example, with SLATE, a user would not have to remind others of a mission component or describe a location before providing the new information. The message itself would be visually integrated with that mission component or location. Further, it is easy for users to review prior topics simply by selecting old annotations and causing them to reappear on the display. All collaboration stages should benefit from the ability to simplify the reference to shared context. It should improve and hasten the development of shared understanding and the convergence of mental models.

Coordinating representations

In addition to supporting a natural visual discourse, SLATE seeks to support the maintenance of shared situation awareness and the coordinated action of team members. SLATE draws on the concept of 'coordinating representations' to accomplish this support. Coordinating representations explore the idea that human conventions and representations can guide and structure collaborative interactions and discourse (Alterman and Garland 2001). For example, the convention of passing on the left facilitates traffic flow on highways and even airport moving sidewalks, where people stand on the right and walk on the left. Similarly, the use of 'over' in radio communications coordinates turn-taking. The conventions save team members from having to renegotiate each interaction or fumble against one another. Much like conversational context, coordinating representations can facilitate communication and collaboration by guiding team members to interact in conventional, expected and efficient ways.

This research suggests that users communicate best, with minimal co-referencing errors, when they share a common workspace, in our case mission representations such as maps and timelines, that can be pointed to and annotated. SLATE incorporates this concept by providing a shared set of visual 'canvases' on which the shared annotations are exchanged. Additionally, the infobs (see below) provide an overview, and they control the sequencing and presentation of messages. Together, the canvases and infobs provide the discourse structures that are the coordinating representations that keep SOF team's 'on the same slate'. Coordinating representations simplify team interactions by making them more standardized, and they should therefore support all stages of collaboration.

Information encapsulation

The concept of information encapsulation derives from the electronic card wall or EWall project (Keel, Chapter 11, this volume; Fleming, Chapter 12, this volume). An idea or fact is encoded onto a 'card' in the form of one or more key words, iconic graphic, and amplifying text – it becomes an information object, or 'infob'. The cards are then arranged on a wall to connote the organization of the concepts. The encapsulation of ideas onto cards or infobs provides an overview of the ideas and facilitates collaboration since all participants can view the cards and quickly grasp their topic and organization. SLATE incorporates this concept by using graphically encapsulated infobs to provide an overview of message content for each new message or annotation.

SLATE innovates on the infob concept to transform it into a useful visual message exchange system. Whereas conventional e-mail provides a subject line, SLATE provides infobs to convey message topic, sender and when it was composed. Infobs for messages are organized in chronological order in a tray along the bottom of the display. This arrangement allows the sequence of messages to be seen easily (see Figure 17.1). The laptop/desktop version of SLATE for command center use supports multiple conversation and mission threads by arranging infobs from each thread into separate trays. This arrangement visually segregates and coordinates the messages within each thread.

When new messages arrive, they are initially displayed as infobs in the appropriate tray. This organization is analogous to the way that the EWall News View (Keel, Chapter 11, this volume) displays news stories as cards that visually encapsulate the content of the stories. This capability overcomes a common dilemma in message and alerting systems: users should be alerted to new information, but not inadvertently distracted from important tasks by low priority messages. SLATE gives users discretion over their attention and helps them to integrate new information in a meaningful, context-sensitive way rather then being driven by the strict order in which messages arrive (McFarlane 2002).

Further, the infob icons avoid a common pitfall of email tools: the frequent vagueness and obsolescence of subject headers. Email threads frequently use the original message's subject header for a series of exchanges that may transform into entirely new topics, rendering the subject header obsolete and non-descriptive.

SLATE's infob icons, on the other hand, are inherently descriptive of their content because they are created automatically for each message.

Once a user decides to read a message, tapping the infob causes the message to 'play'. The correct canvas for the message is displayed, such as a map, photograph, or sketchpad; the spatial annotations appear; and any text or verbal message plays. Tapping the infob again removes the annotations. Through this simple toggle feature, users have control of the presentation of annotations and other message information. Users can retain critical annotations, remove less important or expired annotations and then recall them if needed.

The infobs also support the process of regaining situation awareness following interruptions. Upon returning from an interruption, team members will see infobs for any new messages that arrived during the interruption period. They can then use the infobs to overview and prioritize their access to the new messages. For example, users could quickly search icons containing red symbols for news of hostile forces, or search for other topics of significant and immediate interest. A similar graphical display of critical situation updates, in the time-pressured domain of air warfare, substantially improved both the speed and accuracy of situation awareness recovery compared with alternative message representations (Smallman and St. John 2003; St. John, Smallman and Manes 2005). On the other hand, contemporary chat and shared whiteboard collaboration systems focus on synchronous interactions and offer little support for maintaining or recovering situation awareness across interruptions. When an interrupt ends, and a team member attempts to recover awareness for an ongoing mission and collaboration, whiteboard tools offer a busy tangle of new and old annotations devoid of order and any amplifying text or voice information. SLATE's infob design supports better interruption recovery (St. John 2008).

This capability should primarily benefit the collaboration stage of team problem-solving when collaboration exchanges are more likely to be asynchronous, and interruptions are more common.

Grounded transactions

The collaboration concepts reviewed above provide many innovative ideas for the design of a distributed team communication tool. A final step, however, was needed to ensure that the design facilitated collaboration at the level of exchanging individual messages. We developed a simple framework, called 'grounded transactions', to guide our design and integration of the collaboration concepts described above into the message exchange process. The term 'grounded' refers to the embedding or grounding of messages within their mission or discourse context. The term 'transactions' connotes the focus of the framework on the detailed communication exchanges that occur among team members. Although more micro-than macrocognitive in emphasis, we believe the framework has value for other designers who have to instantiate collaboration tools for macrocognitive processes. The grounded transaction framework divides the message exchange process into five familiar and tractable stages. Consideration of and designing for each stage should facilitate collaboration exchanges overall. In fact, many collaboration tools can be viewed as methods for facilitating one or more of these stages of transaction. The grounded transaction stages are:

1. determining the need for the message,
2. selecting recipients,
3. compositing the message,
4. receiving and interpreting the message by recipients and,
5. responding to the sender.

The first transaction stage is to determine that a message is required in the first place. Although often obvious and implicit, there are occasions when determining that a message is required is a major hurdle to collaboration. For example, there are cases in which a potential message sender holds information that would be useful to others but does not realize the importance of the information or who might benefit from it. One key to crossing this hurdle is knowledge of ongoing missions or other work. Often this knowledge comes in the form of prior collaboration or dialog, but it may also come from access to mission information via briefings, documents, or continuously updated web portals. Accordingly, better representations of mission intent and status may facilitate important collaborations from otherwise disconnected sources. SLATE is designed to support this situation awareness need by keeping team members aware of the evolving situation through the sharing of annotations and the canvases on which they reside.

The second transaction stage is to select the recipients for the message. Again, although selecting recipients is trivial for one-on-one transactions, recipient selection can be tricky, especially when selecting appropriate recipients for broad requests for information and situation updates. For example, a question may be sent to other team members, other organizational elements, outside experts, or broadcast to a broad community of interest. Significant work may go into identifying appropriate organizational elements or experts that may have relevant information. This task is a significant focus of research into transactive memory. Many companies, for instance, have developed elaborate systems for facilitating the sharing of information across organizations by identifying experts and placing questioners in contact with them (Stewart 1997). SLATE simplifies this stage by simply assuming in the context of SOF operations that all team members should be aware of all mission-relevant messages. The distribution of messages to higher command elements, however, is controlled at the operational command center.

The third grounded transaction stage is to compose the message. As indicated by discourse theory, above, a substantial portion of the message may involve establishing context in order to ground the message. The amount and type of explicit reference to context will vary for different recipients, with team members and other frequent collaborators requiring less explicit reference to context than more distant and less frequent collaborators. A key issue for this stage of transactions is how to design an interface to make grounding automatic or implicit so that senders can concentrate on new information, the 'comment', and recipients can correctly interpret the new information in the appropriate context quickly. One design approach is to allow the sender and recipient to share a situation display that provides the context for most transactions. This is the approach taken by SLATE.

The fourth stage is the receipt of the message by the recipient. Ideally, there should be two representations for this stage, though not all collaboration technologies provide both. One representation displays all incoming messages to notify the recipient of their arrival and to help the recipient prioritize the messages for urgency and relevance. Standard e-mail programs provide one example of this type of notification representation. Each incoming message is represented as a row in a table, with a brief text description of the subject, the sender and sender's opinion of the message's importance. Messages also can be sorted automatically by general topic, such as project A and project B, though typically at the cost of reduced visibility of incoming messages. Finally, the arrival of new messages is typically signaled to the user by some type of alert.

The second representation required for the message reception stage is the actual content of the message. Again, the goal is to provide context in such a way that the recipient can interpret the new information of the message as quickly and easily as possible. SLATE's ability to control the presentation of multimedia annotation on mission maps and other documents serves this need (see Figure 17.3).

The fifth and final stage of transactions is the response by the recipient back to the sender. Some transactions do not require responses, and other responses may be simple. However, more complex responses must again be couched in sufficient context for thc original sender to be able to interpret the response. Again, the goal for

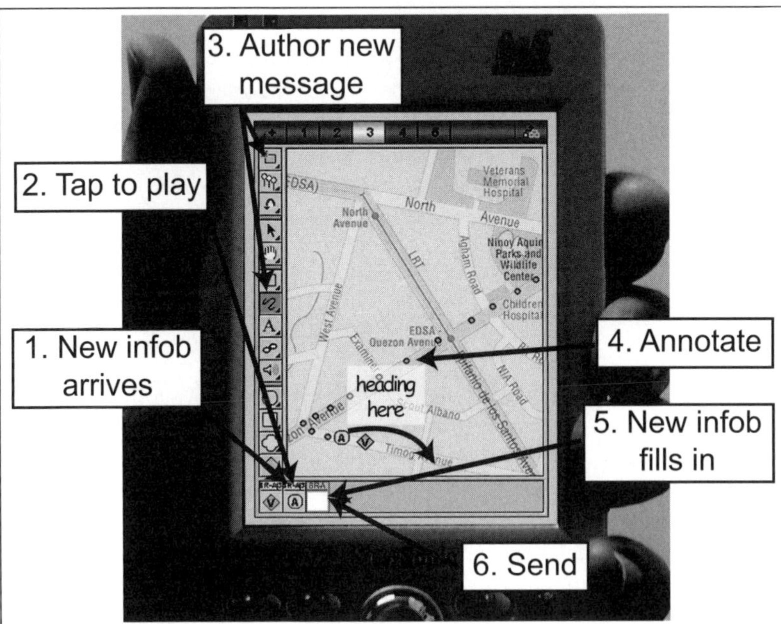

Figure 17.3 Sequence of steps for receiving and viewing an infob, and responding by making a new infob

this representation is to be effective and efficient both for the recipient composing the response and for the original sender receiving the response.

Additionally, from the original sender's point of view, the response must be evaluated not only for its content, but for its pedigree, as well. The original sender will typically want to evaluate the expertise, and possibly the trustworthiness, of the recipient, who in some cases may be unknown to the original sender. Therefore, both the trustworthiness of the information and the responder must be evaluated. The goal for the original sender's representation must be to facilitate this evaluation process.

Beyond these stages of individual transactions, many collaborations require ongoing sequences of transactions. Further, an individual may be involved with several collaborations at any given moment. Consequently, there is also a need to understand and facilitate both the context and content of multiple ongoing collaborations. Additionally, transactions may occur in a synchronous or asynchronous fashion. Queries and responses, for example, may be tightly coupled in a group planning session or decoupled by hours or days. In fact, synchrony or timing is better viewed as a continuum from short intertransaction intervals to longer intervals. Support for transactions must be tailored according to the dominant interval length. For example, chat messaging systems are well suited to synchronous and informal conversations, primarily due to the low interface overhead in sending and receiving messages. E-mail, on the other hand, is better suited to transactions with longer intervals. It has a higher interface overhead, since message windows must be opened, recipients must be selected for each individual message and subject lines must be composed. The selection of recipients, however, is useful for asynchronous transactions which tend to have more varied recipients, and the subject line included with each message is useful for re-establishing context across longer intervals.

A critical focus for SLATE is to provide context for asynchronous collaboration in an automatic and implicit way that keeps the interface overhead of message composition as low as possible. In this respect, grounded transactions should primarily benefit the team problem-solving stage of collaboration, again, because of the greater likelihood of interruptions during this stage. However, designing SLATE to facilitate the grounding of message exchanges to the mission context should improve collaboration across stages.

Summary

SLATE is an innovative collaboration tool designed to address the needs of SOF operations. It integrates several cutting edge collaboration concepts into an effective spatial communication tool to improve team situation awareness and tactical decision-making. In particular, SLATE offers two key innovations over conventional shared whiteboard and chat communication technologies. These innovations are contextually embedded spatial annotations and attention management concepts. These innovations are combined into an intuitive and easy-to-use interface that should enhance sense-making, agile replanning and both synchronous and asynchronous collaboration. The SEAL personnel who have been introduced to SLATE have been universally enthusiastic about its capabilities and potential for use at many command levels, from platoon to tactical and joint operations centers, as well as reach-back assets.

We are currently in the process of developing a functioning SLATE demonstration prototype. As the prototype matures, we will begin testing and refining the design under increasingly realistic conditions, beginning with one-on-one discussions with SOF experts, and then moving to multi-player laboratory scenarios at the Cognitive and Automation Research Laboratory (CARL) at Naval Air System Command, and finally to operational demonstrations at Navy exercises, such as the Trident Warrior/Silent Hammer series. These demonstrations will enable us to evaluate the performance improvements afforded by SLATE's design features that we have outlined here.

Acknowledgments

The authors would like to thank Commander Bradley D. Voigt for his subject matter expertise. Both he and Daniel I. Manes of Pacific Science and Engineering have made important contributions to the design of SLATE. Work described in this chapter was supported by Contract N00014-06-C-0359 from the Office of Naval Research.

References

Alterman, R. and Garland, A. (2001). 'Convention in Joint Activity', *Cognitive Science* 25, 1–23.

Anderson, R.J., Hoyer, C., Wolfman, S.A. and Anderson, R. (2004). 'A Study of Digital Ink in Lecture Presentation', in *Proceedings of the SIGCHI Conference on Human Factors in Computing Systems,* pp. 567–574, Vienna, Austria, 24–29 April 2004. (New York, NY: ACM).

Cadiz, J.J., Gupta, A. and Grudin, J. (2000). 'Using Web Annotations for Asynchronous Collaboration around Documents', in *Proceedings of the 2000 QACM Conference on Computer Supported Cooperative Work*, pp. 309–318, Philadelphia, PA. (New York, NY: ACM).

Clark, H.H. (1985). 'Language Use and Language Users', in G. Lindzey and E. Aronson, (eds), *The Handbook of Social Psychology, 2*, 3rd edn, pp. 179–231. (New York: Harper and Row).

Gernsbacher, M.A. (ed.) (1994). *Handbook of Psycholinguistics.* (San Diego, CA: Academic Press).

Heiser, J., Tversky, B. and Silverman, M. (2004). 'Sketches for and from Collaboration', in J.S. Gero, B. Tversky and T. Knight, (eds), *Visual and Spatial Reasoning in Design III*, pp. 69–78. (Sydney: Key Centre for Design Research).

McFarlane, D.C. (2002). 'Comparison of Four Primary Methods for Coordinating the Interruption of People in Human–computer Interaction', *Human–Computer Interaction* 17, 63–139.

Moreland, R.L. (1999). 'Transactive Memory: Learning Who Knows What in Work Groups and Organizations', in L.L. Thompson, J.M. Levine and D.M. Messick, (eds), *Shared Cognition in Organizations: The Management of Knowledge*, pp. 3–31. (Mahwah, NJ: Lawrence Erlbaum Associates).

Searle, J.R. (1976). 'A Classification of Illocutionary Acts', *Language in Society* 5, 1–23.

Smallman, H.S. and St. John, M. (2003). 'CHEX (Change History EXplicit): New HCI Concepts for Change Awareness', in *Proceedings of the 46th Annual Meeting of the Human Factors and Ergonomics Society*, pp. 528–532. (Santa Monica, CA: Human Factors and Ergonomics Society).

St. John, M., Smallman, H.S. and Manes, D.I. (2005). 'Recovery from Interruptions to a Dynamic Monitoring Task: The Beguiling Utility of Instant Replay', in *Proceedings of the Human Factors and Ergonomics Society 49th Annual Meeting*, pp. 473–477. (Santa Monica, CA: Human Factors and Ergonomics Society).

St. John, M. (2008). 'Desiging a Better Shared Whiteboard: Interruption, Recovery, Message Prioritization, and Decluttering', in *Proceedings of the Human Factors and Ergonomics Society 52nd Annual Meeting.* (Mahwah, NJ: Lawrence Erlbaum Associates).

Stasser, G., Stewart, D.D. and Wittenbaum, G.M. (1995). 'Expert Roles and Information Exchange During Discussion: The Importance of Knowing who Knows What', *Journal of Experimental Social Psychology* 31, 244–265.

Stewart, T.A. (1997). 'Does Anyone Around Here Know…?', *Fortune*, 29 September, pp. 279–280.

Warner, N., Letsky, M. and Cowen, M. (2005). 'Cognitive Model of Team Collaboration: Macro-cognitive Focus', in *Proceedings of the 49th Annual Meeting of the Human Factors and Ergonomics Society*, pp. 269–273. (Santa Monica, CA: Human Factors and Ergonomics Society).

Wegner, D.M. (1986). 'Transactive Memory: A Contemporary Analysis of the Group Mind', in B. Mullen and G.R. Goethals, (eds), *Theories of Group Behavior*, pp. 185–208. (New York: Springer-Verlag).

Chapter 18

JIGSAW – Joint Intelligence Graphical Situation Awareness Web for Collaborative Intelligence Analysis

Harvey S. Smallman

Overview

Intelligence analysts in the US military have never been under such pressure as today. In the wake of 9/11, the most widely publicized intelligence failure since Pearl Harbor, the business of intelligence analysis has received close public scrutiny and repeated calls for reform (e.g., The 9/11 Commission 2004). Further, the two wars that followed 9/11 have been prosecuted against dispersed and rapidly mutating foes – different enemies indeed from the monolithic, slower-moving and more predictable Cold War opponent that much of the intelligence infrastructure was set up to analyze. These two wars have placed an enormous premium on good, actionable intelligence, and analysts have been under pressure to deliver it. Yet the barriers to obtaining good intelligence are extremely high. Systemic problems with intelligence analysis exist in at least three areas related to:

1. the business processes employed to structure it,
2. the characteristics of the technology used to execute it, and
3. the cognitive processes required to conduct it.

Here, we review a project called JIGSAW (Joint Intelligence Graphical Situation Awareness Web) that is designed to solve these problems. By taking a macrocognitive perspective that emphasizes team-level collaboration, we have been able to suggest different interventions than those that have emerged from previous and more microcognitive analyses. JIGSAW represents the application and integration of new collaborative science and technology concepts, many of which were researched and developed in the Collaborative Knowledge Interoperability (CKI) research program of the Office of Naval Research (ONR). JIGSAW was developed with a user-centered design process that incorporated requirements and feedback from analysts serving as subject-matter experts (SMEs) and it has been validated in four separate empirical studies. JIGSAW has the potential to transform the way analysis is performed, making it more systematic and responsive, and a more satisfying group activity, too.

The intelligence cycle: problems with the process

Although cloaked in an aura of mystery and complexity, intelligence analysis is, fundamentally, simply about answering questions. Questions are generated by operational commanders or other senior military or government decision-makers, normally as they plan for or conduct operations. These questions are termed requests for information (or 'RFIs') if they are processed formally, or information requests (henceforth, simply 'requests') if they can be handled informally, usually by locally available analysts. The RFI/request answering process is roughly organized into a cycle, 'the intelligence cycle', which is illustrated in Figure 18.1.

Proceeding around the cycle, an analyst is assigned primary responsibility for answering an RFI/request because of their own or their branch's specialty. Depending on the size and scope of the RFI, other analysts may be assigned to assist and collaborate with the primary analyst on working the problem. The primary analyst then tasks a collections manager to find information relevant to the RFI/request, and the primary analyst may forage for relevant information themself. The information that they find (the 'collections') may come from a variety of sources and are generally classified into types that have specific meaning to the intelligence community.[1] Analysts then integrate and evaluate these collections to try and make sense of the potentially conflicting and ambiguous picture they paint. They then produce a report, in customer-specified format, that hopefully provides a reasoned set of answers to the RFI. This product is returned to the customer for their feedback and their potential follow-on request.

The intelligence cycle is in ubiquitous use across different agencies (government and military), for problems of different scope (from tactical through strategic) and timeline (from next day, through months, years, or with no end date, so-called 'standing RFIs'). The cycle is essentially a service model, with the operational commanders as consumers who are served products by the intelligence community in response to the requests that they generate.

At least three separate cognitive task analyses (CTAs) have been conducted to determine what analysts they do and what tools they need to support them (Badalamente and Greitzer 2005; Hutchins, Pirolli and Card 2004; Patterson, Woods, Tinapple and Roth 2001). All have focused on the detailed microcognitive tasks analysts engage in when processing requests, and have highlighted such issues as analysts' foraging difficulties, time-pressure and information overload. In our own CTA, we conducted structured interviews with 14 San Diego-based current and former analysts to broaden the scope to include issues at the process and team level. By taking a more macro approach, we were able to reveal two new problems related to the collaborative processes required to make analysis both responsive and successful.

The cycle's implied team collaborative processes are not well served by its organization. One problem relates to intra-team collaboration among the community of analysts tasked to respond to an RFI. That team may be composed of a primary

1 The so-called 'INTs', including open-source intelligence (OSINT), signals intelligence (SIGINT) and at least six other types (see Heuer 1999, for review).

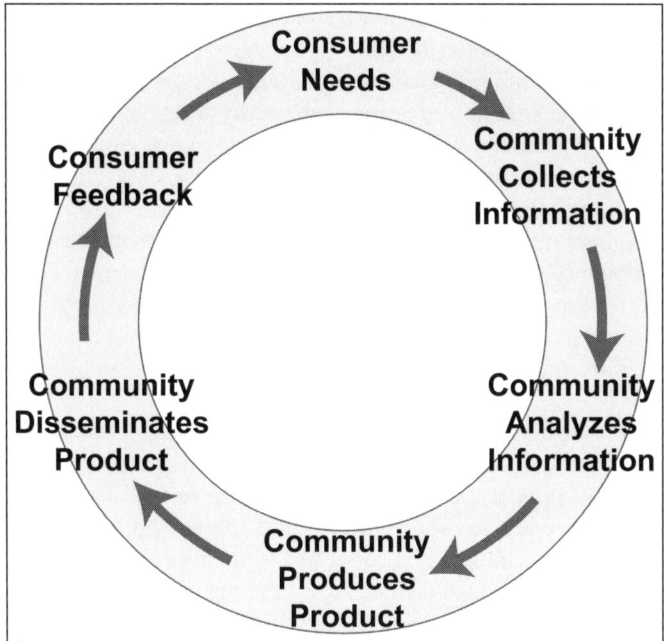

Figure 18.1 The intelligence cycle: The generic business process underlying analysis (modified from Heuer 1999)

analyst assigned responsibility for the analysis, perhaps one or two additional analysts assigned to assist for larger problems, and a collections manager assigned to find and retrieve relevant information. For the intra-team collaboration to work, the primary analyst must articulate what collections are needed to answer the RFI and the collections manager must correctly understand that request and deliver relevant, diagnostic information related to the problem. SMEs reported that this process was flawed, with collections often returning a deluge of material that may be wide of the mark.

The second problem also relates to collaboratively resolving requirements but is inter-team. Because customer feedback is postponed until the end of the cycle, it is critical for analysts to correctly interpret the RFI and its requirements when the request is first passed to the analyst team. Misinterpreting the request or delivering an answer that doesn't meet the customer's needs can lead to time-consuming and wasteful iterations of the cycle. One apocryphal RFI that was independently related by a couple of SMEs was 'how long is the runway on X?' where X represents the name of any small, fictitious island. The apparently appropriate analyst response of '4,600 ft' was greeted with frustration by the customer who replied that he wanted to know which platforms could land on the runway. In this case, he cared a whit about its physical length. Current Navy doctrine requires RFIs to be written as single sentence questions, with elaborating information intended primarily to justify elevating a request to a formal RFI, not to clarify it for the analyst assigned to work it. Thus, problems of misinterpretation are common. A wealth of psycholinguistics has documented the difficulties inherent in correctly interpreting text, particularly when stripped of context (e.g., Gernsbacher 1994). Although SMEs reported some back and forth with a customer for the most obviously ambiguous RFIs, it was hit and miss, focusing on clarifying wording to understand the question – not clarification on the form of acceptable answer. These problems are only likely to worsen in the future with the move to a full network-centric warfare model. Such a model will offload capabilities such as intelligence analysis to geographically dispersed analysts, stripping context that analysts might use to disambiguate RFIs if they were co-located in-theater with the customer, as they often are today.

To share or not to share, that is the question

There are other severe problems related to promoting shared situation awareness and effective collaboration in intelligence analysis that are beyond the scope of what the JIGSAW project can address, but they are relayed here to provide key context to understand the environment that JIGSAW had to be developed in. The 9/11 commission report highlighted the need for intelligence sharing – specifically inter-agency sharing – to 'connect the dots' to prevent future intelligence failures (The 9/11 Commission 2004). But despite a widespread desire for analysts to share and collaborate, there are delicate personal trade-offs, domain-specific clearance and security compartmentalization issues and often severe institutional barriers to full sharing and collaborative analysis.

Personally, analysts may realize that they may benefit from everyone working towards a common goal and leveraging the knowledge, expertise and assistance of others in different organizations. Professionally, though, the currency of analysts' worth is the information they hold, and their value decreases if they 'give away' what they hold close. Of course, each analyst would like to know what each other holds, resulting in a likely posture of, 'Please, after you!' The situation is almost the reverse of Gareth Hardin's Tragedy of the Commons, his famous 1968 game theory thought experiment, because instead of over-grazing the common area, no one wants to even enter the shared commons in the first place. Interestingly, during the drafting of this chapter, news reports surfaced that several three-letter Government intelligence agencies are experimenting with an Intellipedia, their own internal classified version of Wikipedia, the famous online shared knowledge repository, in an effort to begin to address the inter-agency sharing problem (*New York Times* 2006).

On the other hand, the very nature of intelligence, with its multiple classifications and limited releasability, of course factors into whether it can be shared with others due to the risk of compromise. By definition, compromise is inevitable if the circle of people that a piece of information is shared with grows broad enough. When faced with a potentially marginal benefit for sharing versus a potentially catastrophic cost of compromise, the reluctance to share makes perfect sense. However, these factors place additional important constraints on collaboration in this domain.

Some of the contradictions inherent in trying to make sharing and more effective collaboration with classified material work in a strict hierarchical organization are discussed usefully in Hall (1998).

The backdrop of a rapidly changing information landscape: problems with the technology

As a group, intelligence analysts are avid and savvy consumers of information technology (IT). But their IT picture is decidedly mixed. On the one hand, there are a host of extremely capable tools designed for solving specialized analysis problems, such as visualizing and extracting patterns in bank accounts and telephone records, say, in Analyst Workbench. Other systems exist for tracking the assignment of RFIs to different intelligence agencies and for tracking the production of the reports that result. On the other hand, IT support for collaboration is restricted to a conventional array of tools such as e-mail, chat, shared whiteboards and bundles of such components in suites of tools such as InfoWorkspace. However, there are currently no tools developed to support the specific requirements of collaborative intelligence analysis, although there are a couple of related projects, currently in development, that are discussed later.

At least two problems stem from these generic tools related to the complete flexibility they provide. First, the collections that analysts get back are often in a bewildering variety of formats. For example, an analyst may get a mix of soft and hard copies of documents, books, maps, pictures, MS PowerPoint briefs and e-mails – almost as many formats as collection items. This makes interpreting and integrating available intelligence across formats a formidable task. Second, the generic tools provide no common representation to structure analysis and to help foster a shared goal among collaborating analysts working a problem. Again, the future move to a full network-centric warfare model is likely to exacerbate these problems. Maintaining a shared goal and staying coordinated when connected only through networked tools are likely to be significant challenges.

Biases in decision-making: problems in the head

Analysis involves carefully evaluating, integrating, weighing and forming a balanced opinion from often conflicting and fragmentary information obtained from a variety of different sources, often of questionable credibility. Of course, if questions were straightforward to answer, then customers would answer them themselves in the first place and not engage analysts. Complicating matters is that humans show consistent biases in integrating information and decision-making. For example, they may become inappropriately 'anchored' to the information they have at the start of any decision-making task, tend to discount (under-weigh) new and especially conflicting information, and act as if seeking to confirm their initial thinking (the so-called 'confirmation bias'). The problem is reinforced by other biases, such as the positivity bias, which is a general tendency to answer all questions in the affirmative (see Nickerson 1998, for an excellent review).

These heuristics that tend to discount conflicting information are likely to be exacerbated by the *ad hoc* nature and sheer volume of the products that collections deliver and the time-pressure to meet operational deadlines, making analysts especially prone to biases. The CIA, for one, is an agency that is aware of these problems and it attempts to train its analysts about cognitive decision biases (Heuer

1999). However, training may not be sufficient to counter the biases, and the extent of training varies.

JIGSAW: Joint Intelligence Graphical Situation Awareness Web

JIGSAW is intended to support the collective brainstorming of geographically dispersed analysts to allow them to produce timely, unbiased analysis with responsive answers that meet the needs of operational consumers.

Overview of a new concept for collaborative intelligence analysis

The overall aim of the JIGSAW project is to apply advances in collaboration science and technology and seamlessly integrate them together into the design of a new prototype collaborative intelligence analysis system that addresses the six problems identified above.

JIGSAW provides an Analysis Plan Generator (APG) to help analysts interpret an RFI. An analyst uses the APG to break down an RFI to demonstrate their understanding of it, the requirements of an answer and the hypotheses that they need to test. The analyst then sends the plan to the customer for their review and approval. The problem of clearly articulating collections is addressed by laying out these hypotheses in a shared graphical evaluation space called an Intelligence Landscape or I-Scape. This space also addresses the collections understanding problem by creating a structured, labeled visual space for analysts and collection managers to populate. The inconsistent collections format problem is addressed by posting consistently formatted, graphical Intelligence Posts or I-Posts, directly onto the I-Scape. The I-Scapes also help address the collections integration problem by creating an intuitive graphical landscape of all the evidence laid out together for the looking. The decision bias problem is addressed both by making refuting evidence visually salient on the graphical I-Scape and also by retaining the interpretation of the analyst who contributed that information in the spatial location of the I-Post on the I-Scape.

Overall, JIGSAW creates a new coordinating representation (Alterman and Garland 2001) for collaborating analysts that structures and clarifies their discourse and interaction and that serves as a glue to resolve and clarify the intra- and inter-team requirements. Discourse is structured through analysts posting I-Posts to each other, and is clarified through posting onto the shared visual I-Scape.

Below, we first review how JIGSAW was designed through often microcognitive analysis and instantiated design support and then discuss how JIGSAW supports macrocognitive team level collaborative processes.

Approach and target domain JIGSAW's formulation and development followed a user-centered design process. We conducted structured interviews with a group of 14 San Diego-based intelligence analyst subject matter experts (SMEs). These interviews identified the requirements and challenges of collaborative analysis. Next, storyboards for JIGSAW design elements were developed and refined progressively with feedback from the SMEs. In particular, a systematic array of different I-Post layouts and different I-Scape schemes were shown to the SMEs for their feedback at the beginning of the project.

Several JIGSAW design features draw on collaborative research concepts from the Navy's Office of Naval Research Collaborative Knowledge Interoperability (CKI) research program, such as I-Posts which were inspired by EWall cards (see Keel, Chapter 11, this volume). Others, such as the Analysis Plan Generator (APG), were innovations that we struck upon only after conducting a close analysis of the information transactions in our chosen domain. We settled on an Operational Intelligence Cell (OIC) as a domain in which to explore these ideas as it was small and tractable. Finally, a fully functioning web-based demonstration prototype of JIGSAW was developed and refined progressively for both demonstration and evaluation purposes.

Main design features and how they work JIGSAW is composed of three main interconnected software modules which afford end-to-end processing of RFIs for collaborating analysts: see Figure 18.2 (from left to right: the RFI manager, Hypothesis manager and I-Scapes; overall information flow between users is shown with orange arrows). These modules are discussed in turn.

RFI manager

The RFI manager supports receiving, logging, assigning, tracking and filtering the RFIs/requests that the OIC is currently working on. The manager is somewhat akin to a sortable MS Excel table that is used to track the progress of RFIs through the

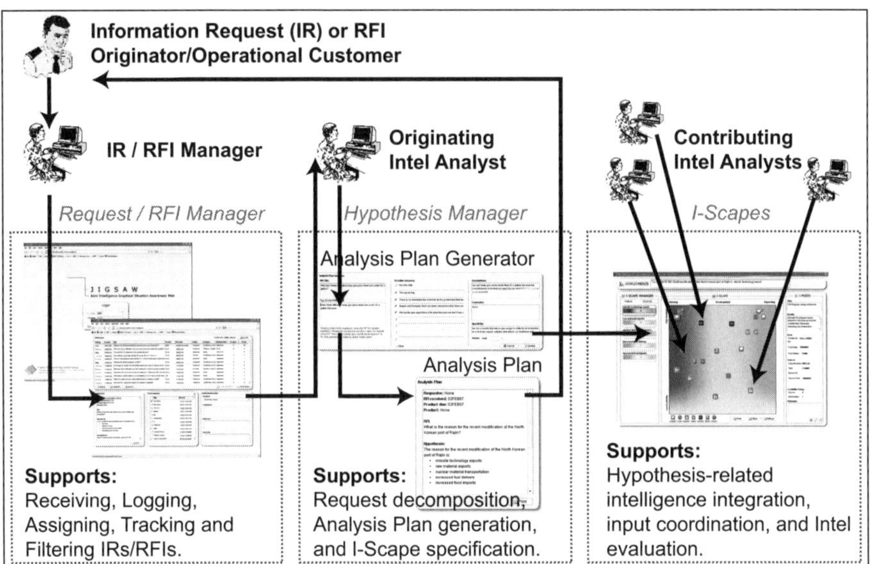

Figure 18.2 Illustration of JIGSAW's three main modules

system. The columns in the table are arranged from left to right in an order that reflects RFI processing, from receipt, to analyst assignment, through I-Scape population, through output product generation. Ordering the columns this way afforded the table a simple task management capability. Although not the focus of the JIGSAW project, the RFI manager was created as a 'glue' to create a complete solution for demonstrating JIGSAW's collaborative capabilities to OIC users. Further, it served as a hub representation to link all of the I-Scapes in an organized fashion.

Hypothesis manager

The hypothesis manager represents a key JIGSAW innovation. It stemmed from the realization that although analysts are often instructed to think in terms of testing hypotheses (Heuer 1999; Badalamente and Greitzer 2005), they receive no explicit assistance and negligible training in the process of turning the RFIs they are assigned, which are questions, into suitable hypotheses to test. This problem of delayed consumer feedback and RFI misinterpretation, combined with the problems discussed above, motivated us to try to solve all at a stroke with a hypothesis manager which is a key nexus in the JIGSAW prototype and concept of operations.

The main feature of the manager is the APG, which supports, decomposing the RFI/request into a series of testable hypotheses necessary to answer it, generating a formatted analysis plan for the consumer which makes explicit the analyst's understanding of the RFI, any assumptions made when interpreting it, and the form the answer to the RFI will likely take, and specifying the number and format of the graphical I-Scapes necessitated by the plan.

The APG is implemented as a structured form that an analyst fills out when first assigned an RFI. The form consists of a series of text and check boxes that lead an analyst through the decomposition in logical order. First, the analyst rewords the RFI to a testable hypothesis, using minimal rewording (e.g., the RFI 'What would the local militia reaction be to a downed pilot south of Kirkuk?' would be reworded to 'The local militia reaction to a downed pilot south of Kirkuk would be:'). Next, the analyst checks a box to indicate whether the RFI is binary (yes/no answerable). If not, they then generate some possible answers that complete the reworded statement (e.g., 'offering shelter', 'holding for ransom', 'execution', etc., continuing the downed pilot RFI example). This externalization helps to focus the analyst on what form an answer might take and how specific that answer may be, and to generate testable statements of fact to be supported or refuted by upcoming collections. At this stage, the number of likely answers and their prioritization, if more than one, can be indicated. Next, the analyst fills in a section of the form to clarify the assumptions they are making when decomposing the RFI. In the downed pilot RFI example, an analyst might clarify their interpretation of 'local' to mean local to Kirkuk, not to their location, the main operation to the south, or the location of the customer posing the RFI. The analyst can also add elaborating comments on the decomposition for the customer and, finally, they state the likely specificity of the answer (e.g., location defined as region, city, or street).

The analysis plan is then sent to the RFI's originator (usually, an operational customer) for their review and approval to establish their buy-in and to ensure that the plan is in line with the intent of their RFI. In the case of the infamous runway RFI, mentioned above, review of an analysis plan listing the likely response as a length to the nearest 100 feet could be redirected to an aircraft platform of a given type by the customer before painstaking and irrelevant detailed satellite imagery collections are ordered up by an analyst. Likewise, in the case of the downed pilot RFI, a customer could clarify that the 'local' referred to Baghdad, the focus of an upcoming operation, and not to Kirkuk, before an analyst headed off down a blind alley by ordering up a detailed analysis from a Kurdish specialist.

The design of the APG was mainly an exercise in applied microcognitive psycholinguistics. Because RFIs are single sentence questions, we were able to apply Art Graesser's taxonomy of inquiry (Graesser and Person 1994) which breaks down inquiries into 1 of 18 categories. For each category, we identified the minimal number of information requests necessary to make of an analyst decomposing an RFI in order to determine what form the answer to the RFI might take. This number turned out to be less than half a dozen, and therefore lent itself to a small series of user queries which we then further rationalized to keep the APG as simple as possible. We also ensured that our testable hypothesis rewording technique was sufficiently general to accommodate exemplars from all 18 categories.

Collaborative intelligence landscapes (I-Scapes)

The I-Scape module is JIGSAW's second main innovation. I-Scapes are graphical coordinating representations that support hypothesis-centered collections, fusion and evaluation. The five main features of I-Scapes are called out in Figure 18.3 and are discussed in turn below (the I-Post symbols on the I-Scape are icons for the different intelligence types but the symbology was user-tailorable).

I-Scapes The actual I-Scape is a shared visual workspace populated with formatted I-Posts for knowledge exchange, akin to an EWall populated with graphical cards, or 'information objects' (see Keel, Chapter 11, this volume). I-Scapes can be roughly conceived as shared whiteboards upon which collaborating analysts and collections can post information to each other about a particular hypothesis component from the AP. However, unlike conventional whiteboards, and even the unstructured EWall, I-Scapes are structured to coordinate shared understanding and collaboration. The graphical space has explicit, task-relevant axes to coordinate both the posting and interpretation of intelligence. Intelligence is multi-attribute data, so we made the two axes the two attributes rated most important by our analyst SMEs. The x-axis is the extent to which an I-Post supports (left) or refutes the hypothesis (right). The background horizontal hue gradient of the I-Scape from red (refuting) through gray (for gray, uncategorized region) to green (supporting) reinforces the meaning of this axis. The y-axis is the credibility of the intelligence from high (top) to low credibility (bottom). The background vertical saturation gradient from highly saturated (credible) to low saturation (less credible) reinforces the meaning of this axis for users.

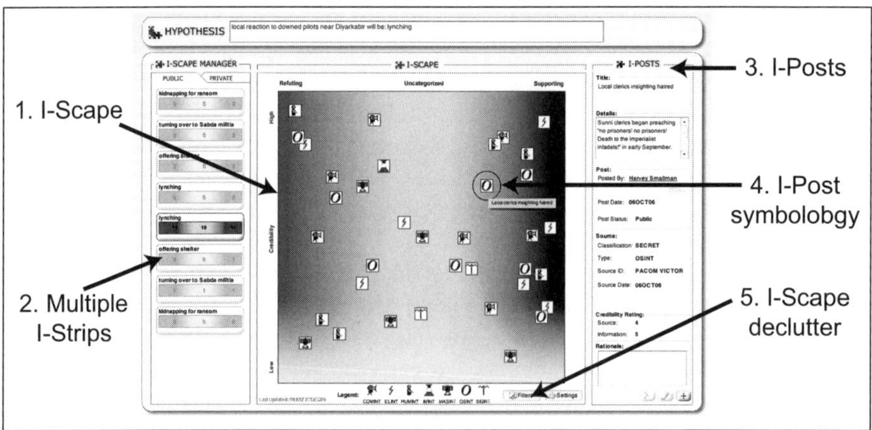

Figure 18.3 I-Scapes and their five main features

Analysts' interpretations of the intelligence they contribute are preserved in order to facilitate sense-making and to debias interpretation. Fleming's DCODE[2] project (Chapter 12, this volume) highlights the value of preserving and graphically encoding interpretations. JIGSAW employs a different graphical encoding scheme from that of DCODE, however. Whereas DCODE employed tags that are emblazoned on, and embedded within each EWall card, here the position of the I-Post on the I-Scape indicates the two most important subjective attributes: the perceived degree of support of that intelligence for the hypothesis in question and its credibility. In this way, multiple I-Posts can be assessed at once, at a glance, without either integrating all of the tags or digging into and integrating each I-Post's contents.

I-Strips Analysis plans will often involve testing multiple component hypotheses, each of which has its associated I-Scape. I-Strips were created to facilitate comparing the evidence distributions across competing I-Scapes associated with the same RFI and for navigating between them. I-Strips are lozenge-shaped summaries of an I-Scape that has had the vertical (credibility) dimension collapsed. The I-Post distribution across the I-Scape is summarized on the I-Strip into three numbers: the number of I-Posts supporting, uncategorized or refuting the hypothesis, see Figure 18.3. By stacking the I-Strips vertically, users are afforded a way to easily compare the evidence distributions across the I-Scapes in order to determine which hypothesis is the best supported, for example. Navigation across I-Scapes is supported by selecting an I-Strip, which populates and foregrounds the selected I-Scape.

I-Posts I-Posts are consistently formatted intelligence posts that transform collected information into hypothesis-relevant knowledge. I-Posts have dual representations. On the I-Scape, I-Posts are shown as graphically encapsulated information objects akin to symbols on a map. Just to the right of the I-Scape a formatted readout for the

2 Decision-making COnstructs in a Distributed Environment.

selected I-Post that functions much like a character readout (CRO) for a geospatial tactical display enabling drill-down into the I-Post's content. The layout of the I-Post readout was derived from the SME cognitive task analysis and places the information rated as more important by the SMEs nearer the top of the record.

Giving the I-Scape a 'hook'n'look' tactical display feel exploited operational users' familiarity and comfort with that display metaphor. Using the word 'landscape' in the I-Scape title was to deliberately reinforce that metaphor in users' minds. Further, the tactical display metaphor for the I-Scape enabled us to integrate HCI research concepts from the Navy's successful TADMUS[3] and successor projects into the I-Scape design for CROs (Kelly, Morrison and Hutchins 1996), variable-coded symbology (Osga and Keating 1994) and declutter (St. John, Smallman, Manes, Feher and Morrison 2005).

Intelligence symbology Creating a new symbology for intelligence was another JIGSAW design innovation. SMEs informed us that a given RFI/request might entail piecing together anywhere from a handful of I-Posts to possibly a couple of hundred of them. Therefore, a user can change the symbols for the I-Posts on the I-Scape to a variety of different forms (a so-called variable-coded symbology, after Osga and Keating 1994), to assist them in the different tasks they put the I-Scape to. I-Posts can be shown as small icons for larger problems, with those icons for the different intelligence types (different icons for signals intelligence – SIGINT, open source intelligence – OSINT, etc., see Figure 18.3). Alternately, the I-Post symbols can show the national flags of the contributing coalition analyst, if the analysis domain were coalition, say. Finally, I-Posts can be shown as a bigger symbol with the national flag and a text string of the intelligence type, more akin to an EWall card. When integrating and evaluating the intelligence, analysts can change the symbology to code it as they desire to begin to see how it relates together, for example, initially beginning by intelligence type and then seeing which country it is from, etc.

I-Scape declutter To further facilitate interpretation and analysis in busy I-Scapes, some simple filtering controls were provided. Instead of turning off certain symbol types completely, we implemented an 'intelligent' declutter that decreases the visual salience of all except the symbol type of interest. For example, in order to find recent I-Posts, an analyst could filter the I-Scape by the time it was posted to simply highlight all I-Posts made within a recent time window, with all the others visually de-emphasized on the I-Scape. The utility of this decluttering technique has recently been documented in the tactical domain (St. John *et al.* 2005).

JIGSAW's support for team collaboration: A macrocognitive perspective

Many of JIGSAW's innovations were derived from close, microcognitive design and analysis. How well do these innovations translate into changes at the level of macrocognitive teamwork? Warner, Letsky and Cowen (2005) have proposed a

---

3 Tactical Decision Making Under Stress.

Structured Collaboration Team Model to characterize the macrocognitive processes a team engages in as it solves problems. Figure 18.4, below, illustrates how JIGSAW's concept of operations fits the four stages of this model, concentrating on the I-Scape's role as a coordinating representation for the intra-team analyst collaboration. First, team knowledge is built by authoring collections as I-Posts. I-Posts transform data into hypotheses-related knowledge by encapsulating them into consistently formatted objects. Second, collaborative problem-solving is facilitated both by the interoperability afforded by I-Posts being in a common currency and by the team shared understanding promoted by the graphical I-Scape representation. Third, the state of team consensus is visualized by the evolution of the dispersion versus clumping of I-Posts on the I-Scape. Fourth, outcome evaluation is promoted by the integrated, graphical evidence landscape (I-Scape) for visualizing and judging the likely veracity of the explicit hypotheses developed using the APG. Book-ending these four stages were inter-team requirements rationalizations accomplished with the design of the new APG that helps establish that the team was even working towards an answer that would be acceptable to the operational customers.

Structured collaboration team model

Given that macrocognition refers to the higher, team-level naturalistic artifact-dependant cognitive processes (Klein *et al.* 2003), it is notable how deeply JIGSAW affects both the tone and currency of those processes. JIGSAW helps foster a shared goal between analysts who now think of themselves as part of a team engaged in principled group hypothesis testing, instead of individuals on the hook for guessing the right answers to short-fuse ambiguous questions. These macrocognitive changes are motivated by a new fabric that JIGSAW weaves to bind analysts together. The weave of that fabric was created stitch by individual stitch through task analysis and attendant design decisions made at a lower, more microcognitive level of analysis necessary to actually design and make JIGSAW. One of the ways we accomplished this harmonized mesh across different levels of analysis was to operationalize and ground every possible information transaction a user could engage in with the system and with another collaborator (see St. John and Smallman, Chapter 17, this volume). The next section reviews our empirical assessments of how well this design process achieved its goals.

Assessing JIGSAW's performance impact

We have conducted four empirical studies, to date, to evaluate how well JIGSAW works, to provide a feedback for design improvements and to quantify its likely performance benefits. The four studies targeted those steps and collaborative interactions at the beginning of the intelligence cycle that JIGSAW was designed to support. The studies employed Naval trainee intelligence analysts, reservists and postgraduate students as participants and either used JIGSAW *in vivo* for extensive periods of time to work realistic analysis problems collaboratively (e.g., Risser and Smallman 2008), or design elements of JIGSAW stripped-out and studied in isolation with carefully constructed analysis vignettes (e.g., Cook and Smallman 2007).

Very briefly, first, the APG was shown to improve the inter-team processes at the customer–analyst interface of rationalizing RFI requirements. Naval reservists role-playing both customers authoring RFIs and analysts breaking them down rated the process and products of the APG superior to baseline, existing processes. Second,

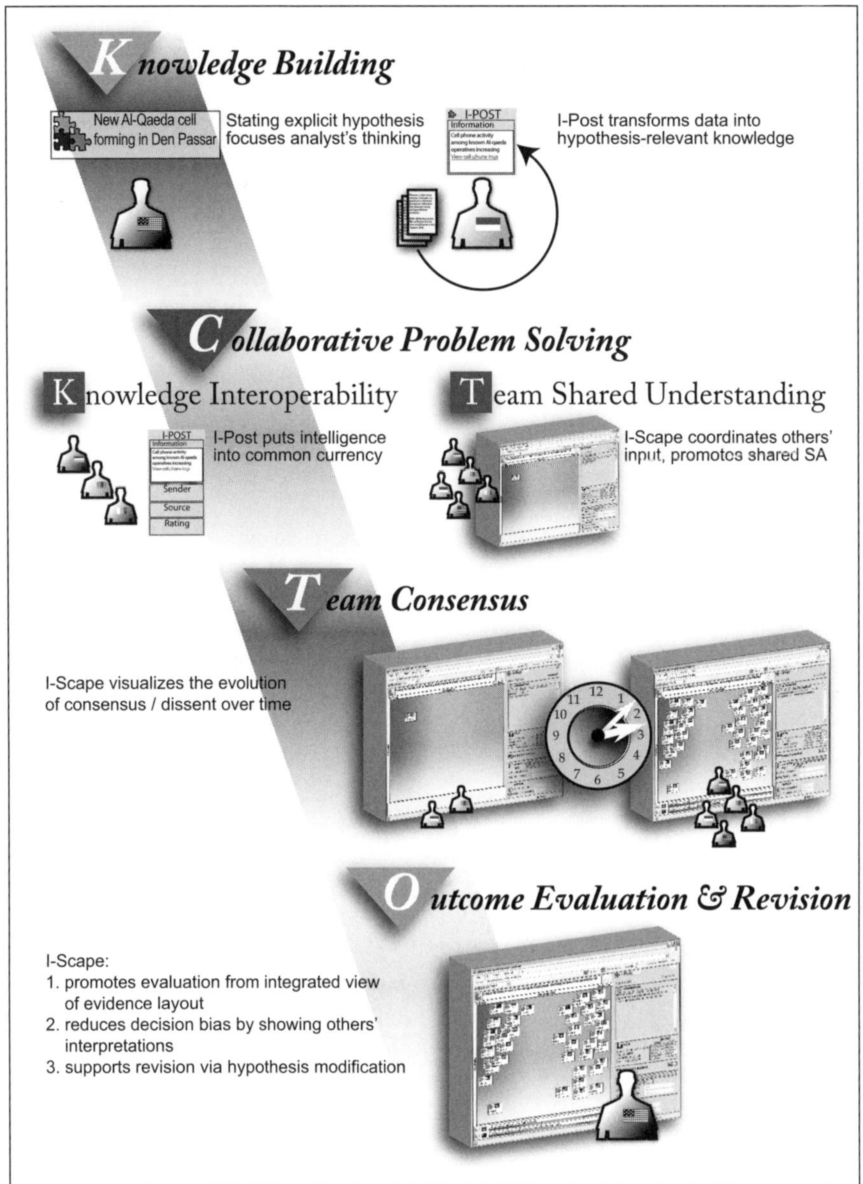

Figure 18.4 Illustration of JIGSAW's support for the four macrocognitive stages of the Warner, Letsky and Cowen (2005) model

consistently formatted I-Posts were shown to speed the interpretation of collections up to 60 percent over informationally equivalent collections shown in today's conventional unformatted email format. Third, we were able to both demonstrate the presence of the confirmation bias in working quasi-realistic analysis vignettes and mitigate it with the graphical I-Scape (Cook and Smallman 2007). Fourth and finally, we were able to investigate how the discussion architecture among collaborators using JIGSAW is likely to influence their re-evaluation of intelligence when it is used realistically over an extended period of time (Risser and Smallman 2008). We showed that allowing all collaborating analysts to view and discuss their I-Posts together promoted more rapid consensus (reduction in the dispersion of I-Posts locations on an I-Scape) than when that discussion was routed through a controlling analyst. Therefore, intriguingly, JIGSAW's collaborative architecture may well require tailoring to the different operational or strategic analysis environments in which it is deployed.

Summary

Specific support for collaborative analysis, reasoning and hypothesis exploration is lacking in analysts' existing IT infrastructures. Currently only a few systems for such exploration even exist. For example, to promote science education, collaborative systems such as Belvedere have been developed (Suthers 1999) so that high school students might explore and visualize evidence–hypothesis relationships. Other representations are being developed for capturing reasoned discourses, such as Toulmin structures, and for visualizing design rationales and argumentation. But the field is still young (for review see Kirschner, Buckingham-Shum and Carr 2003).

Notwithstanding, at least two research groups are developing collaborative tool suites for intelligence analysts. Both are centered around the same framework originally proposed by Heuer (1999) as a solo pencil and paper technique exercise. Heuer developed an alternate competing hypotheses diagram for analysts (the so-called ACH diagram) in which alternate hypotheses for a phenomenon are listed in the columns of a table and relevant evidence points, either pro or con, are listed as pluses or minuses in the rows. Heuer's intent with the ACH diagram was to encourage analysts to search out and exploit diagnostic evidence that discriminates between competing hypotheses. Cluxton and Eick (2005) have incorporated ACH diagrams into a suite of tools called DECIDE for hypothesis visualization. Similarly, Billman *et al.* (2006), have developed a collaborative ACH tool, called CACHE.[4]

Although undeniably useful, ACH diagrams differ from I-Scapes in two key ways. They lack any explicit support for coordinating collaboration between analysts, and they are tabular – not graphical – and therefore may afford a less immediate appreciation of how different types of intelligence relate. To date, no systematic comparison of JIGSAW to DECIDE or CACHE has been conducted, although the time is approaching when such a comparison could and should be made.

4 Collaborative Analysis of Competing Hypotheses Environment.

With emphasis on aiding collaborating analysts by actually showing each of them what they and the others are contributing and what their various individual opinions on their contributions are, JIGSAW has the potential to radically alter intelligence analysts' macrocognitive processes. JIGSAW is more than a tool, then; it represents an entirely new way of approaching the job of intelligence analysis. Web technologies have been radically democratizing for the population in general. It remains to be seen whether they will similarly affect what has been until now one the most individual of all human endeavors. The analysts that we interviewed are certainly conflicted over their desire for JIGSAW for just this reason, although the user-centered design process has ensured that we have delivered largely what they asked for. As we begin transitioning JIGSAW to operational users, we are increasingly aware that we may be just at the beginning of an interesting, larger experiment in the application of collaborative science and technology.

Acknowledgments

Work described in this chapter was supported by Contract N00014-05-C-0356 from the Office of Naval Research. The work represents the efforts and insights of a wider project team of researchers and developers at PSE, to whom the author is deeply indebted. The team included Dr. Maia B. Cook, Chris Griffith, Dr. Nancy J. Heacox, J. Marty Linville, Daniel I. Mancs, Dr. Heather M. Oonk, Dr. Matthew R. Risser and Geoff Williams. Thanks also to Dr. Mark St. John for helpful comments on a previous draft of this chapter.

References

Alterman, R. and Garland, A. (2001). 'Convention in Joint Activity', *Cognitive Science,* 25(4), 1–23.

Badalamente, R.V. and Greitzer, F.L. (2005). 'Top Ten Needs for Intelligence Analysis Tool Development'. Paper presented at the 2005 International Conference on Intelligence Analysis, MITRE Corp., McLean, VA, 2–4 May 2005.

Billman, D., Convertino, G., Pirolli, P., Massar, J.P. and Shrager, J. (2006). 'The CACHE Study: Supporting Collaborative Intelligence'. Paper presented at the 2006 HCIC Winter Workshop on Collaboration, Cooperation, Coordination, Fraser, CO, 1–5 February.

Cluxton, D. and Eick, S.G. (2005). 'DECIDE – A Hypothesis Visualization Tool'. Paper presented at the 2005 International Conference on Intelligence Analysis, MITRE Corp., McLean, VA, 2–4 May 2005.

Cook, M.B. and Smallman, H.S. (2007). 'Visual Evidence Landscapes: Reducing Bias in collaborative Intelligence Analysis', *Proceedings of the 51st Annual Meeting of the Human Factors and Ergonomics Society*, pp. 303–307. (Santa Monica, CA: Human Factors and Ergonomics Society).

Gernsbacher, M.A. (ed.) (1994). *Handbook of Psycholinguistics.* (San Diego, CA: Academic Press).

Graesser, A. and Person, N. (1994). 'Question Asking During Tutoring', *American Educational Research Journal* 31(1), 104–137.

Hall, T.H. (1998). 'Collaborative Computing Technologies and Hierarchical Command and Control Structures'. Paper presented to the 3rd International Command and Control Research and Technology Symposium. Naval Postgraduate School, Monterey, CA, 29 June–1 July.

Hardin, G. (1968). 'The Tragedy of the Commons', *Science* 162, 1243–1248.

Heuer, R.J. Jr. (1999). *The Psychology of Intelligence Analysis.* (Washington, DC: Center for the Study of Intelligence).

Hutchins, S.G., Pirolli, P.L. and Card, S K. (2004). 'A New Perspective on Use of the Critical Decision Method with Intelligence Analysts'. Paper presented at the 9th International Command and Control Research and Technology Symposium. Space and Naval Warfare Systems Center, 15–17 June, San Diego, CA.

Kelly, R.T., Morrison, J.G. and Hutchins, S.G. (1996). 'Impact of Naturalistic Decision Support on Tactical Situation Awareness', in *Proceedings of the 40th Annual Meeting of the Human Factors and Ergonomics Society*, pp. 199–203. (Santa Monica, CA: Human Factors and Ergonomics Society,).

Kirschner, P.A., Buckingham Shum, S.J. and Carr, C.S. (eds) (2003). *Visualizing Argumentation: Software Tools for Collaborative and Educational Sense-Making.* (London: Springer-Verlag).

Klein, G., Ross, K.G., Moon, B.M., Klein, D.E., Hoffman, R.R. and Hollnagel, E. (2003). 'Macrocognition', *IEEE Intelligent Systems* 18 (3), 81–85.

New York Times (2006). 'Open-Source Spying', 3 December, pp. 54–60.

Nickerson, R.S. (1998). 'Confirmation Bias: A Ubiquitous Phenomenon in Many Guises', *Review of General Psychology* 2, 175–220.

Osga, G. and Keating, R. (1994). *Usability Study of Variable Coding Methods for Tactical Information Display Visual Filtering.* Naval Command, Control and Ocean Surveillance Center, San Diego, CA. Technical Document 2628, March 1994.

Patterson, E.S., Woods, D.D., Tinapple, D. and Roth, E.M. (2001). 'Using Cognitive Task Analysis (CTA) to Seed Design Concepts for Intelligence Analysts under Data Overload', in *Proceedings of the 45th Annual Meeting of the Human Factors and Ergonomics Society*, pp. 439–443. (Santa Monica, CA: Human Factors and Ergonomics Society).

Risser, M.R. and Smallman, H.S. (2008). 'Networked Collaborative Intelligence Assessment'. Paper presented to the 13th International Command and Control Research Technology Symposium, Bellevue, WA, 17–19 June.

St. John, M., Smallman, H.S., Manes, D.I., Feher, B.A. and Morrison, J.G. (2005). 'Heuristic Automation for Decluttering Tactical Displays', *Human Factors* 47(3), 509–525.

Suthers, D.D. (1999). 'Effects of Alternate Representations of Evidential Relations on Collaborative Learning Discourse', in C.M. Hoadley and J. Roschelle, (eds), *Proceedings of the Computer Support for Collaborative Learning (CSCL) 1999 Conference*, pp. 611–620. (Palo Alto, CA: Stanford University).

The 9/11 Commission (2004). *The 9/11 Commission Report: Final Report of the National Commission on Terrorist Attacks Upon the United States.* (New York, NY: W.W. Norton and Company) .

Warner, N.W., Letsky, M.P. and Cowen, M.B. (2005). 'Cognitive Model of Team Collaboration: Macro-cognitive Focus', in *Proceedings of the 49th Annual Meeting of the Human Factors and Ergonomics Society*, pp. 269–273. (Santa Monica, CA: Human Factors and Ergonomics Society).

Chapter 19

The Collaboration Advizor Tool: A Tool to Diagnose and Fix Team Cognitive Problems

David Noble

Introduction

The Collaboration Advizor™ Tool (CAT) is an expert system that helps teams diagnose and fix cognitive problems that can impede effective teamwork. The premise of CAT is that team members need specific kinds of knowledge in order to work together effectively, that teams whose members lack this knowledge often encounter certain predictable kinds of problems (Wegner 1987), and that once diagnosed there are effective ways to correct the knowledge deficiencies responsible for the problem and thus improve team effectiveness.

As part of the Office of Naval Research (ONR) research program to support collaboration in quick-response *ad-hoc* teams, CAT is designed to support teams in time compressed, high stress, dynamically changing environments supported by uncertain data. It is also designed to address the kinds of teams becoming increasingly important to the military: geographically distributed, asynchronous and culturally diverse.

Though originally designed to be focused on a specific military environment, crisis action planning in the Pacific area of operation, CAT has been tested with several different kinds of teams. These include a multicultural student crisis action team at the Naval PostGraduate School, two human factors research groups in the United Kingdom and a product development team and corporate support team, both at Evidence Based Research. In these cases, the team used the tool to diagnose problems and to identify candidate solutions. Several of the teams implemented the suggested remedies, and according to retests corrected the problem.

CAT addresses only the cognitive causes of poor teamwork. Cognitive problems can cause serious team shortfalls, being the cause, for example, of the Bay of Pigs fiasco in 1962 and the Vincennes Iranian airliner shootdown in 1988. There is an extensive literature of the cognitive basis of collaboration (e.g., Argote and Ingram 2000; Beach and Mitchel 1987; Canon-Bowers and Converse 1993; Janis 1972; Kirkman and Shapiro 1997; Liang *et al.* 1995; Mathieu *et al.* 2000; Noble 2002, 2003a, Noble *et al.* 2004) that CAT builds on. However, cognitive problems are not of course the only reason why teams can encounter problems. Teams can also experience difficulties because they lack adequate resources, because they have poor leadership, because team members don't like one another, or because they do not agree with the team objectives. There are extensive resources that address the social aspects of teamwork (Katzenbach and Smith 1993, 2001; Brounstein 2002; Herbelin 2000; Maxwell 2001), while planning procedures address resource requirements.

This chapter has two major sections. The next section reviews the cognitive framework that CAT uses to make its diagnoses and recommendations. That section describes CAT's twelve categories of team knowledge. It also describes the framework that CAT uses to relate knowledge to impediments of obtaining needed knowledge and to symptoms of knowledge shortfalls. The third section briefly reviews the tool itself. It describes how team members give CAT information about team characteristics and behaviors. It then shows the kinds of output CAT provides to help teams diagnose and fix cognitive problems.

CAT cognitive framework

Twelve knowledge enablers

CAT organizes the knowledge that teams need to work together effectively into twelve categories. These are CAT's twelve knowledge enablers. This is the knowledge that teams draw on in each of the phases described in the ONR team collaboration model described in Chapter 1: knowledge construction, collaborative problem-solving, team consensus and outcome evaluation and revision. Generally, all types of knowledge are important in each of these phases, though in some cases certain kinds of knowledge are of greater importance in some phase than another. For example, the enabler 'plan assessment', which is understanding whether a plan will still enable the team to reach its objectives, is most prominent during the outcome evaluation and revision stage.

There are many different kinds of knowledge that can contribute to effective teamwork and which can be critical to team effectiveness under various conditions. This knowledge includes, for example, knowing the team's goals and plans, knowing how one task impacts another, knowing where team members agree and disagree and the knowing the team's rules for sharing information. In order to help people be aware of all of the different types of knowledge important to team success, CAT organizes the needed team knowledge into twelve 'knowledge enablers'. The first six, 'team preparation', comprise the foundational knowledge that tends to build and change slowly over time. This is the knowledge that accounts for team members being able to work together more effectively as they get more experience working together as a team. The second six, status assessment and decision-making, are the real-time knowledge and understandings that can change dramatically instant to instant. This is the knowledge and understandings that enable people to react quickly to changing circumstances.

Of course, these twelve enablers are not the only way to organize needed team knowledge (Mathieu *et al.* 2000). These twelve classes of knowledge are based on numerous case studies and extensive literature review. In addition, they have additional attributes that make them well suited to CAT's needs, they are:

1. provide a level of diagnosis that points to concrete actions able to improve team performance;
2. are easy to understand;
3. map reasonably cleanly onto knowledge acquisition risks and behavioral symptoms of team problems, and
4. as a set, account for key team behaviors, such as team agility, team member backup, accountability, and coordination.

The Collaboration Advizor™ Tool, the expert system that helps teams diagnose and fix knowledge-based collaboration problems, builds on this organization of knowledge. This organization has worked well for this tool. All of the teams that tested the tool during its development found this knowledge organization useful.

The twelve enablers are:

1. *Goal understanding* encompasses understanding team mission, the goals of the client, the criteria for evaluating team success and the criteria for evaluating task progress. Understanding of team objectives includes understanding both the explicit and implied goals of the team, taking into account the cultural norms of the tasking authority. Goal understanding may be the most important of the knowledge enablers, for teams that do not know their goals are unlikely to achieve them.
2. *Understanding that plan,* including roles, tasks, and schedule is the 'surface' understanding of the plan. Project plans usually decompose the team's work into separate tasks, assign these tasks to individuals or groups of people and then specify a schedule. The plans may specify team member responsibilities, to include both fixed and context dependent leadership roles, principal task performers and task back-ups.
3. *Understanding of relationships and dependencies* is the 'deeper' understanding required to project success and make adjustments between tasks, resources, time, information and the situation. The dependencies that are important to understand are the temporal, spatial and causal (logical) relationships between separate tasks and between tasks and goals, information, resources and the external situation. This understanding enables team members to predict the consequences of resource or information shortfalls, or of inadequately performed or delayed tasks.
4. *Understanding of other team members' backgrounds and capabilities* includes knowing other team members' values and decision criteria, their capabilities and knowledge and their level of interest and engagement. This knowledge enables other team members to predict what people can and will do under various circumstances. It is the cognitive basis of trust.
5. *Understanding of team 'business rules'* includes both formal and unspoken rules by which team members work together. These are the rules for:

   (a) talking, listening, brainstorming and hearing outside perspectives at meetings;
   (b) critiquing and editing;

(c) offering/asking for help and information;
(d) providing performance feedback;
(e) setting up meetings (how to schedule, who to invite); and
(f) cc'ing and broadcasting.

Poor business rules contributed significantly to the Bay of Pigs failure described earlier.

6. *Task knowledge* is the knowledge team members need to do their individual tasks. No matter how well they know how to work together, teams cannot be successful if the individual team members lack the skills and knowledge to carry out their parts of the job. Task knowledge includes knowing how to perform assigned tasks, how to find and access documented information, how to use support tools and how to find and access people with needed knowledge.
7. *Activity awareness* is knowing what others are doing, how busy they are, their level of engagement, if they are getting behind or over their heads and if they need help with their workload. Activity awareness, sometimes called 'team transparency', is essential for catching team problems quickly and for enforcing individual accountability.
8. *Understanding of the external situation* is appreciation of everything outside of the team that can impact its work. In military operations it includes the actions of the adversary. In business it may include the actions of competitors and the preferences of customers. Understanding the external situation includes knowing who the significant players are and knowing their status, capabilities, strengths, weaknesses, behaviors objectives and plans.
9. *Task assessment* is determination of what tasks are being worked on and by whom, the status of these tasks, comparison of tasks status with the status called for by the plan and judgment of the adequacy of available information and resources. It includes an assessment of progress and prospects for task success, including an estimate of whether a task needs help and whether required resources and information are available. Task assessment allows everyone on the team to dynamically adjust their work when other tasks progress either better or worse than planned.
10. *Mutual understanding* addresses the extent to which team members know how well they understand each other. It includes the extent to which team members are aware of where and why they agree or disagree about team goals, team progress, the external situation and all the other team knowledge enablers. It does not include everyone agreeing on everything. There is no requirement that all team members always agree. There is a requirement, however, that they know when they do not agree.
11. *Plan assessment* is an estimate of whether the current team, processes, plans and resources will still enable the team to achieve its objectives. It builds on and integrates assessments of team activities, task progress, the external situation and degree of mutual understanding. Unlike a task assessment, which focuses on how well individual tasks are progressing, plan assessment

considers all current factors and projections into the future to estimate the need for plan adjustments. Plan assessment is essential to teams adapting to changing circumstances.

12. *Understanding of decision drivers* includes grasping all of the factors that must be considered when making a decision. These include knowing what can impact the effectiveness of a decision, knowing the decision deadlines, picking the right strategy for decision-making under uncertainty, and knowing who should be consulted when deciding what to do.

The full description of this needed team knowledge (Noble 2004a) discusses these enablers in much greater depth. This reference also provides for each a more complete description of the knowledge elements, the possible consequences to a team should the knowledge be missing, impediments for obtaining the knowledge and symptoms for not having it.

However, even the simplified description in this chapter can help teams diagnose and fix problems. Just being aware of the fact that a team's problems may have cognitive causes and then considering which of these twelve enablers might be responsible can trigger ideas about how to fix the problems. If the team does not understand its goals, it can have a meeting to discuss its objectives and the constraints that must be observed. If two people do the same work unnecessarily and the team thinks the problem was a poor understanding of the plan, then the team can meet to review the plan, post the plan in a prominent place, or draw a diagram summarizing team members' responsibilities. If the team thinks that the cause of the redundant work was poor activity awareness, then it can arrange to have team members make periodic progress reports to each other. It can also set up a central repository for team members to post their products.

Risks, symptoms and diagnoses

CAT builds on a simple framework that relates team knowledge, team behaviors and the quality of team products or action effectiveness. In this model (Figure 19.1) new information augments team members' current understandings, impacting the knowledge that team members need to work together effectively ('needed team member knowledge' in the figure) which in CAT is organized into the twelve enabler categories. This knowledge mediates effective team behaviors, which then impacts team acquisition of additional knowledge and the quality of team output. When the team product is a subset of team knowledge (e.g., the product is an intelligence analysis), then the team member knowledge also contributes directly to the product.

In this framework, a knowledge risk is anything that makes it harder for a team to acquire the needed knowledge. For example, being geographically distributed hinders team members understanding each other (enabler 4) and activity awareness (enabler 7). Behavioral symptoms are behaviors that indicate a knowledge lack. For example, asking the wrong team member for information indicates incomplete knowledge about that person's capabilities and responsibilities, reflecting possible deficiencies in understanding each other (enabler 4) and/or in enabler 2 (knowing the plan). Symptoms are often ambiguous, ambiguity which CAT is designed to handle.

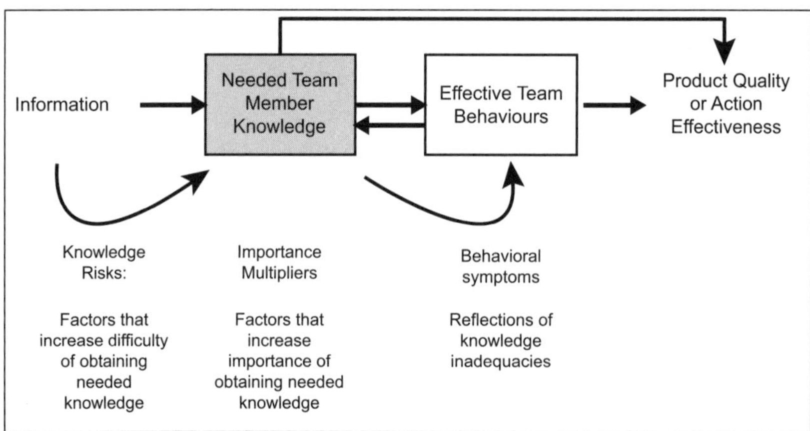

Figure 19.1 Representation of knowledge risks, importance multiplier, and behavioral symptoms in CAT

'Importance multipliers' modify the inherent weight CAT applies to the importance of a kind of knowledge, given the mission and environment of a team. For example, understanding the external situation (enabler 8) is much more important to military teams whose mission is to understand the situation than it is to some other teams.

CAT knowledge bases

CAT captures this knowledge about knowledge enablers, risks, symptoms and importance multipliers in its knowledge base. This knowledge base specifies for each of its twelve knowledge categories the knowledge risks imposed by various environmental factors and by different kinds of team characteristics. The knowledge base also specifies the extent that specific kinds of team behaviors are symptomatic of a knowledge deficiency in each of the twelve categories. In addition, the knowledge base stores methods that teams can use to correct knowledge deficiencies.

CAT functions

CAT life cycle

The Collaboration Advizor™ Tool is intended to support teams as they develop over months and years, with the team using the tool about every six weeks. Each time the team uses the tool, team members answer a set of questions that CAT uses to diagnose possible team cognitive shortfalls, which team members then discuss to assess the significance of the diagnosis. If team members agree that an issue is significant enough to address, they ask the tool to suggest ways to correct the cognitive problem. The team then discusses CAT's suggestions, and may pick one or more of the suggested remedies to implement. That completes use of the tool for that cycle. After a few weeks, the team will use the tool again to assess the team's progress in correcting previously diagnosed shortfalls and to detect any new possible cognitive issues.

This chapter briefly reviews highlights of CAT features. The DTIC report *Understanding and Applying the Cognitive Foundations of Effective Teamwork* (Noble 2004a) describes CAT features more fully.

Collecting team data

To make its diagnoses, CAT presents a list of statements to team members individually and privately and asks each team member whether they agree or disagree with the statement. These statements concern the nature of the team, the characteristics of its environment, and the various kinds of team behaviors that can be symptomatic of team cognitive deficiencies. After all team members have specified their agreements or disagreements with the statements, CAT consolidates the answers to create a team diagnosis, drawing on static data in the knowledge base and on its diagnosis algorithms.

Overall team diagnosis

Figure 19.2 is a screen shot of the overall CAT team diagnosis. The bar graph in the middle of the diagram indicates the diagnosed strengths and weaknesses of each of the twelve knowledge categories. Black bars to the left indicate potentially serious deficiencies in the knowledge areas. Gray bars to the right represent knowledge areas that are doing well. Light gray bars represent knowledge areas with potential problems.

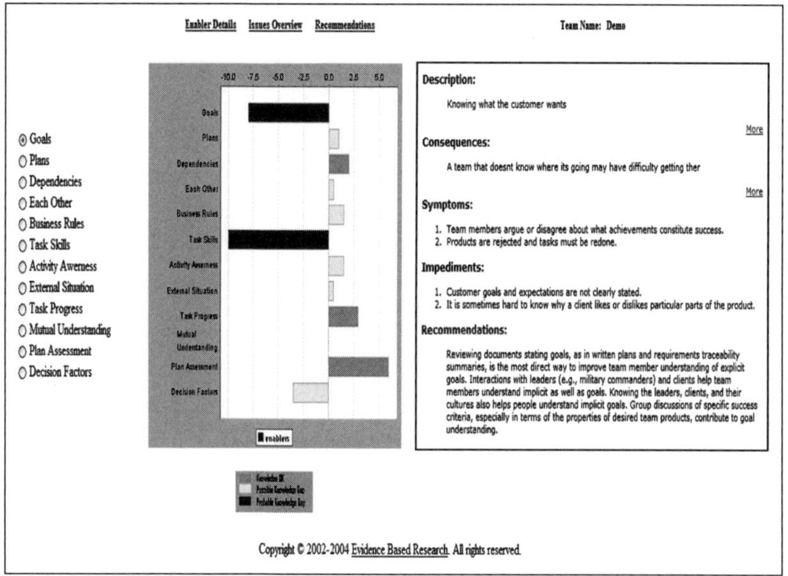

Figure 19.2 Overall team diagnosis

The left side of the chart lists the twelve knowledge enablers, associating each with a radio button. Pressing the radio button for any knowledge area generates a brief top-level summary of that enabler as it applies to that team. In this case, the chart summarizes the enabler 'goals'. The chart briefly describes what the enabler is about ('knowing what the customer wants') and the consequences of a team's not understanding its goals ('a team that doesn't know where it's going may have trouble getting there'). In both cases, clicking on 'more' elaborates on the description of consequences. The 'symptoms' are the team behaviors that caused CAT to flag goal understanding as a possible knowledge deficiency. The 'impediments' are the characteristics of the team, task and work environment which CAT believes would impede the team from understanding its goals. In this case, it is the teams' perception that it is hard to know what the client actually wants.

Diagnosis drill-down

Figure 19.3, generated for another team, illustrates the details CAT provides about a specific enabler, in this case goals. This chart lists the statements associated with a particular knowledge area, and shows the number of team member who agreed or disagreed with each statement. The pull down box under the label 'questions' indicates that in this case the team has limited the list to include only statements related to behavior symptoms of knowledge problems. For each statement CAT shows the number of team members who agreed or disagreed with the statement, thus revealing both what fraction of the team believes there's a problem connected to the statement and also showing the areas of agreement and disagreement within the team. CAT's ability to reveal areas of agreement and disagreement can highlight

Questions for: Goals

High Agreement : There is a problem.
Poor Agreement : There may be a problem.
High Agreement : There is not a problem.

Rank	Questions (Symptoms)	Agree	Disagree	Associated Enablers	Recommendation
1	Team members argue or disagree about what achievements constitute success.	3	0	Goals	Discuss, articulate, and write down success criteria ⊞
2	Team members are sometimes surprised by a client's feedback.	4	10	Task Skills,Goals	Clarify client's goals and expectations ⊞
3	Some team members feel that others do not fully pull their own weight.	2	1	Goals,Activity Awerness,Task Progress,Each Other	Discuss team member responsibilies and performance standards ⊞
4	Team members are uncertain of the criteria for a job done well.	1	2	Goals,Plans,Task Progress,Plan Assessment,Decision Factors	⊞
5	People do not know when they have successfully completed a task or project.	1	2	Goals,Plans,Dependencies,Task Progress,Plan Assessment	⊞
6	People sometimes do much more work on a task than is necessary to support team goals.	1	2	Goals,Plans,Dependencies,Task Progress	⊞
7	People carry out actions that are no longer needed.	1	13	Goals,Plans,Dependencies,Activity Awerness	⊞
8	Products are rejected and tasks must be redone.	0	3	Goals,Task Skills	Clarify criteria for product quality and task success ⊞

Figure 19.3 Team responses to team questions about goals

areas where team members were not aware they disagreed and can often generate discussions within the team which helps the team reach consensus on an issue.

In this example, the team had fourteen members. Four of these team members agreed with the statement 'team members are sometimes surprised by a client's feedback', and ten members disagreed with the statement. This level of disagreement can generate a discussion of the specific instances in which team members were surprised. This discussion can then help the four team members agreeing with the statement to understand the client's requirements better. Note that because team members may see different statements, the total number of people agreeing and disagreeing with a statement may vary from question to question.

The column labeled 'associated enablers' lists all of the knowledge areas that could be associated with each statement. Hence, being surprised by client's feedback might indicate that a team member is unfamiliar with a task. Alternatively, it could also indicate that the team member does not understand a client's goals. The 'recommendations' column lists one possible way to address the problem associated with a statement. Clicking on the '+' sign generates CAT's full list of possible remedies.

Prioritized issues

Often team members wish to quickly focus on the overall key issues facing the team, rather than examining the issues associated with each knowledge area specifically. Figure 19.4 lists the statements which CAT estimates to be the most important for the team to consider, considering the inherent importance of an issue to teams in general,

High Agreement : There is a problem.
Poor Agreement : There may be a problem.
High Agreement : There is not a problem.

Rank	Questions	Agree	Disagree	Associated Enablers	Recommendation
	Discussion				
1	Some team members feel that others do not fully pull their own weight.	2	1	Goals,Activity Awerness,Task Progress,Each Other	Discuss team member responsibilies and performance standards ⊞
2	Team members argue or disagree about what achievements constitute success.	3	0	Goals	Discuss, articulate, and write down success criteria ⊞
3	The team has multiple competing/conflicting goals.	8	6	Goals,Decision Factors	⊞
4	There are no specified criteria for mission success.	7	7	Goals,Plans	Review mission goals and desired outcomes ⊞
5	It is sometimes hard to know why a client likes or dislikes particular parts of the product.	10	4	Goals	Review with customer how he/she wants to use the product ⊞
6	Customer goals and expectations are not clearly stated.	9	5	Goals	Discuss, articulate, and write down customers goals and expectations ⊞
7	Team members are uncertain of the criteria for a job done well.	1	2	Goals,Plans,Task Progress,Plan Assessment,Decision Factors	⊞
8	People do not know when they have successfully completed a task or project.	1	2	Goals,Plans,Dependencies,Task Progress,Plan Assessment	⊞
9	Team members are sometimes surprised by a client's feedback.	4	10	Task Skills,Goals	Clarify client's goals and expectations ⊞
10	People sometimes do much more work on a task than is necessary to support team goals.	1	2	Goals,Plans,Dependencies,Task Progress	⊞

Figure 19.4 Key team issues

the specific problems the team is encountering and how team members responded to the statement when CAT was gathering the diagnostic information from team members.

In the example of Figure 19.4, the team wished to order the issues by the importance of discussing them. The top-rated discussion issue was 'some team members feel that others do not fully pull their own weight'. Only three team members saw this statement during the information-gathering phase, and two of those three agreed with it. This ability of CAT to flag sensitive issues which team members might otherwise be reluctant to raise is one of the most important ways that CAT can help solve team problems.

Recommendations

As in Figure 19.4, the statements listed in Figure 19.3 are linked to recommendations for methods of dealing with the issue. Clicking the '+' sign next to 'discuss team member responsibilities and performance standards' generates the recommendation list shown in Figure 19.5. These recommendations are drawn from multiple sources, including collaboration and teamwork self-help books, military doctrine, theories of decision-making and command and control. Each of the items in this recommendation list suggests a way for the team to address this issue. In this case, some of them suggest direct team discussions while others suggest ways for team members to be more aware of each other's contributions.

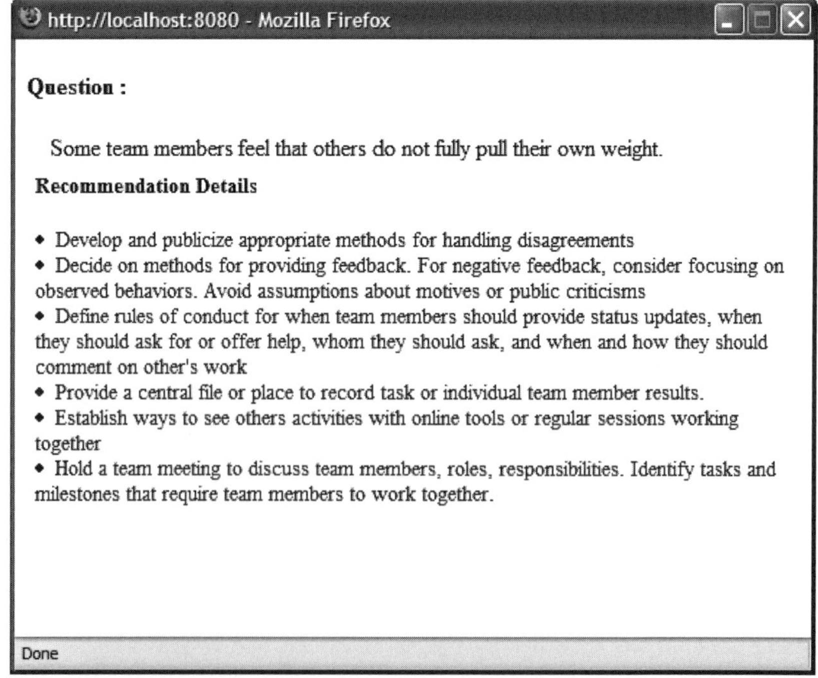

Figure 19.5 Recommendations

Summary

By enabling team members to leverage each others' perspectives, experience, expertise and imagination, collaboration helps teams make better situation assessments, plans and decisions. While collaboration is potentially highly beneficial, not all teams are effective. Sometimes when they are not, the problem is cognitive. Team members do not know what they need to know in order to work together effectively.

The Collaboration Advizor Tool (CAT) is an expert system that helps team diagnose and fix collaboration issues with cognitive causes. CAT's framework for understanding the knowledge foundations for effective collaboration describes the specific kinds of knowledge that teams need for effective teamwork, organizing this knowledge into twelve knowledge enablers, and specifying for each the team, environment and task impediments to acquiring the knowledge needed for effective teamwork.

Teams use CAT over several months to monitor their team's cognitive health, to identify cognitive problems and to identify possible remedies to these problems. In this way, teams can identify cognitive-based problems early, so that they can address these problems before they seriously undermine the team's accomplishment of its goals.

References

Argote, L. and Ingram, P. (2000). 'Knowledge Transfer: A Basis for Competitive Advantage in Firms', *Organizational Behavior and Human Decision Processes* 82(1), 150–169.

Beach, L.R. and Mitchell, T.R. (1987). 'Image Theory: Principles, Goals, and Plans in Decision Making', *ACTA Psychologica* 66(3).

Brounstein, M. (2002). *Managing Teams for Dummies*. (Indianapolis, IN: Wiley Publishing, Inc.).

Canon-Bowers, J.A. Salas, and Converse, S.A. (1993). 'Shared Mental Models in Expert System Team Decision-Making', in M.J. Castellan, (ed.), *Individual and Group Decision Making: Current Issues*. (Hillsdale, NJ: Lawrence Erlbaum Associates, Inc.).

Herbelin, S. (2000). *Work Team Coaching*. (Riverbank, CA: Riverbank Book).

Janis, I.L. (1972). *Victims of Groupthink*. (Boston, MA: Houghton Mifflin Company).

Katzenbach, J.R. and Smith D.K. (1993). 'The Discipline of Teams', *Harvard Business Review* 71(2), 111–120.

Katzenbach, J.R. and Smith, D.K. (1993). *The Wisdom of Teams*. (New York, NY: HarperCollins).

Katzenbach, J.R. and Smith, D.K. (2001). *The Discipline of Teams*. (New York, NY: John Wiley and Sons). New York. NY.

Kirkman, B. and Shapiro, D. (1997). 'The Impact of Cultural Values on Employee Resistance to Teams: Toward a Model of Globalized Self-Managing Work Team Effectiveness', *Academy of Management Review* 22(3), pp. 730–57.

Liang, D., Moreland, R. and Argote, L. (1995). 'Group Versus Individual Training and Group Performance: The Mediating Role of Transactive Memory', *Personality and Social Psychology Bulletin* 21(4), 384–393.

Mathieu, J., Goodwin, G., Heffner, T., Salas, E. and Cannon-Bowers, J. (2000). 'The Influence of Shared Mental Models on Team Process and Performance', *Journal of Applied Psychology* 85(2), 273–283.

Maxwell, J.C. (2001). *The 17 Indisputable Laws of Teamwork.* (Nashville, TN: Thomas Nelson Publishers).

Noble, D. (2002). 'A Cognitive Description of Collaboration and Coordination to Help Teams Identify and Fix Problems', in *Proceedings of the 2002 International Command and Control Research Technology Symposium*, http://www.dodccrp.org/files/7th_ICCRTS.zip (pdf 089), Quebec, Canada.

Noble, D. (2003a). 'Understanding and Applying the Cognitive Foundation of Effective Collaboration', in *Proceedings of the 15th International Conference on: Systems Research, Informatics and Cybernetics: Collaborative Decision-Support Systems Focus Symposium, Baden-Baden, Germany*.

Noble, D. (2003b). 'Understanding and Applying the Cognitive Foundation of Effective Collaboration', in *Proceedings of the 5th ONR Conference on Collaboration*, *Quantico, VA*.

Noble, D., Shaker, S. and Letsky, M. (2004). The Cognitive Path to Team Success, *Government Executive* 36(13).

Wegner, D.M. (1987). 'Transactive Memory: A Contemporary Analysis of Group Mind', in B. Mullen and G.R. Goethals, (eds), *Theories of Group Behavior*, pp. 185–206. (New York: Springer-Verlag).

Chapter 20

Collaborative Operational and Research Environment (CORE): A Collaborative Testbed and Tool Suite for Asynchronous Collaboration

Elizabeth M. Wroblewski and Norman W. Warner

In recent years, advances in communication and information technology have changed the face of team collaboration. Collaborative technology is developing at an unprecedented, astonishing rate. The vast majority of current development has stemmed from the fields of business, management and information technology. Without a doubt, input from each of these fields is critical to the development of effective collaboration tools. The problem, however, is that inputs from these fields typically stem from intuition rather than relevant empirical research (DeFranco-Tommarello and Deek 2004). To date, there is little understanding of the fundamental mechanisms used by teams to collaborate effectively (Stahl 2006). This lack of understanding represents a serious gap in our scientific knowledge that must be addressed. Key to successful tool development is empirical research in team collaboration and the supporting cognitive processes rather than relying solely on intuition.

But is empirical research enough to ensure a comprehensive understanding of team collaboration, and, in turn, ensure effective tool development? Controlled laboratory experimentation focuses on testing and formalization, allowing researchers to study team collaboration by manipulating key variables. It would, however, be impossible to replicate the 'real' world (e.g., stress, environmental factors, high stakes) in a controlled setting. On the other hand, supporters of Naturalistic Decision-Making (NDM) stress the importance of observing expert decision-making in a natural environment (Klein 1998; Klein *et al.* 2003; Zsambok and Klein 1997). While direct observation provides critical insight, it fails to identify specific variables which influence behavior.

The solution is to find a balance between NDM and laboratory experimentation. NDM and controlled laboratory experiments should be used to complement each other. This can be done by *progressively* increasing the fidelity of a collaboration study through a spectrum of experimentation and observation. Laboratory testing can provide a small window to the world in a controlled environment while field observations can provide critical insight into expert behavior. The Cognitive and Automation Research Lab (CARL) at the Naval Air Systems Command (NAVAIR) in Patuxent River, Maryland was designed to encompass both empirical and NDM research through the Collaborative Operational and Research Environment (CORE) testbed. Although in the early stages of development, the CORE testbed has proven to be an effective research environment. This chapter provides a detail discussion of the iterative process of conducting empirical experimentation with natural observation to obtain a better understanding of the cognitive aspects of team collaboration. In addition, a description is provided of how the CORE testbed and tools are used in support of this spectrum of research.

Collaboration in today's military

Recent advances in communication and information technology have resulted in a dramatic shift in military collaboration. The shift to network-centric operations at both the tactical and strategic levels provides the warfighter with improved abilities for sharing and leveraging information (Garstka 2004). This shift to a knowledge-based force will have a direct impact on shared situation awareness, intelligence analysis and decision-making.

Despite these improvements, however, information technology in the military is not without its problems. Because of the tremendous amount of data available, collaboration teams are often faced with information overload. Much of that information comes from open sources such as the Internet. As a result, knowledge uncertainty becomes a primary concern. In addition, military intelligence is dynamic in nature, and therefore is constantly changing. Military strategists and operational personnel must continually monitor the flow of information to ensure accurate and timely mission planning and execution (Warner, Letsky and Cowen 2005; Warner and Wroblewski 2004).

Geographically distributed collaboration teams face additional burdens. Co-located teams have the advantage of *real time* collaboration. On the other hand, distributed teams often receive information asynchronously. In addition, the widening realm of contributors often results in teams that represent a disparity of experience, knowledge and cultural backgrounds. To complicate the collaborative effort further, accessibility of enhanced commercial and military technology has resulted in a market flooded with diverse, often incompatible collaboration tools.

To ensure continued effectiveness, it is imperative that this advancing technology enhance the collaborative effort rather than impede it. Without focusing on the principles of cognitive psychology, team dynamics, human factors and knowledge management, with validation through supporting experimentation, collaboration tools will not develop to their full potential.

An empirical model of team collaboration

Distributed collaborative teams often have the ability to successfully reach consensus and accomplish their assigned tasks. How effectively those tasks are accomplished, however, usually depends more on the dynamics of the specific teams than on the collaboration tools they are using. As DeFranco-Tommarello and Deek (2004) suggest, 'there is more emphasis on the technology for collaboration and less focus on the methodology. Current models need to consider the psychology and sociology associated with collaborative problem solving'. From an empirical perspective, Warner, Letsky and Cowen's (2005) Model of Team Collaboration provides an initial framework for collaborative study. The model, derived from extensive literature reviews and supported by empirical research (Warner and Wroblewski 2004, 2005, 2006), emphasizes the macrocognitive aspects of the collaboration process with a focus on identifying the collaboration stages and macrocognitive processes that occur during collaborative team problem-solving, together with a specific emphasis on the unique characteristics of military collaboration.

Macrocognition is defined as the high-level mental processes, both external and internal, used by team members while engaged in collaborative problem-solving. External processes are those processes associated with observable and measurable actions in a consistent, reliable, repeatable manner or through standardized conventions of a specific domain. Examples would include writing, gesturing and speaking. Internal processes are those processes that are not expressed externally, and can only be measured indirectly. Developing mental models and understanding concepts are examples of internal processes.

As discussed in Chapter 2, this model identifies four unique, but interdependent stages of team collaboration: knowledge construction, collaborative team problem-solving, team consensus and outcome evaluation and revision. Knowledge construction occurs when team members build individual task knowledge and team knowledge. In collaborative team problem-solving, team members communicate data, information and knowledge to develop solution options to the problem. The majority of collaboration occurs in this stage. Team members negotiate solution options and ultimately reach final agreement in the team consensus stage. Evaluation of the solution option against the problem-solving goal is done in the outcome evaluation and revision stage. Although the model includes a feedback loop for revising team solutions, the stages are not necessarily sequential. Because team communication is very dynamic, the flow of collaboration can follow virtually any path. In addition, current research has identified fourteen internal and external macrocognitive processes that contribute to the success or failure of team collaboration. Examples of these macrocognitive processes include individual task knowledge development, team shared understanding development, team negotiation, critical thinking and the convergence of individual mental models to team mental models. Chapter 2 provides a full description of each of these processes.

Naturalistic decision-making versus laboratory research

Zsambok and Klein (1997) define NDM as the study of how experienced people, working as individuals or groups in dynamic, uncertain, and often fast-paced environments, identify and asses their situation, make decisions and take actions whose consequences are meaningful to them and to the larger organizations in which they operate.

More simply put, NDM is the study of how experienced people make decisions in demanding, real-world situations.

Several constraints are key to defining a naturalistic decision making setting (Klein 2003, 1998; Zsambok and Klein 1997). Decisions are often complex, involving high stakes and high risk. Goals can be ill-defined and conflicting. Information is often inadequate, ambiguous, or erroneous. Organizational procedures are often poorly defined. Conditions are often dynamic, with constantly changing information and goals. These constraints can have a distinct effect on real-world decision-making and can be difficult, if not impossible, to replicate in a laboratory setting. Proponents of NDM ascertain that real-time field observation, critical to NDM research (Lipshitz *et al.* 2001; Klein 1998), provides a solution. Observation allows the researcher to gain insight into expert behavior during real-time decision making. Introspection techniques also play a critical role. Structured and unstructured interviews, cognitive task analyses, critical incident reviews, think-aloud protocol analyses and other techniques are often used for knowledge elicitation during an NDM study.

By definition, NDM seems to be the antithesis of traditional laboratory research. Proponents of NDM maintain that controlled laboratory findings are not applicable in a real-world setting (Zsambok and Klein 1997; Klein 1998; Klein *et al.* 2001). Furthermore, they assert that models developed as the result of empirical research fail to explain or predict real-world decision-making behavior, and are, in fact, 'useless in natural settings' (Azar 1999). On the other hand, proponents of a more traditional empirical approach to research claim that theories based on NDM research 'are consistent with expert decision making but tend to be very general, referring to broad categories of cognitive processes such as recognition, pattern matching, and mental simulation, without explaining exactly how these processes are performed' (Bryant 2002). As Klein (1998) states, 'People who study naturalistic decision making must worry about their inability to control many of the conditions in their research. People who use well-controlled laboratory paradigms must worry about whether their findings generalize outside the laboratory.' Without the manipulation of carefully controlled variables, it would be impossible to identify those processes which influence behavior.

From a researcher's perspective, which methodology should be implemented as the most effective strategy to understand the macrocognitive processes of team collaboration? To capitalize on the strengths and compensate for the shortcomings of both NDM and laboratory research, a viable solution would be to develop a series of studies that incorporate both strategies.

The team collaboration research spectrum

Ideally, NDM and traditional experimental research should be used, not at the exclusion of each other, but rather as a means to compliment each other. While this seems to be a daunting task, it can be accomplished through a series of closely integrated laboratory and field studies. Conducting research through this spectrum of progressive validation synthesizes NDM and traditional research, and as a result, effectively increases the fidelity of the research findings.

The team collaboration research spectrum consists of four tightly interwoven research venues (see Figure 20.1):

1. a traditional, controlled research study in a laboratory setting;
2. a controlled field experiment;
3. a controlled experiment in a real world setting; and
4. field operations in which the researcher closely observes experts responding to a real world event.

Each stage of the spectrum builds upon a common scenario, while gradually decreasing the amount of researcher control.

Phase 1: Controlled laboratory research In this traditional research venue, the researcher uses a rich, outcome-based scenario, manipulating predefined variables to gain insight into the collaborative process. The result of this research is statistical data to be used for quantitative analysis. Qualitative data, such as post-experiment interviews, may also be used. Critical to this phase of the spectrum is the use of an effective, domain-specific scenario to elicit information about the internal and external cognitive processes of collaborative decision-making. To ensure the collaborative process is replicated as closely as possible in a laboratory setting, the scenario used should be multidimensional, with several complex solutions possible. For further information on scenario development see Chapter 22.

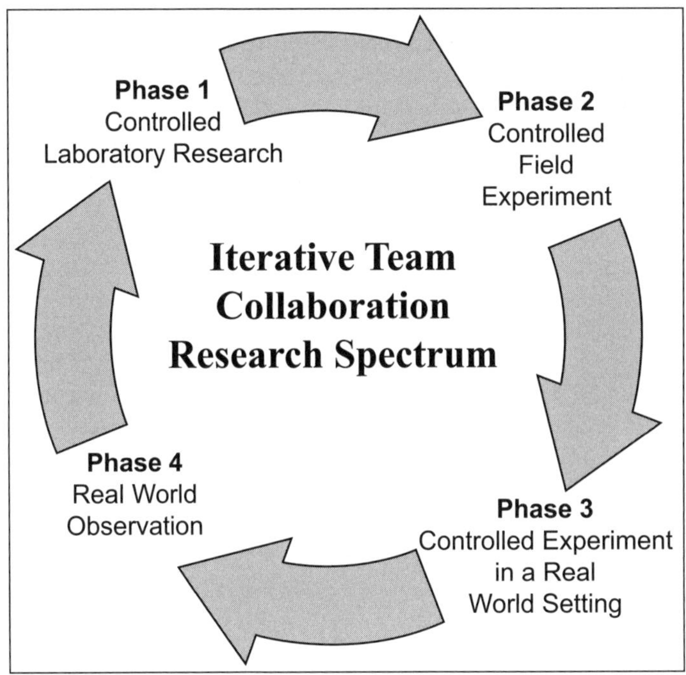

Figure 20.1 Iterative team collaboration research spectrum

Investigators at CARL at the Naval Air Systems Command (NAVAIR) designed a Non-combatant Evacuation Operation (NEO) scenario (Warner, Wroblewski and Shuck 2004) to investigate Warner, Letsky and Cowen's (2005) model of team collaboration as it pertains to this domain. This complex scenario required subjects to develop a plan to rescue three Red Cross workers trapped in a church basement on the remote South Pacific island of Drapo. Required elements of the final plan included a list of US forces to be used in the rescue, modes of transportation, weapons to be used, a specific timeline of events and a detailed explanation of the plan. Subjects were randomly assigned as team members in a face-to-face environment, or asynchronous teams using MIT's Electronic Card Wall (EWall) (Keel 2006). (For more information on EWall see Chapter 11.) Because of limited military availability, civilians were used as subjects. Each team member was given the General Information section of the NEO scenario, which included a situation report, topographical maps, a limited explanation of US military assets available and information on hostile forces in the area. In addition, each team member was assigned as an expert in either local intelligence, available weapons or local environmental issues and provided with information pertinent to their area of expertise. By combining the general information with the expertise information, teams could develop a realistic plan to rescue the workers. The effectiveness of the final plan submitted was rated using a scoring matrix developed with input from military operational personnel.

Through the use of this NEO scenario, several controlled experiments (Warner and Wroblewski 2004, 2005) were conducted to investigate the team collaboration model. Obviously, as NDM supporters would suggest, carefully controlled experiments such as this would fail to replicate the constraints and pressures found in a real-world NEO. It did, however, provide critical insight into the collaborative process, as it pertains to this specific domain, and serves as a basis for research conducted in the next three phases of the team collaboration research spectrum.

Phase 2: Controlled field experiment Using the same scenario developed in Phase 1, a controlled experiment would be conducted in a simulated operational setting. The concept of these controlled field experiments is similar to the field exercises (e.g., Trident Warrior) used by the military to train for specific operations. Conducting the research in this quasi-realistic environment allows the researcher to examine the impact of extraneous factors on cognitive processes. In other words, do factors like physical stress, environmental factors, time pressure, and dynamic conditions influence task-specific behaviors, such as time on task or solution outcome? Through this process, both quantitative and qualitative data can be collected for analysis.

In our NEO example, future controlled field experiments will be conducted at the NAVAIR's Advanced Maritime Technology Center (AMTC). This facility, a joint initiative between NAVAIR and the Naval Surface Warfare Center (NAVSEA), is located at the Patuxent River Naval Air Station and serves as an RDTandE facility for experimentation and demonstration. Using military personnel as subjects, they will be required to not only develop a rescue plan for the Red Cross workers trapped on Drapo, but actually carry out the evacuation plan. Participants posing as the trapped workers will be hiding in a simulated church on a nearby island or inlet. Once the subjects develop a course of action, they will instruct rescuers to carry out the mission. Quantitative data, such as time to completion, verbal protocol analysis and quality of decision would be captured. Qualitative measures would include questionnaires and subjective workload measures.

Phase 3: Controlled experiment in a real-world setting Again, using the same scenario, the next phase of the team collaboration spectrum would be to conduct a similar experiment to the controlled field experiment, but conduct the experiment in an actual operational setting. Typically, the infrastructure of a military operational environment is much more extensive and dynamic than could be replicated in a simulated environment. Extensive databases, joint multi-agency resources, the availability of diverse expertise would be virtually impossible to replicate. The key question to be answered in this phase of the process is, 'Does a real-time operational infrastructure influence the collaborative process?'

Again, using our NEO scenario as an example, the Mission Support Center (MSC) under US Special Operations Command (USSOCOM), would be an ideal venue to conduct this phase of the research spectrum. MSC maintains an extensive, networked infrastructure and information management system in an effort to provide reach-back capabilities for Special Operations forces. Using MSC capabilities and personnel, mission planners would develop a course of action for rescuing the Red Cross workers. NDM observations, as well as empirical data, would be used for analysis.

Although many of the constraints that influence collaborative behavior would be represented in this environment, it would be naive to claim that this is a true representation of an actual event. High levels of stress could, potentially, have a serious impact on the collaboration and the decision-making. Undoubtedly, it would be impossible to replicate the intense stress present in a real-world NEO mission planning situation in this controlled setting. Simply put, there are no life or death consequences if mission planners fail to devise an adequate plan to rescue the Red Cross workers. It does, however, serve to validate the findings of the earlier studies and identify any previously unidentified influences on the collaborative process.

Phase 4: Real-world observations To qualify as a true NDM study, it is necessary to observe experts making decisions in a real-world situation. This final stage of the research spectrum requires the investigator to step away from the scripted scenario and see if the findings from the previous studies hold up in the real world. This final validation should represent an event as close to the scripted scenario as possible. Obviously it would be impossible to find a NEO situation that identically reflected the Drapo scenario. The task, then, would be to find a team collaboration response to a real NEO event. The evacuation of New Orleans during Hurricane Katrina or the response to the 2004 Indonesian tsunami would serve as comparable examples. Unlike the initial controlled laboratory research, no variables would be manipulated and the result of these NDM observations would be strictly qualitative.

The challenge is to understand how collaboration teams behave throughout the entire research spectrum. Traditional laboratory research fails to replicate the dynamics of the real world. NDM, on the other hand, fails to identify the specific causes of behavior. By gradually decreasing experimenter control through a series of studies, and synthesizing both empirical and NDM studies, we can systematically increase the fidelity of the research and achieve a better understanding of the cognitive mechanisms of team collaboration.

CORE testbed

The Cognitive and Automation Research Lab (CARL) at NAVAIR was designed to incorporate the entire collaborative research spectrum, thereby incorporating a series of studies around both NDM and traditional empirical research. The testbed was designed to serve several functions:

1. to provide empirical investigations into the Model of Team Collaboration (Warner, Letsky and Cowen 2005);
2. to support, test and integrate mature products developed by the Office of Naval Research (ONR) Collaboration and Knowledge Interoperability (CKI) program; and
3. to serve as a liaison between the military and CKI investigators to demonstrate CKI team collaboration products.

How does this testbed help us understand macrocognition? Through an integrated, networked collaborative environment, coupled with appropriate metrics and measures, we can identify and quantify the impact of macrocognitive processes on team collaboration. Critical components of the CORE testbed include hardware and software applications, as well as carefully defined metrics for data collection. (see Figure 20.2).

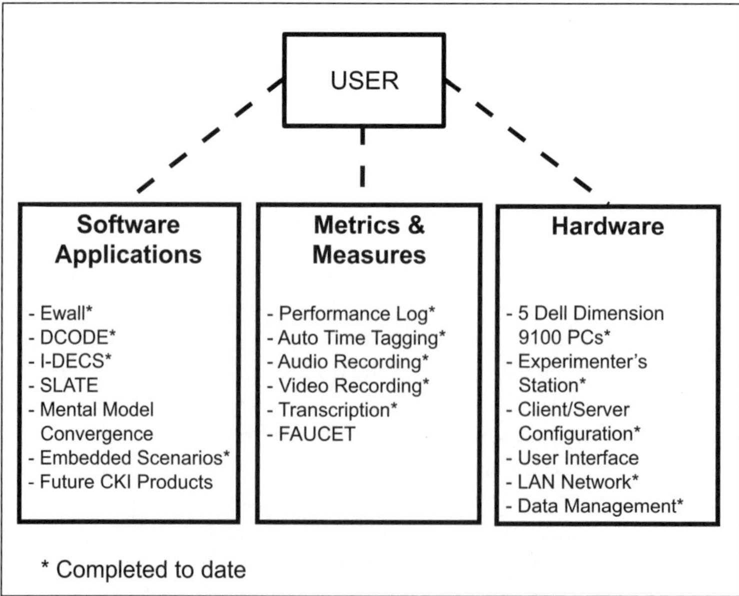

Figure 20.2 CORE tool suite and testbed

CORE testbed metrics Key to the success of any testbed is the metrics and measures incorporated into its foundation. The CORE testbed is designed to identify the macrocognitive processes associated with collaboration on both an individual and team level. To date, the CORE testbed has only been validated in a Phase 1 study of the research spectrum. Metrics for the remaining three phases have not been identified as of yet.

By definition, macrocognition involves both internal and external cognitive processes. Because external processes are associated with observable behaviors, the metrics used to measure these processes are more readily defined. Time-stamped digital video and audio recordings capture both individual and team behaviors during the experimental session for later analysis. In our NEO example, time to task completion and quality of decision (determined by the team's score on their final plan) was also used for statistical analysis.

The internal cognitive processes, which occur inside the head, can only be measured indirectly. As a result, they present a unique challenge to an investigator. As Stahl (2006) states, 'discourse is the embodiment of group cognitive processes, and the analysis of that discourse makes the group's learning and meaning … visible and comprehensible'. With that in mind, verbal analysis serves as a cornerstone of the CORE testbed. In this technique, the cognitive processes are carefully defined, and each utterance is coded for its applicability to the definitions. This technique relies, not only on accurate transcription, but also on clear, concise definitions. Ambiguous definitions lead to ambiguity in the coding. As a result, inter-rater reliability would be adversely effected. To ensure accuracy, raters must be carefully trained, not only in the applicable coding scheme and definitions, but to remain objective while coding. It is critical that the analysts have an appropriate background to understand the culture, dialect and language of the subjects (Stahl 2006). To date, CORE transcript coding is done by hand, a tedious and time-consuming endeavor. In the near future, Flow Analysis of Utterance Communication Events in Teams (FAUCET) (Cooke and Kiekel 2006) and a latent semantic analysis (LSA) technique developed by Bolstad and Foltz (2006) will be used to automate the coding process. These techniques are explained in detail in Chapters 4 and 15, respectively.

For asynchronous teams an automated performance log is critical for future analysis. The performance log used in the CORE testbed has been designed to capture all keystrokes, events and interactions. It is configurable to meet a variety of analytical needs and can easily be sorted to be time sequential or following the actions of a specific user.

While a variety of statistical analyses are conducted using the data from CORE studies, common analyses include frequency of occurrence of the individual macrocognitive processes, mean percentage of time spent in each process and the transition probabilities between processes. Other internal measures used in the CORE testbed include subjective workload assessments, pre- and post-experiment surveys, concept maps, and structured interviews. As Stahl (2006) states:

> If learning takes place in an interaction, we should be able to observe it by analyzing an adequate record of the interaction. It should show in how participants use words,

> in the ways they build on each others' utterances, in their expressions, gazes, postures, expressive noises, and in their interactions in similar circumstances later.

A similar testbed, the Global Information Grid Testbed, has been developed at the Naval Postgraduate School (NPS) in Monterey, CA. The NPS testbed has been designed to provide a plug-and-play test environment for mature CKI products, with a focus on network-centric, tactical command and control experimentation (Bordetsky and Hutchins 2006). (See Chapter 22 for further details.) The CORE testbed, while also designed to test and integrate CKI products, focuses more on product testing and tool development from conception through transition, with an emphasis on intelligence analysis, mission planning, and mission execution.

CORE tool suite

With current technology, developing a single collaboration tool to address all collaboration needs would be a daunting, if not impossible task. A more practical solution is to develop a suite of integrated tools to support the different types of collaboration problem domains and their associated cognitive processes. For example, problems in today's military include information overload, knowledge uncertainty, dynamic intelligence, distributed teams and a disparity of experience, knowledge and cultural backgrounds. To facilitate effective collaboration, a suite of integrated collaboration tools should be designed to address each of these specific problem areas.

The ideal collaboration tool suite would be one that guides a distributed team through each collaboration stage (knowledge construction, collaborative team problem0solving, team consensus and outcome evaluation and revision), while ensuring the cognitive processes (e.g., individual mental model construction, knowledge interoperability development, team shared understanding, team negotiation) are addressed. The design architecture for the CORE tool suite will provide this type of functionality.

The CORE Tool Suite incorporates mature CKI products into an integrated set of collaboration tools for testing in all phases of the research spectrum. A prototype of the tool suite will ultimately be developed for testing, using configurable, but domain-specific, embedded scenarios. As each CKI product is completed, the product is incorporated into the tool suite and tested for usability, configurability and functionality. Since products mature at different rates, the testing process is iterative, being repeated with the inclusion of each new product. The ultimate goal is to transition a complete, multifaceted tool to the warfighters to aid collaborative decision-making.

Fundamental to the development of the CORE tool suite is the design of the integrated tool's interface. It is essential to ensure the design of the interface is consistent with the needs of the domain in which the tool will be used. Although the *functionality* of a collaboration tool used by a mission planner may be similar to a tool used by Special Operations forces on the ground, the *interface* must be designed to meet their very specific, individual needs. For example, the mission planner might require multiple displays and unlimited bandwidth, unrealistic luxuries for the warfighter, who requires a tool capable of stealth and rapid deployment. While the initial configuration of the CORE tool suite is geared more towards the needs of the mission planner, future adaptations will be made with the individual warfighter in mind.

To date, several mature CKI products have been integrated into the CORE Tool Suite for laboratory testing. MIT's EWall (see Chapter 11) (Keel 2006), with its embedded Decision-Making Constructs in a Distributed Environment (DCODE) (see Chapter 11) (Fleming 2006), serves as the cornerstone of the tool suite. In addition, Pacific Science and Engineering's Integrative Decision Space (I-DecS) Interface (see Chapter 12) (Handley and Heacox 2006) and Evidenced Based Research's Collaboration Advizor Tool (CAT) (see Chapter 18) (Noble *et al.* 2004) have also been tested for usability and functionality. Other CKI tools will be implemented as they mature.

Conclusion

Military communication and information technology is rapidly evolving at both the tactical and operational levels. Reliance on effective collaboration tools for distributed teams has become critical to the military's efforts in mission planning and decision making. To ensure continued effectiveness, it is imperative that this advancing technology enhance the collaborative effort rather than impede it. As former Secretary of Defense Donald Rumsfeld stated in the 2003 *Transformation Planning Guidance*, 'we must achieve: fundamentally joint, network-centric, distributed forces capable of rapid decision superiority and massed effects across the battlespace. Realizing these capabilities will require transforming our people, processes, and military force.' It is towards this end that the NAVAIR CORE testbed and tool suite have been developed. The CORE tool suite has far-reaching impact on enhancing team collaboration, both on the battlefield and off.

References

Azar, B. (1999). 'Decision Researchers Split, but Prolific', *APA Monitor Online* 30 http://www.apa.org/monitor/may99/split.html, accessed 21 December 2006.

Bolstad, C. and Foltz, P. (2006). 'Automated Communication Analysis for Interactive Situation Awareness Assessment', *Collaboration and Knowledge Management Workshop Proceedings*, pp. 212–232. (Arlington, VA: Office of Naval Research, Human Systems Department).

Bordetsky, A. and Hutchins, S. (2006). 'NPS–CKM Testbed: Architecture and Cognitive Aspects', *Collaboration and Knowledge Management Workshop Proceedings*, pp. 147–169. (Arlington, VA: Office of Naval Research, Human Systems Department).

Bryant, D. (2002). 'Making Naturalistic Decision Making "Fast and Frugal"', *Proceedings of the 7th International Command and Control Research and Technology Symposium, Quebec City*, pp. 105–125. (Washington, DC: United States Department of Defense).

Cooke, N. and Keikel, P. (2006). 'Automatic Tagging of Macrocognitive Collaboration through Communication Analyses', *Collaboration and Knowledge Management Workshop Proceedings*, pp. 182–195. (Arlington, VA: Office of Naval Research, Human Systems Department).

DeFranco-Tommarello, J. and Deek, F. (2004). 'Collaborative Problem Solving and Groupware for Software Development', *Journal of Information Systems Management* 21(1), 67–80.

Fleming, R. (2006). 'Decision Making Constructs in a Distributed Environment (DCODE)', *Collaboration and Knowledge Management Workshop Proceedings*, pp. 80–94. (Arlington, VA: Office of Naval Research).

Garstka, J. (2006). *Network Centric Warfare: An Overview of Emerging Theory.* http://www.mors.org/publications/phalanx/dec00/feature.htm, accessed 30 November 2006.

Handley, H. and Heacox, N. (2006). 'Modeling and Simulation of Cultural Differences in Human Decision Making', *Collaboration and Knowledge Management Workshop Proceedings*, pp. 259–273. (Arlington, VA: Office of Naval Research, Human Systems Department).

Keel, P. (2006). 'EWall: Electronic Card Wall: Computation Support for Individual and Collaborative Sense-Making Activities', *Collaboration and Knowledge Management Workshop Proceedings*, pp. 47–59. (Arlington, VA: Office of Naval Research, Human Systems Department).

Klein, G., Ross, K., Moon, B., Klein, D., Hoffman, R. and Hollnagel, E. (2003). 'Macrocognition', *Human Centered Computing* 18(3), 81–85.

Klein, G. (1998). *Sources of Power: How People Make Decisions* (Cambridge, MA: The MIT Press).

Lipshitz, R., Klein, G., Orasanu, J. and Salas, E. (2001). 'Focus Article: Taking Stock of Naturalistic Decision Making', *Journal of Behavioral Decision Making* 14, 331–352.

Noble, D., Christie, C., Yeargain, J., Lakind, S., Richardson, J. and Shaker, S. (2004). 'Cognitive Guidelines to Support Effective Collaboration', *Collaboration and Knowledge Management Workshop Proceedings*, pp. 167–179. (Arlington, VA: Office of Naval Research, Human Systems Department).

Stahl, G. (2006). *Group Cognition: Computer Support for Building Collaborative Knowledge* (Cambridge: The MIT Press).

U.S. Department of Defense (2003). *Transformation Planning Guidance.* (Washington, DC: Department of Defense).

Warner, N., Letsky, M. and Cowen, M. (2005). 'Cognitive Model of Team Collaboration: Macro-Cognitive Focus', *Proceedings of the Human Factors and Ergonomics Society 49th Annual Meeting*, pp. 269–273. (Orlando, FL: Human Factors and Ergonomic Society).

Warner, N. and Wroblewski, E. (2004). 'Achieving Collaborative Knowledge in Asynchronous Collaboration', *Collaboration and Knowledge Management Workshop Proceedings*, pp. 63–72. (Arlington, VA: Office of Naval Research, Human Systems Department).

Warner, N. and Wroblewski, E. (2005). 'Achieving Collaborative Knowledge in Asynchronous Collaboration', *Collaboration and Knowledge Management Workshop Proceedings*, pp. 95–122. (Arlington, VA: Office of Naval Research, Human Systems Department).

Warner, N. and Wroblewski, E. (2006). 'Macro Cognition in Team Collaboration (MCTC)', *Collaboration and Knowledge Management Workshop Proceedings*, pp. 64–83. (Arlington, VA: Office of Naval Research, Human Systems Department).

Warner, N., Wroblewski, E. and Shuck, K. (2004). 'Noncombatant Evacuation Operation: Red Cross Rescue Scenario', *Office of Naval Research Collaboration and Knowledge Management Workshop Proceedings*, pp. 290–325. (Arlington, VA: Office of Naval Research, Human Systems).

Zsambok, C. and Klein, G. (1997). *Naturalistic Decision Making.* (Mahwah, NJ: Lawrence Erlbaum Associates).

Chapter 21

Plug-and-Play Testbed for Collaboration in the Global Information Grid

Alex Bordetsky and Susan Hutchins

Military tasks have grown in complexity over the past decade so that most require a team effort to address the problem. Moreover, these problems are often addressed at the international level, requiring a coalition effort. An emerging Department of Defense (DoD) platform of ubiquitous end-to-end information services, referred to as the Global Information Grid (GIG), provides warfighters and first responders with a seamless geographically distributed collaboration capability around the world. While developments such as the GIG have widened the accessibility of participants to information and provided greater communication between contributors, at the same time this increased access to information can create problems of information overload. In line with the theme of this book, the objective of this chapter is to describe a unique testbed, operated by the Center for Network Innovation and Experimentation (CENETIX), which enables researchers to experimentally explore team decision-making in complex, data-rich situations. Team collaboration data captured in this testbed, collected to validate a model of team collaboration, is also discussed.

Global information grid and collaborative technology

The GIG, as defined by DoD directive (USJFC, GIG, US Joint Froces Command 2001), is the DoD enterprise network of globally interconnected, end-to-end set of information capabilities. It is evolving as a single secure grid providing seamless information sharing capabilities to all warfighting, national security and support users, supporting DoD and intelligence community requirements from peace-time business support through all levels of conflict. From the command and control (C2) perspective (Sturk and Painter 2006), the GIG represents a network-centric service platform that enables:

1. joint, high-capacity netted operations,
2. supporting strategic, operational, tactical and base/post/camp/station levels of operation,
3. plug-and-play interoperability for US and Coalition users; and
4. tactical and functional data fusion, information and bandwidth on demand and defense in depth against all threats.

Many, if not all, of the benefits of participating in a face-to-face meeting can be gained using collaborative tools: information flows quickly, outstanding issues are raised and a certain amount of brainstorming can occur to arrive at a decision. Additionally, all relevant users, or providers of information, reach a fuller understanding of the issues because they have seen other viewpoints and received a freer flow of information (Truver 2001). From a military perspective, advantages of using a collaborative environment include:

1. fewer personnel have to be located in the area of conflict,
2. enhanced opportunities to share information among planners and decision-makers,
3. experts in remote locations can participate in all phases of the planning, decision-making and assessment process,
4. increased access to many additional sources of information that previously were not possible, and
5. reduced time required for the planning, decision-making and assessment process.

Collaboration across layers of the GIG and C2 architectures is crucial for achieving missions as described in network-centric warfare doctrine. Proliferation of GIG services is expected to enable data sharing and collaboration between operational and tactical levels. The NPS plug-and-play testbed for collaboration in the GIG represents a unique research extension of the GIG, which provides a plug-and-play environment for research teams and field operators to explore emerging models of collaboration, self-synchronization and data fusion through experiments with Special Forces, coast guard, emergency response units, and coalition partners.

Plug-and-play testbed with reachback to GIG

In accordance with the DoD vision, the GIG involves more than just technology (Richard and Roth 2005). It incorporates a collaborative environment of people who, in conjunction with processors and sensors, perform the roles of information producers and information consumers, depending on the situation, by publishing and retrieving information from shared information spaces. Users seeking experimentation with this tactical extension of the GIG can connect their network, sensors, unmanned vehicles, collaboration and situational awareness tools and other tools for investigating team collaboration.

Based on the tactical network topology (TNT) backbone (Bordetsky, Dougan, Foo and Kihlberg 2006) the NPS testbed for collaboration in the GIG provides several layers for integrating models, tools and experimentation procedures for research teams. Users can connect their remote local area network, including command and operation centers, via the virtual private network (VPN) client, and sensors and unmanned vehicles can be added via the situational awareness (SA) agent. Sensors can also provide their feed to the C2 center via the EWall poster, described by Keel in Chapter 11, this volume. Human operators (both remote and local) can access the testbed collaborative environment via the Groove client, SA agent, video conferencing room and video client. In addition to Groove workspaces, researchers can join the testbed environment via the tactical extension of EWall.

Maritime interdiction operation experiments

The Naval Postgraduate School (NPS), Monterey, CA, is conducting a series of experiments that focus on detecting, identifying and interdicting nuclear materials in open waters. A Maritime Interdiction Operation (MIO) involves a Coast Guard operation to search a suspect ship for contraband material and suspect persons where the basic scenario entails detecting a moving vessel emitting signs of ionizing radiation. Coast Guard officers and laboratory researchers (role-playing Coast Guard officers) board the vessel and take in-depth readings with portable radiation-detection instruments. These readings are quickly electronically relayed to scientific experts at remote sites and results of data analysis are transmitted back to the boarding vessel for use by first responders who are continuing their search on the intercepted vessel.

First responders working in marine enforcement are confronted with the immense challenge of attempting to screen cargo inside the containers carried on the continuous flow of container ships that enter major ports such as the Port of Oakland, the fourth largest container port in the United States. Successful interdiction requires a combination of modern technology, excellent intelligence, robust communications and collaboration between an array of people who possess different types of expertise, scattered across the nation, or across several continents.

The testbed described in this chapter provides the capability to conduct experiments in San Francisco Bay, between Yerba Buena Island and the Port of Oakland. The target, drive-by detection and boarding vessels, augmented by unmanned surface and aerial vehicles with a self-forming adaptive broadband network, provide a sensor-networking platform for sending, sharing and synchronizing in real time, large data files across collaborative work spaces.

Maritime interdiction operations

This series of experiments is being conducted to test the technical and operational challenges of developing a global Maritime Domain Security testbed. One goal is to test the applicability of using a wireless network for data sharing during a MIO scenario to facilitate ‘reach back’ to experts for radiation source analysis and biometric data analysis. This technology is being tested and refined to provide networking solutions for MIOs where subject matter experts at geographically distributed command centers collaborate with a boarding party in near real time to facilitate situational understanding and course of action selection.

The objective of these experiments is to evaluate the use of networks, advanced sensors and collaborative technology for conducting rapid MIOs. Specifically, the ability of a boarding party to rapidly set up ship-to-ship communications that permit them to search for radiation and explosive sources while maintaining contact with the mother ship, C2 organizations and collaborating with remotely located sensor experts. The boarding team boards the suspect vessel and establishes a collaborative network and then begins their respective inspections and data collection processes. The boarding officer boards the vessel with his laptop so he can collaborate with all other members of the team, including those who are located on the ship, but physically dispersed around different areas of the ship while searching for contraband material and obtaining fingerprints of crew members, as well as the virtual members of the boarding team – the experts who are located at the different reach-back centers. Since there are numerous commercial uses for certain radioactive sources, positive identification of the source in a short time is imperative.

Testbed networking layer: Self-forming wireless tactical network topology

Figure 21.1 shows the MIO TNT used in San Francisco Bay for the MIO experiments where each node plays a specific, unique role in the scenario. VPN is used to provide access between the testbed and individual participants via software clients from Lawrence Livermore National Labs (LLNL) among other places. The VPN architecture is essentially a hub-and-spoke model with NPS in the center, and the San Francisco Bay Area, San Diego, Army Biometrics Fusion Center, in West Virginia, the Stiletto in San Diego, and researchers in Austria, Sweden, and Singapore as spokes.

Figure 21.2 depicts the broadband wireless 100-mile stretch of the testbed to Camp Roberts, CA. This wireless tactical network provides real-time access to unmanned aerial vehicles (UAVs), unmanned ground vehicles (UGVs), unmanned surface

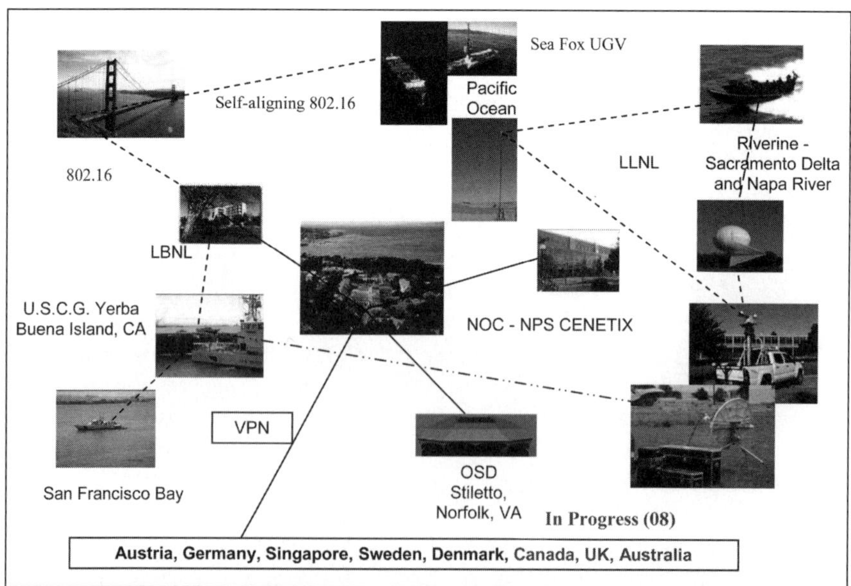

Figure 21.1 NPS testbed network with unmanned vehicles and sensors in San Francisco Bay

vehicles (USVs) and mobile Special Operations Command (SOCOM)/Marine Corps units for exploring collaboration for high-value target tracking operations. This part of the testbed will also be used for remote integration of UAVs/USVs in MIO or other scenarios, such as non-combatant evacuation experiments, where flying UAVs or operating UGVs is not feasible.

Collaborative network tools

Groove Virtual Office provides the majority of the collaborative tools for these experiments. A tactical operations center (TOC) and associated networking workspace was established to resolve network malfunctions and other associated issues necessary to optimize network performance and to coordinate logistics and issues which are not part of the scenario. A boarding party (BP) workspace provides a venue for scenario participants (including remote experts) to perform the following tasks: analyze and compare spectrum files, track radiation materials, share atmospheric modeling and predictions, consider emergency medical actions, post biometrics matches, radiation files/photos, responses and expert evaluations and make recommendations regarding additional actions to be taken by the BP when additional search information is required. The District 11 workspace provides access to C2 assets, including agencies and assets not directly under the commander's authority, and provides C2 to ensure appropriate actions are taken by the BP.

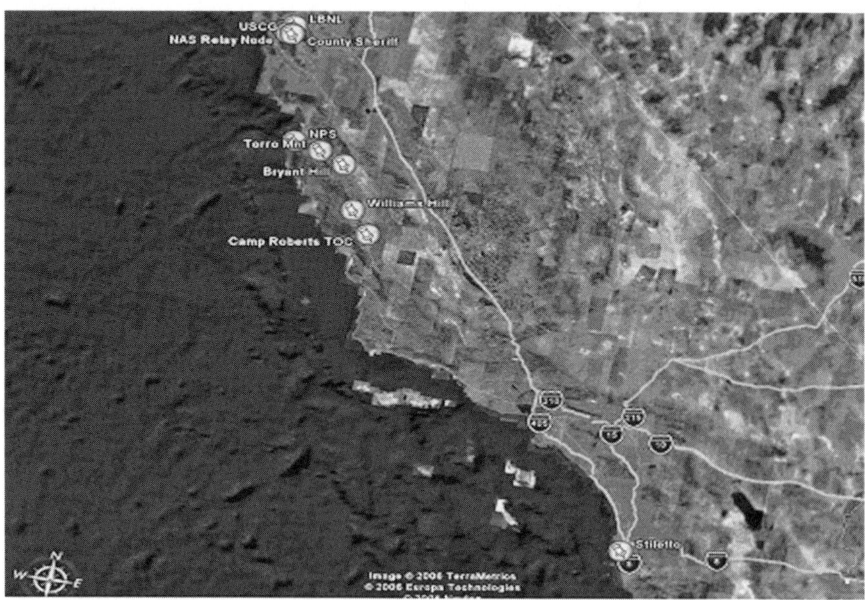

Figure 21.2 Area included in the testbed broadband wireless link to unmanned systems

Each of the collaborative network nodes can use any subset of the collaborative tools. These tools provide the ability to remotely monitor both the status of operational assets (such as boats/vessels) and the progress of scenario events. Groove provides a *Discussion board and chat* which are used for text communication between nodes. *File transfer* is primarily used for transferring data files between the BP and the reach back facilities. *Task manager* is used by experiment control in the TOC and the networking workspace to provide participants a way to monitor the progress of the scenario. *EWall* was used to monitor information alerts and video streams were monitored from various nodes.

The primary collaboration tools used on the boarding vessel were Groove and VOIP phone. The boarding vessel served as a coordination entity that provided a link between the TOC, District 11 and the boarding party. The boarding vessel also provided a video feed and the physical network link between the target vessel and Yerba Buena Island 802.16 node. The NPS CENETIX network operations center (NOC) used all of the available collaborative tools for the experiment: Groove, EWall, SA agent, video conferencing and audio conferencing. Groove was used for file sharing, messaging, chat, discussion board, pictures and web links. Teams from Sweden, Austria and Singapore used Groove, SA agent, a live video link and occasional cell phone communication. In addition, the Swedish team provided feedback to EWall. Figure 21.3 depicts the different teams who participated in the MIO experiment collaborative network.

MIO scenario

Based on intelligence, a commercial vessel of foreign origin is suspected of transporting uranium-enriching equipment. The Coast Guard has ordered one of its cutters to stop, board and search the commercial vessel. This involves a boarding party team who board the suspect ship to search for contraband cargo

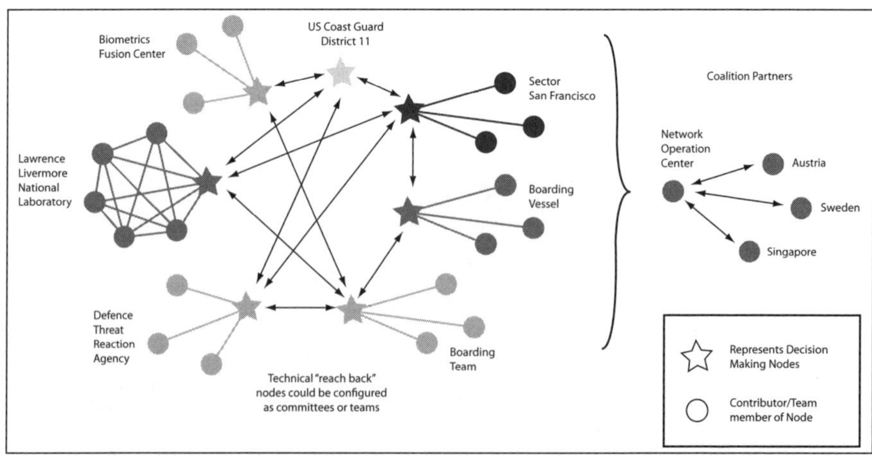

Figure 21.3 A team of teams works within the MIO collaborative network

(e.g., explosives, machinery) and possible terrorist suspects. The boarding party brings radiation detection and biometric gear, drawings of dangerous equipment and people and video recording capability. Data collected on suspicious material, equipment and people is sent to experts at distributed reach-back centers. Digital prints of the crew, taken by the biometric team, are compared to known criminal prints and latent prints from terrorist and crime scenes. A network extension capability is used from the Coast Guard cutter to the boarding team to enable the network to reach back to LLNL and the Defense Threat Reduction Agency (DTRA) to assist in identification of suspect cargo. Support from the National Biometric Fusion Center is used to quickly and accurately discriminate between actual vessel crew members and non-crew suspect persons.

The Groove collaborative workspace brings expert services into the boarding party team's tool set and facilitates voice and text communications between all members of the virtual boarding party and physical boarding party. Remote sites are able to quickly receive and open posted files to begin their analysis. For example, expert services provided at LLNL quickly determine the need for additional data capture of longer length spectrums and different angles of approach. Requests for additional data are transmitted by text message and taken for action, and radiation source spectrum captures are made of suspect containers that were detected to have a radiation signature presence. Analysis of this data may lead the boarding officer to recommend that the vessel be quarantined for further inspection. The biometric team takes digital prints of the crew to be compared to databases of known terrorist suspects.

Participating units and role player

Each quarterly experiment has new elements added to make the scenario different for each experiment. This includes the 'cover story', new data files and sources of radiological material and other possible suspect material, new role players, new technology and other new elements of the scenario that add to the technology being tested (such as boarding the boat while moving) and so on. Table 21.1 lists the participants and roles performed by this large and diverse group that is comprised of real-world subject matter experts and role-players (graduate students at NPS and interns at LLNL). A brief description of the roles performed by experiment participants follows.

San Francisco Police Department, Marine Unit, conducts the initial drive-by sensing of the suspect vessel and the Alameda County Sheriff's Office Marine Patrol unit is called to the scene. US Coast Guard, District 11 and Pacific Area, provide watch officers and an analyst. The Maritime Intelligence Fusion Center (MIFC) provides the initial intelligence injects. The Coast Guard provides the level-two capable boarding team who work in conjunction with LLNL using portable radiation detection devices and reach-back to various experts for remote analysis of data captured aboard the target vessel. LLNL provides source material, source security and data files for detection as well as a remote analysis cell and assistance from Livermore, CA, via Groove.

Networking and technical advisors at the NPS NOC and TOC provide expertise to the boarding party to deal with complex networking problems that might arise at any point during the boarding operation and which might prevent them from exploiting the unique capabilities offered by the network. LLNL provides weapons of mass destruction (WMD) expertise on radiological material detection, identification and categorization. This includes tracing the origins of the materials, correlation with other related findings worldwide, evaluation of the associated dangers and implications of their discovery. Identification of certain materials that can lead to the development of WMD, such as machinery and other non-proliferation material, is the focus of the MIO scenario.

Maritime Intelligence Fusion Center (MIFC) provides maritime traffic information such as ships' registries, cargo and crew manifests, ports of call and shipping schedules. This type of information is helpful in designating a vessel as suspect, locating the vessel, making its interdiction possible and confirming discrepancies onboard, such as fake documentation. Command Centers at NPS and District 11 provide both tactical and operational commanders, who are the primary decision-makers concerned with decisions before, during and after the boarding with particular emphasis on post-sensor analysis of the suspect vessel. The boarding officer makes decisions at the lowest level, following standard operating procedures, including how to organize and conduct the search using the best resources at his or

Table 21.1 Maritime interdiction operations scenario participants and roles

Participating Unit/ Command	Role Performed
San Francisco Police Dept. Marine Unit	Conducts initial drive-by sensing
Alameda County Sheriff's Dept. Marine Unit	Provides boarding vessel
SF Police 175' Clear Bay II Cargo Vessel	Provides target vessel
US Coast Guard	Watch Officer leads boarding party
Marine Intelligence Fusion Center Alameda	Watch Officer and analyst
Lawrence Livermore National Laboratory	Boarding party members collect data
DoE Radiation Assistance Program	Provides input on material handling
Army Biometric Fusion Center	Provides data files for detection teams
NPS Network Operations Center	Network support team and experiment control
Meteorology Experts	Provide alerts on visibility, radar detection limits
Coast Guard Boarding Team	Provide source security, data files for detection
Swedish Team	Provide video surveillance and radar detection
Austrian Team	Provide video surveillance and radar detection
Singapore Team	Simulate drive by radiation detection

her disposal. Decision-making of the tactical or operational commanders may differ depending on the details of each boarding case, their parent organizations, relevant policies and standing orders. The primary purpose of the collaborative network is to support information exchange and situational awareness among the various participants.

The boarding party physically enters the vessel and conducts a thorough visual and sensor inspection to locate the source of radiation detected by the drive-by sensor. The boarding party passes sensor data to LLNL for analysis and evaluation of the type and severity of the source. Singapore provides initial information to maritime intelligence law enforcement agencies such as MIFC (such as, to explain the presence of specific materials and/or personnel onboard the suspect vessel). Sweden acts as a counterpart MIO agency, exchanging real-time information that might be useful to them or to the TNT operation.

Cognitive complexity of scenarios

Scenarios used for this research focus on detecting, identifying and interdicting nuclear materials in open waters. The critical task involves the cognitively complex issue of discrimination – that is, how to determine the presence of contraband radiological material against a background containing multiple benign radiation sources. A variety of benign sources, such as smoke detectors, radiant signs, a container load of bananas all share the ability to be moved in commercial vessels and all three can cause radiation detectors to alarm (Schwoegler 2006). For example, 'smoke detectors contain small amounts of americium, radiant signs glow because they contain tritium, a radioactive hydrogen isotope, and bananas, contain a small fraction of potassium-40 which emits ionizing radiation' (ibid, p. 4). A container full of bananas will cause a Geiger counter to tick.

Technical expertise, provided by remotely located experts, is required to interpret the radiological signals emitted from complex detectors to enable on-site personnel to make the fine discriminations required. Performing these complex discriminations is made possible by the collaborative capability provided by the collaborative workspace in terms of bringing remote expertise to the vessel undergoing the search and the ability to rapidly send and receive communications between a diverse team of experts who all bring their respective expertise to bear to deal with a potentially high-threat situation.

A large effort has been invested in increasing the cognitive complexity of the scenario for each follow-on experiment. This includes meetings with the various role players and participants, as well as weekly telephone conferences to discuss ways to increase the fidelity of the scenarios in order to produce more realistic cognitive activity and discussion among the various participants. The last several experiments extended the number of participating organizations to include two international teams in Sweden and Austria, and San Francisco Police Department and Alameda County Marine Units. The C2 part of the scenario was designed to be more cognitively challenging and the chain of command is beginning to reflect more of the actual organizations involved in a real-world MIO.

Conceptual model of team collaboration

The model of team collaboration, described by Warner, Letsky and Wroblewski in Chapter 1 (this volume) defines macrocognitive processes that guide team collaboration. These macrocognitive processes include:

1. individual conversion of data to knowledge,
2. team integration of individual knowledge for common understanding,
3. team agreement on a common solution, and
4. solution adjustment to fit goals and exit criteria.

The model defines information processing components the team performs to achieve each collaborative stage, and emphasizes cognitive aspects of the collaboration process. Major cognitive processes that underlie this type of communication include:

1. individual knowledge construction,
2. collaborative team problem-solving,
3. team shared understanding/consensus,
4. outcome evaluation and revision.

Validating the model of team collaboration

Verbatim transcripts were analyzed from two series of experiments and one real-world situation where teams collaborated to solve a complex problem. Decision-making domains included MIOs, air warfare decision-making scenarios, and one real-world large-scale firefighting situation. In all three problem-solving tasks assessment is particularly difficult due to time pressure, high workload and because the available information is often incomplete or ambiguous. Transcripts included communications that transpired between all team members as well as with decision-makers at the distributed sites.

Types of problem-solving situations

The model of team collaboration is a structural model that emphasizes the cognitive processes employed by teams engaged in the types of problem-solving situations that are characterized as one-of-a-kind, time-compressed, high-stakes problems where decision-makers make use of uncertain, open-source data. For the MIO task, time pressure manifests itself in the urgency to conduct the boarding and visual inspection as quickly as possible so as to not detain the ship any longer than necessary. Uncertain data refers to the overlapping parameters between different types of radioactive materials which makes identifying the source particularly challenging. An additional source of ambiguity is the imperfect nature of the sensors. Many sensors have different degrees of accuracy that can vary depending on environmental conditions (such as weather and atmospherics) and other factors.

The dynamic situation also creates ambiguity, in that the situation is ill-structured and ill-defined: therefore the level of threat presented by the contraband materials may change quickly as new information is discovered. High cognitive workload refers to a situation where there are many tasks to be performed under time pressure and they entail a large amount of highly specialized knowledge. Operational tasks include team data processing, developing shared situational awareness, team decision-making and course of action selection.

Team types

Team types described by the model include teams who operate asynchronously, whose members are distributed, and culturally diverse, as the scenario involved a multinational team, where members possess heterogeneous knowledge, due to the unique roles played by each team member, and who operate in a hierarchical organizational command structure, and in some situations involve rotating team members (Warner *et al.* 2004). Members of the boarding party team each have distinct roles and bring their respective expertise (such as radiological detection) to bear, and combine their heterogeneous knowledge to arrive at a final solution and course of action. A large body of heterogeneous knowledge is involved as team members bring their respective areas of expertise to bear to solve the problem.

Approach to measuring team collaboration

Macrocognition refers to ‘a level of description of the cognitive functions that are performed in natural (versus artificial) decision-making settings’ (Klein, Ross, Moon, Klein, Hoffman and Hollnage 2003, p. 81). To gain insight into the role of these cognitive processes, verbatim transcripts were analyzed using the cognitive process definitions developed by Warner *et al.* (2004). Transcripts included communications that occurred between all team members as well as with decision-makers at the distributed sites. The focus of the collaboration model is on knowledge-building among the team members, developing team consensus and selecting a course of action. This research builds on previous work to validate this model (Warner *et al.* 2004). The current effort uses a similar methodology applied to three different decision-making scenarios (Hutchins, Bordetsky, Kendall and Garrity 2007).

Coding process

Cognitive process coding definitions developed by Warner *et al.* (2004) were used to code all speech turns. The coders attempted to develop criteria for applying the coding schema as some coding categories appear to have similar meanings. This codification of the coding process is part of the overall validation of the model, in that one goal is to have high inter-rater reliability between coders. The first time a team member discussed a track the speech turn was coded as a two – *individual mental model* (imm) construction – where an individual team member, using available information, develops their mental picture of the problem situation. After three speech turns that discussed the same track (typically involving at least four of the team members) it was coded as a 4 – *team knowledge development* (tkm) – where all team members participate in clarifying information to build team knowledge. Once five–six team members had discussed a track it was coded as a 10 – team shared understanding development – which includes discussion among all team members on a particular topic or data item. As more team members get involved in discussing a contact (i.e., more reports and/or updates have been shared among team members, the cognitive process coding category reflects a higher level of team understanding of the situation regarding that particular track.

Results

Table 21.2 presents the number of speech turns for air warfare, MIO, and the 9-11 firefighting tasks coded as representing each of the macrocognitive processes included in the model. The large number of speech turns coded as categories 1–4 reflects the huge emphasis on individual *knowledge construction* that is required for all three tasks.

Evidence was found for all seven cognitive processes that occur during the *collaborative team problem-solving* phase (7–13), where teams integrate individual knowledge to develop a team common understanding, indicating the role these cognitive processes play for teams who engaged in all three tasks. Far fewer speech turns were coded as representing processes included in the *team consensus* phase of collaboration (14–18).

Significant macrocognitive processes

Macrocognitive processes frequently employed for MIO problem-solving, based upon data analyzed thus far, emphasize cognitive processes used to manage uncertainty; these include development of individual task knowledge, individual mental models and team knowledge. The macrocognitive process employed most frequently – across all decision-making domains – is individual task knowledge development. Individual task knowledge development (#3, itk) is defined as a team member asking for clarification to data or information, or a response to a request for clarification. The large number of speech turns coded as itk (3) reflects the high degree of uncertainty inherent in these decision-making tasks.

During the collaborative team problem-solving stage the most frequently used process varied between all three decision-making domains. The most frequently used macrocognitive processes for the air warfare, MIO and firefighting domains, respectively, were team shared understanding, iterative collection and analysis and convergence of mental models. These differences reflect differences in the tasks performed for these three domains. Air warfare teams focus on developing team shared situation awareness regarding the status (threat/non-threat) of the various tracks (air and surface) in the vicinity of the ship. During MIOs teams focus on iterative collection of data, and receiving feedback on results of analysis of this data which drives additional data collection until they are certain regarding the types of

Table 21.2 Cognitive process coding frequencies for air warfare, MIO and 9-11 firefighters scenarios

		Air Warfare Scenarios				MIO Scenarios			Firefighting
	Macrocognitive Process Coding Categories	**Scen D-Run A**	**Scen D-Run B**	**CG-59**	**DDG-54**	**Nov 06**	**June 06**	**Sept 06**	**Firefighters 9-11**
	Knowledge Construction								
1.	Data to information (dti)	1	4	-	37	2	5	-	2
2.	Individual mental model (imm)	8	11	18	25	1	7	8	14
3.	Individual task knowledge development (itk)	25	30	31	29	35	7	47	325
4.	Team knowledge development (tk)	11	5	18	1	3	5	8	210
5.	Knowledge object development (ko)	-	-	-	-	-	2	8	-
6.	Visualization and representation (vrm)	-	-	-	-	-	-	-	-
	Collaborative Team Problem Solving								
7.	Common understanding (cu)	-	6	-	-	2	6	7	16
8.	Knowledge interoperability (kio)	-	5	-	1	2	-	10	8
9.	Iterative collection and analysis (ica)	1	11	-	-	6	4	14	-
10.	Team shared understanding (tsu)	1	17	28	34	3	2	3	6
11.	Solution alternatives (sa)	-	3	-	-	6	-	-	13
12.	Convergence of mental models (cmm)	1	-	-	-	1	-	-	22
13.	Agreement on Common solution (cs)	-	2	-	-	-	-	-	1

Table 21.2 (continued)

		Air Warfare Scenarios				MIO Scenarios			Firefighting
	Macrocognitive Process Coding Categories	**Scen D- Run A**	**Scen D- Run B**	**CG-59**	**DDG-54**	**Nov 06**	**June 06**	**Sept 06**	**Firefighters 9-11**
	Knowledge Construction								
	Team Consensus								
14.	Team negotiation (tn)	-	-	-	-	4	-	-	1
15.	Team pattern recognition (tpr)	-	-	-	-	-	-	-	3
16.	Critical thinking (ct)	-	-	-	-	-	-	-	3
17.	Sharing hidden knowledge (shk)	-	2	-	-	-	-	-	5
18.	Solution adjustment against goal (sag)	-	-	-	-	-	-	-	-
	Outcome Evaluation and Revision								
19.	Compare solution options against goals (csg)	-	1	-	-	-	-	-	2
20.	Analyze, revise solutions (aro)	-	-	-	-	-	-	-	1
21.	Miscellaneous (misc)	38	27	57	61	6	-	-	-
22.	Issue order regarding course of action (coa)	7	5	17	37	-	-	2	92
23.	Request take action (rta)	3	2	18	8	1	2	11	53
	Totals	96	131	187	233	73	40	118	777

materials found on the vessel. During the 9-11 firefighting task, many completely unanticipated events occurred which made it necessary for the firefighters to develop a new mental model of the situation they were confronted with in response to each new event: a terrorist attack, the collapse of World Trade Center (WTC) 1, collapse of WTC 2, and not being able to rescue people due to the collapse of the two buildings.

Firefighters from 9-11

Following the collaboration stages and cognitive processes icluded in the model of team collaboration, the first problem for the firefighters on 9-11 was to figure out what happened and develop a mental model within which the Fire Department of New York could work. One striking difference between the 9-11 data and the two sets of coded experiments is the difference in the percentage of speech turns coded *individual task knowledge* development, itk (team member asking for clarification to data or information). Twenty-six percent of the total number of speech turns in experiments 1 and 2 were coded itk compared to a significantly larger percentage, i.e., 41 percent, for 9-11 data ($p < 0.0001$, $z = 6.5$, combined $n = 1555$). The second largest coded difference was tk, or team knowledge development, which comprised 7 percent for the combined experimental groups and 27 percent for the 9-11 data ($p < 0.0001$, $z = 11.03$, $n = 1555$). Itk and tk combined represent 32 percent of the experimental teams' speech turns and a majority of speech turns for the 9-11 team (69 percent). A majority of the communications for 9-11 involved *knowledge construction* and significantly less team communications for the other stages: collaborative team problem-solving, team consensus and outcome evaluation and revision.

Coding examples of collaborative cognitive processes

Table 21.3 presents an excerpt of the communications coding from a MIO scenario where the team is developing solution alternatives by using data to justify a solution. First (1), individual team members (TMs) are developing team shared understanding by clarifying data regarding the degree of danger inherent in the material discovered and using data to justify a solution (2) and exchanging knowledge among each other, i.e., the material needs to be confiscated (3), based on information provided by one of the remote centers (the material needs to be handled carefully). An individual exchanges knowledge with other TMs (4) to develop knowledge interoperability regarding whether the Coast Guard ship has a suitable storage area for the confiscated material (5). Finally, TMs combine individual pieces of knowledge to achieve a common understanding (6) regarding the next action to be taken.

Table 21.4 presents an excerpt that represents the team developing knowledge interoperability and agreement on a final plan.

Table 21.3 Excerpt from MIO scenario: communications coding for developing solution alternatives

	MIO Team Communications		Cognitive Process Coding	
	Speaker		Code	
1	DTRA	Cesium 137 can be used to make an RDD. If there are no explosives, then it is not configured as a weapon yet. Recommend material be confiscated.	tsu sa	*Team shared understanding*: Discussion among all team members on a particular topic or data item. *Solution alternatives*: Develop, ration-alize and visualize solution alternatives; using data to justify a solution
2	BO	Roger will confiscate.	itk	*Individual task knowledge* development; individual TM clarifying data.
3	BO	Make sure you handle carefully. Cs-137 is an external gamma hazard.	kio	*Knowledge interoperability*: TMs exchanging *knowledge* among each other.
4	BO	Roger. Will take precautions.	kio	*Knowledge interoperability*: TMs exchanging *knowledge* among each other.
5	SOCOM	Does CG ship have proper storage area for material confiscated?	itk	*Individual task knowledge* development: individual TM clarifying data, asking for clarification.
6	SOCOM	Search team will report size of material and its current containment condition; then make recommendations.	cu	Team integration of individual TM knowledge for *common understanding*; one or more TMs combine individual pieces of knowledge to achieve common understanding.

Table 21.4 MIO scenario: knowledge interoperability development and agreement on a final plan

	MIO Team Communications		Cognitive Process Coding	
	Speaker		Code	
1	BO	Negative for explosives Station 2.	kio	*Knowledge interoperability*: TMs exchanging *knowledge* among each other.
2	LLNL	Finally received RAD data from station 2.	kio	*Knowledge interoperability*: TMs exchanging *knowledge* among each other.
3	SOCOM	Will need to resolve RAD containment hazard if it exists.	cu	Team integration of individual TM knowledge for *common understanding*; one or more TMs combine individual pieces of knowledge to achieve common understanding.
4	DTRA	If you have plutonium, you need to confiscate. It's an alpha hazard, but still must be handled carefully.	ica	Iterative *information collection and analysis;* collecting and analyzing information to come up with a solution but <u>no specific solution exists</u>.
5	BO	Roger.	Misc	Acknowledge report.
6	DTRA	By the way, if plutonium is in solid metal form, your team can handle safely with rubber gloves and a dental face mask, depending on how much is there.	ica	Team shared understanding development – discussion among <u>all</u> team members on a particular topic or data item.
7	BO	Talking to search team to see if this is within their capabilities or if we will need outside assets.	kio	*Iterative information collection and analysis;* collecting and analyzing information to come up with a solution but <u>no specific solution exists</u>.
8	LLNL	Hazard is probably minimal, can isolate and confiscate.	cs	Team agreement on a *common solution* – all team members agree on the <u>final plan</u>.

New coding categories

Several new coding categories were added to the set of macrocognitive process definitions that are included in the model. The first involves a person in command issuing an order regarding a course of action, that is, a person with higher rank tells them to take some specific action against a potential threat track. This could include issuing verbal warnings, illuminating or locking-on with radar, developing a firing solution, covering with missiles, etc. A second new coding category is *Request a team member take an action* which involves asking a team member to do something. The difference between these two categories is that the latter does not involve a direct action against a threat track, e.g., 'Can you try and change 7006 and 7005 to assumed hostile?'

Discussion

Communications involved in all three tasks did not include many speech turns involving the cognitive processes included under the team consensus or outcome evaluation and revision phases included in the model (processes 14–20). Course of action selection for air warfare tasks – and based on the team communications analyzed thus far, for the MIO and firefighting tasks – tends to be done less collaboratively, due to the inherent time pressure to make decisions and take actions. Klein (1993) found that when decision-makers use a recognition-primed decision-making strategy to perform decision-making tasks, usually the situation itself either determines or constrains the response options and that experienced decision-makers make up to 90 percent of all decisions without considering alternatives.

Very few of the cognitive processes included in the team consensus stage were evident in the team communications that transpired during these three collaborative tasks. All three tasks are tactical level tasks; that is, these operations are conducted within a very short time period. Due to the speed of a potentially hostile aircraft, political/economic pressure to conduct the MIO quickly, and for obvious reasons for the firefighters, these decisions need to be made within minutes, or even seconds, as the situation unfolds. In contrast, operational and strategic level operations occur over longer time intervals, e.g., weeks or months. More of the macrocognitive processes included in the collaborative team problem-solving and team consensus stages may occur in collaborative problem-solving tasks that span a longer time period.

Anther explanation for the lack of evidence for the macrocognitive processes associated with the team consensus and outcome evaluation and revision stages is that the MIO experiments have not focused on the C2 part of the overall task. The focus for the MIO experiments has been on assessing new technologies and not on human decision-making. In future MIO experiments the human decision-making part of MIOs will be included. We expect to see evidence for more of the macrocognitive processes in follow-on MIO experiments.

Challenges

Major gains regarding including increased cognitive complexity have been made, yet the scenarios do not always play out to completion, i.e., selection of final course of action. Because the emphasis is on technological capabilities, once sufficient data files have been exchanged, the experiment is determined to be complete. For example, when a team member requested a worst case analysis based on atmospheric modeling, the estimated effects of plume analysis indicated the ship was in a 'bad location' (i.e., near a highly populated area where people might die and there would be potentially severe medical effects). The team considered a solution alternative: if the boat could be moved outside San Francisco Bay, the impact on the public could be reduced. However, they were told by higher command to continue with boarding and that analysis of worst case does not modify current tasking. Another team member reiterated: 'This file shows the effects, if this material is dispersed', as if trying to get convergence of the individual's mental model to the team mental model, but then the discussion shifted back to collecting more data.

Conclusions

Rapid access to current, accurate and relevant information, and the ability to engage in real-time collaboration with other decision-makers who are geographically distributed, have become indispensable elements of the C2 planning and decision-making process. Describing the cognition involved in performing collaborative decision-making tasks is an essential step prior to developing systems to support these types of tasks. Designing complex cognitive systems based on inaccurate views of cognition often produce systems that degrade performance rather then improve it.

References

Bordetsky, A., Dougan, A., Foo, Yu, C. and Kihlberg, A. (2006). 'TNT Maritime Interdiction Operation Experiments: Enabling Radiation Awareness and Geographically Distributed Collaboration for Network-Centric Maritime Interdiction Operations'. *Proceedings of the 12th International Command and Control Research and Technology Symposium, Newport, RI, 19-21 June.* Published on CD. (Washington, DC: DoD Command and Control Research Program). Also available at http://wed.nps.navy.mil/~brutzman/International/MaritimeProtecxtionSyumposium2006/.

Hutchins, S.G., Bordetsky, A., Kendall, A., Looney, J. and Bourakov, E. (2006). 'Validating a Model of Team Collaboration', in *Proceedings of the 11th International Command and Control Research and Technology Symposium, Cambridge, UK, 26-28 September.* Published on CD. (Washington, DC: DoD Command and Control Research Program).

Hutchins, S.G., Bordetsky, A., Kendall, T. and Garrity, M. (2007). 'Empirical Assessment of a Model of Team Collaboration', in *Proceedings of the 8th International Conference on Naturalistic Decision Making, Asilomar, California, 4–6 June*. Published on CD. (Pacific Grove, CA).

Klein, G.A. (1993). 'A Recognition-Primed Decision (RPD) Model of Rapid Decision Making', in G.A. Klei Orasanu, R. Calderwood and C.E. Zsambok, (eds), *Decision Making in Action: Models and Methods*, pp. 138–147. (Norwood, NJ: Ablex).

Klein, G., Ross, K.G., Moon, B.M., Klein, D.E., Hoffman, R.R. and Hollnagel, E. (2003). 'Macrocognition', *IEEE* May/June.

Richard, M. and Roth, D. (2005). 'How the Global Information Grid is Transforming Communications for the Warfighter', *EDGE Magazine*, MITRE, Fall, available at http://www.mitre.org/news/the_edge_fall_05/richard.html.

Schwoegler, D. (2006). 'Marine Experiment Tests Detection Capability', *NEWSLINE*, Lawrence Livermore National Laboratory, 29 September.

Sturk, B., Col. and Painter, R., Lt Col. (2006). C2 Constellation and Constellation Net Briefing, AFC2ISR Warfighter, Langley Air Force Base. website: http://www.afc2isrc.af.mil.

Truver, S.C. (2001). 'Spearheading Joint Transformation—And Supporting Homeland Defense', *Sea Power*, December, 46–48.

US Joint Forces Command (2001). *Global Information Grid Capstone Requirements Document*. JROCM 134-01, availble at https://jdl.jwfc.jfcom.mil/. (Norfolk, VA: USJFCOM).

Warner, N., Letsky, M. and Cowen, M. (2004). 'Cognitive Model of Team Collaboration: Macro-Cognitive Focus'. *Proceedings of the 49th Human Factors and Ergonomics Society Annual Meeting, September 26-30, Orlando, Florida*. Published on CD. (Santa Monica, CA: Human Factors and Ergonomics Society).

Chapter 22

Naturalistic Decision-making Based Collaboration Scenarios

Elizabeth Wroblewski and Norman W. Warner

Introduction

In recent years, military communication and information technology has rapidly evolved at both the tactical and operational levels. In addition, a shift to network-centric operations has provided the warfighter with an unprecedented ability to share and leverage information (Garstka 2006). Reliance on effective collaboration tools for distributed teams has become critical to military efforts in decision-making, mission planning and intelligence analysis. To ensure continued effectiveness, it is imperative this advancing technology enhance the collaborative effort rather than impede it. This chapter discusses the development and use of naturalistic decision-making-based collaboration scenarios for better understanding the macrocognitive processes of team collaboration. By better understanding the processes of team collaboration through empirical research more effective team collaboration tools can be developed.

Historically, the development of collaboration tools has been the result of intuition. While intuition is, undoubtedly, a precursor to technological advancement, the key to successful collaborative tool development lies in empirical research. The challenge of understanding team collaboration is to study behavior in a controlled, but rich environment, emulating a real-world team collaborative problem-solving setting as closely as possible. Critical to this is the use of an effective scenario. Heuer (1999) defines a scenario as a 'series of events leading to an anticipated outcome and linked together in a narrative script'. In the Collaborative Operational and Research Environment (CORE) testbed at the Naval Air Systems Command (NAVAIR) in Patuxent River, MD, scenarios are used to elicit information about the internal and external macrocognitive processes of collaborative decision-making. As discussed in Chapter 19, the CORE testbed was designed to evaluate the cognitive processes associated with team collaboration. The Team Collaboration Research Spectrum, the foundation of experimental research in the CORE facility, consists of four closely integrated phases:

1. a traditional, controlled research study in a laboratory setting;
2. a controlled field experiment;
3. a controlled experiment in a real-world setting; and
4. field operations in which the researcher closely observes experts responding to a real-world event.

Phases 1, 2, and 3 rely on a common, carefully scripted scenario, while Phase 4 focuses on the observation of experts in a real event that closely parallels the scenario used in the earlier phases. For example, if a scenario used in Phases 1–3 requires participants to develop an emergency evacuation plan for a large city during a natural disaster, Phase 4 observations of developers constructing an evacuation plan for New Orleans during Hurricane Katrina would be a comparable real-world event. Each phase of the spectrum builds upon this common scenario, while gradually decreasing the amount of researcher control. Use of this research spectrum allows for progressive validation by increasing the fidelity of collaboration studies through a series of experimentation and observation.

Carefully crafted scenarios can be used as stimuli to understand the high level internal and external macrocognitive processes of collaborative decision-making. As defined in Chapter 1, and based on Warner, Letsky and Cowen's (2005) conceptual Model of Team Collaboration, internal processes are those macrocognitive collaborative processes that occur inside the head. As a result, they can only be measured indirectly through qualitative metrics (e.g., questionnaires, cognitive mapping) or surrogate quantitative metrics (e.g., pupil size, galvanic skin response). Examples of internal processes would be developing mental models and understanding concepts. External processes are those macrocognitive processes that are associated with observable and measurable actions in a consistent, reliable, repeatable manner or through standardized conventions of a specific domain. Examples would include writing, gesturing and speaking. Sample metrics for external processes include discourse analysis, and automated communication flow analysis.

Naturalistic decision-making scenarios

Careful construction of a scenario is critical to the effectiveness of any collaboration study that relies on the use of these event-based scenarios as a foundation for its experimental design. Not only must the scenario be inclusive enough to elicit the desired behavior, it must accurately represent the domain being studied. Results from a study that focuses on one specific domain would not necessarily be applicable in a different domain. For example, findings from a study using a scenario which focuses on decisions medical staff are required to make would undoubtedly not be applicable to the domain of warfighter decision-making. A critical first step for a developer of an event-based scenario is to not only fully comprehend the intended domain, but to clearly understand how an expert would respond to the given situation. Because the focus of the CORE testbed is to understand the macrocognitive processes, it is essential the scenarios used elicit behavior on the cognitive aspects of decision-making.

Naturalistic decision-making (NDM), the study of how experts make decisions in demanding, real-world situations (Klein 1998; Klein *et al.* 2003; Zsambok and Klein 1997), provides a unique methodology for scenario construction as it pertains to domain relevance and cognitive processes. These demanding situations are marked by a variety of constraints such as time pressure, uncertainty, poorly defined problems, uncertainty, ill-defined goals, organizational constraints, dynamic conditions and varying amounts of experience. The NDM framework uses a variety of research strategies to understand the cognitive functions of real world decision-making. While a variety of research methods are used by NDM supporters, cognitive task analysis, decision requirements table, and critical decision method are particularly effective tools for eliciting and representing knowledge for scenario development. Observation and structured interviews are at the heart of these strategies.

The cornerstone of NDM research is the Cognitive Task Analysis (CTA). The CTA provides a procedural methodology for understanding the cognitive skills used by experts performing a given task (Phillips *et al.* 2001). The intent of the CTA is to allow the researcher to attempt to understand the decision-making process of an expert. The first step of the CTA is to develop a task diagram to provide a broad overview of the task. In addition, it allows the researcher to identify those parts of the task that require specific cognitive skills. The steps required to complete a task are recorded in chronological order. The subject matter expert (SME) is interviewed for insight into those steps which require assessment, judgment, decision-making, problem-solving – any cognitive-based mental process (Klein Associates 2005). This task diagram serves as the foundation for the scenario.

The knowledge elicitation phase of the CTA provides elaboration on the responses to the cognitive tasks identified in the task diagram (Klein Associates 2005; Phillips *et al.* 2001). Using this technique, the researcher focuses on extracting information through in-depth interviews, probes and observations. Extraction methods can be either direct (e.g., interview, teach back, questionnaire, storyboarding) or indirect (e.g., role playing, twenty questions). The method chosen depends on the complexity of the information, as well as the willingness and ability for the SME to articulate the information. Burge (2007) provides a comprehensive list of knowledge elicitation methods.

Another NDM elicitation technique applicable to the scenario development process is the Decision Requirements Table (DRT). The DRT is an analytical method used to break down specific incidents into the key cognitive components which impact decision-making (Klein Associates 2005). Based on a single decision point in the CTA, an SME is asked to elaborate on several key points:

1. Why was this decision a cognitive challenge?
2. What mistakes would a novice make in the same situation?
3. What sensory perceptions (e.g., sight, smell, hearing) were noted during the situation?
4. What kind of knowledge did you bring into the situation?
5. How did you apply the strategies you used in dealing with this situation?
6. What information sources did you use to make your decision?
7. What would you do differently next time you're faced with this decision?

If a scenario developer intends to develop a realistic, rich, cognitive-based scenario, the DRT serves as an invaluable tool.

Critical Decision Method (CDM) is another constructive NDM technique that can easily transition to meet the needs of scenario development. While expert observation, the heart of NDM, is critical to understanding the specific domain, CDM allows the developer to narrow the scope of the scenario. Klein, Calderwood and Macgregor (1989) define CDM as 'a retrospective interview strategy that applies a set of cognitive probes to actual non-routine incidents that required expert judgment or decision-making'. Once a critical incident has been identified, the researcher asks the SME a series of probing questions to delve deeper into the cognitive tasks. These questions focus on incident identification, timeline verification, deepening the story and asking what-if queries to elicit information on alternative strategies (Klein Associates 2005). By gaining a clear understanding of the cognitive tasks, the developer can easily incorporate these tasks into the scenario, ultimately resulting in a richer, more clearly defined scenario.

Collaboration scenarios

Event-based scenarios are used extensively throughout the fields of business, education and research. In collaboration research, four types of scenarios are commonly used: gaming or pictorial scenarios, one dimensional scenarios, multidimensional scenarios, and multidimensional/multidomain scenarios. The type of scenario used is dependent on the intent of the research.

Gaming scenarios These empirically based gaming scenarios rely on strategic, interactive games to serve as their foundation. Kirsch's (2005) 'Passing the Bubble' (see Chapter 10) serves as a representation of a typical gaming scenario. Passing the Bubble is a computer-based game which requires participants to annotate a series of video and still images. The intent is to measure the effects of static versus dynamic annotation on team performance, as well as to determine the quality of shared understanding. The strength of gaming scenarios lies in their simplicity. While the range of complexity varies, these games typically rely on well-defined, but minimally complex tasks. As a result, there is a short learning curve for participants, allowing the researcher to spend less time training and more time on task. In addition, a typical gaming scenario requires a relatively low level of subject matter expertise, thereby increasing the participant pool dramatically. Ironically, the simplicity of gaming scenarios limits their use in collaboration research. While gaming scenarios can provide high-level insight into simplistic decision-making, responses tend to be tactical in nature, providing little insight into a complex domain. Military domains, for example, are complex, dynamic environments. Simplifying the scenario allows for little operational relevance and transference. To gain insight into collaborative decision-making, it is imperative that researchers distinguish between team coordination and team collaboration. Cooke and Kiekel (2006) define collaboration as 'the identification, collection, fusion and presentation of data for attaining a shared understanding and sufficient comprehension of a situation'. They define coordination as 'the timely and adaptive sharing of information among team members/entities'. Typical gaming scenarios focus on coordination rather than collaboration between team members. By nature of their simplicity, true collaboration would be virtually impossible.

One-dimensional scenarios One-dimensional scenarios used in collaboration research usually require team members to solve a relatively simple, single solution task. These scenarios tend to be more collaborative in nature than gaming scenarios. Although they tend to be a bit more complex than gaming scenarios, they still are relatively simplistic, rendering it unnecessary to use highly trained operational personnel. Their limitation, however, lies in the fact that once a consensus is reached, the task is complete. No further negotiation or discussion is required. As a result, the metrics used for one dimensional scenarios tend to be outcome-based (e.g., were the subjects right or wrong) rather than process-based. This, again, provides somewhat limited insight into the macrocognitive processes of team collaboration. In Stasser *et al.*'s (1995) 'The Case of the Fallen Businessman' teams are required to solve a murder mystery by evaluating transcripts from police interviews, newspaper clippings and other forensic evidence. Once they have identified the suspect they believe committed the murder, the collaborative task ends. Warner *et al.* (2005) used this scenario to provide empirical data on the validity of the collaboration stages and macrocognitive processes of their conceptual model of team collaboration. The results supported the existence of knowledge construction, collaborative problem-solving and team consensus stages during collaborative problem-solving, as well as several high-level macrocognitive processes. However, this study failed to elicit evidence of the outcome evaluation and revision stage. Simply put, once a team has reached a decision, there is no need for further discussion – an unrealistic occurrence in military collaboration.

Multidimensional scenarios Similar to one-dimensional scenarios, multidimensional scenarios allow for a collaborative exchange between team members. The difference lies in their complexity. While one-dimensional scenarios require participants to reach a consensus on a single issue, multidimensional scenarios required several, complex solutions. The number of possible solutions is limited only by the team's creativity, resulting in a more realistic, NDM-based scenario. While the simplicity of one-dimensional scenarios limits the collaborative process, multidimensional scenarios allow for a rich, collaborative environment. As a result, the metrics used for empirical studies using these event-based scenarios can not only be outcome based (e.g., percentage correct), but also processed-based (e.g., utterance frequency by stages and process states across the independent variables, transition probabilities). This allows for a more thorough and concise analysis of the macrocognitive processes of team collaboration. The *Non-combatant Evacuation Operation (NEO): Red Cross Rescue Scenario* (Warner, Wroblewski and Shuck 2004) developed for the ONR's Collaboration and Knowledge Interoperability (CKI) program serves as an example of a multidimensional scenario. This scenario, which serves as the cornerstone of NAVAIR's CORE testbed, requires participants to develop a course of action to rescue three Red Cross workers trapped in a church basement on a remote island in the South Pacific. Required elements of the final plan include a list of US forces to be used in the rescue, means of transportation, weapons to be used, a specific timeline of events and a detailed explanation of the plan. Each team member is given the General Information section of the NEO scenario, which includes a situation report, an explanation of US military assets available, information on hostile forces in the area, topographical maps and other descriptors of the island. In addition, each team member is assigned as an expert in local intelligence, available weapons or local environmental issues, and provided with information pertinent to their area of expertise. An appropriate plan could be developed by combining both the general and expertise information. While the NEO scenario emulates a military domain more closely than a one-dimensional scenario, it was designed for non-military participants. Unlike Stasser's murder mystery scenario, there is no right or wrong answer to the NEO scenario. The final plan is scored, however, using a complex scoring matrix developed with input from military operational personnel. Points are deducted for unrealistic components of the solution (e.g., choosing an aircraft with a limited range without addressing refueling needs, failing to address medical treatment for the workers, failing to include all requirements of the final plan). This score serves as the basis for the outcome metrics of the study. The NEO scenario has proven to be adaptable to a variety of research environments. Modified versions of the NEO have been used in a variety of research conducted through ONR's CKI program. It was modified to support usability and functionality studies for Pacific Science and Engineering's Integrative Decision Space (I-DecS) Interface (Handley and Heacox 2006) (see Chapter 13) and Evidenced Based Research's Collaboration Advizor Tool (CAT) (Noble *et al.* 2004) (see Chapter 19). A web-based version was created for MIT's Electronic Card Wall (EWall) (Keel 2006) (see Chapter 11). In addition, it has been used in the Communication and Meaning Analysis Process (C-MAP) research at the University of Tennessee (Rentsch 2006) (see Chapter 8) and Arizona State University communication analysis studies (Cooke and Kiekel 2006) (see Chapter 4).

Multidimensional/multidomain scenarios To ensure a comprehensive empirical investigation into the macrocognitive processes of team collaboration, it is essential the scenario chosen serves as an accurate representation of the intended domain. The complexity of multidimensional/multidomain scenarios ensures a sense of realism and operational relevance as it applies across integrated domains. It requires situational constraints (e.g., time pressure, information uncertainty, dynamic information, cognitive overload) to ensure the realism critical to content validity.

Equally as important, they elicit a high level of team collaboration so critical to the investigation. Because the focus of a multidimensional/multidomain scenario is on operational relevance, it must be carefully constructed, with input from subject matter experts at every stage of its development. Naturalistic decision-making (NDM) elicitation techniques (cognitive task analysis, decision requirements tables, and critical decision methods as describe above) offer a methodology for understanding the cognitive tasks (and expert responses) associated with a given situation. Through this thorough understanding, the scenario developer will be better equipped to develop an effective, rich and clearly defined scenario. While still under development, The Special Operations Reconnaissance Scenario: Intelligence Analysis, Mission Planning, and Mission Execution (Wroblewski, Warner, St. John, Smallman and Voigt 2007) is representative of a multidimensional/multidomain scenario used empirically to investigate the macrocognitive processes of team collaboration. Originally intended as a vehicle for Pacific Science and Engineering's (PSE) Shared Lightweight Annotation Technology (SLATE) tool (see Chapter 16), this scenario is being developed as a joint venture between NAVAIR and PSE, and will ultimately serve as a critical component of the CORE testbed. Modular in its design, this scenario is made up of three distinct, yet carefully intertwined modules: intelligence analysis, mission planning, and mission execution. Researchers can choose to use each module individually or as a whole. While it was designed specifically for operational personnel, the level of complexity can be simplified for non-operational personnel. This fictional scenario requires an intel team, a mission planning team, and operational personnel to complete a series of tasks. The premise of this scenario focuses on the activities of Farah, a coalition-supported warlord in Afghanistan. Module 1 of the scenario requires a team of intelligence analysts to evaluate a collection of historical background, forensic evidence, human intelligence, open source information and satellite imagery to determine if Farah has allegiance to a local terrorist organization. If a tie is established, Module 2 requires mission planners to develop a course of action (COA) for a Special Operations reconnaissance mission to gather further information on Farah's activities. In addition, the mission planners are required to analyze dynamic information and modify the COA as necessary. In Module 3, operational personnel would carry out the mission as defined in Module 2 in a simulated operational setting. This quasi-realistic setting allows the researcher to examine the effects of situational constraints on the collaborative process more thoroughly than through the use of gaming or one-dimensional scenarios.

Generating an effective collaboration scenario

Scenario development is a concise, iterative process that can be broken down into eight steps:

1. define the parameters;
2. research the domain;
3. storyboard the scenario;
4. write the scenario text;
5. validate the scenario;
6. debrief the participants;
7. analyze the data;
8. evaluate and revise the scenario as necessary.

1. Define the parameters More specifically, identify any limitations and expectations. Start with a plausible hypothesis and define the objective. What do you want to accomplish through the use of this scenario? Define the domain that best represents your needs. Define the metrics to measure both internal and external macrocognitive processes. What do you want to measure when the session is over? Should the metrics be outcome or process-based? Identify the appropriate subject pool. Are operational personnel available to serve as participants or do you need to rely on non-military personnel? Establish the time pressure. There should be enough of a time constraint that the participants feel somewhat pressured to make a decision, but not too little time that they feel forced to make a quick decision.

2. Research the domain Carefully research your intended domain. Identify the types of cognitive decisions that need to be made. Conduct a cognitive task analysis or other NDM elicitation techniques to help define a specific cognitive task performed by experts. This critical step serves as the foundation for an NDM-based scenario.

3. Storyboard the scenario Include as many NDM constraints (e.g., time pressure, information uncertainty) as possible to emulate a real environment. Add enough unrelated, ambiguous cues (i.e., red herrings) to add an element of uncertainty. Create scenarios with multiple outcomes to avoid compromise between teams. Establish safety procedures as well as a plan to deal with any contingencies.

4. Write the scenario text Clearly define the mission statement so avoid participant misconceptions of their tasks. Write the scenario as a story. Begin with a general overview, leading to more specific details. Be sure to include all information needed to complete the task. The text should also include clear, concise and carefully scripted facilitator guidelines to ensure repeatability and consistency between researchers.

5. Validate the scenario Progressive validation is critical throughout the developmental process. Continually seek feedback from subject-matter experts to ensure content validity. Once the scenario is written, elicit independent readings to ensure such things as grammatical accuracy and content flow. Conduct a pilot study to identify any problems such as timing, clarity, instructions, procedures and metrics.

6. Debrief the participants Elicit feedback from participants. Ask open-ended questions (e.g., Why did you make the choices you made? What would you do differently?) Explain any results or scores in a positive, non-judgmental manner (Klugiewicz and Manreal 2006).

7. Analyze the data How robust are the findings? Are they dependent on the domain or the collaboration environment (e.g., face-to-face, asynchronous and distributed)?

8. Evaluate and revise the scenario as necessary The development of a scenario is an iterative process. Throughout the process, the developer should constantly ask: Does the scenario tell a realistic story? Does it unintentionally guide the participants towards the right answer? Is there enough uncertainty? Is there more than one solution? Are the allotted time requirements realistic? Are the scoring metrics measuring the appropriate information? Once the need for modification is recognized, it is necessary to repeat each of the steps above.

Common pitfalls

Despite careful preparation, collaboration scenario developers often make basic errors that ultimately will undermine the effectiveness of the scenario:

1. Creating a scenario not representative of the intended domain. In most instances, collaboration research is not transferable across unrelated domains.
2. Failure to obtain SME input throughout the entire development process.
3. Wording the text of the scenario in such a way that participants are guided towards the correct answer. This can be avoided by providing some degree of vagueness in the scenario or using a complex scenario that requires more than one decision.
4. Using inaccurate domain terminology. This can become a major distraction among participants and undermines content validity.
5. Creating a scenario of inappropriate length often results in either rushed decisions or extraneous conversation, either of which will skew the results of any communication analysis.
6. Creating a scenario of inappropriate difficulty. The challenge is to write a scenario that is neither too complex nor not challenging enough. Conducting a careful CTA and obtaining SME input throughout the course of the development process will lessen the likelihood of this occurring.

Conclusion

Writing a simple story is a relatively painless process. Generating an events-based scenario to be used to investigate the macrocognitive processes of team collaboration, however, is a more daunting endeavor. Following the methodology used in naturalistic decision-making research allows the scenario developer to construct a rich, multidimensional/multidomain scenario to be used as a means of understanding the high level internal and external macrocognitive processes of collaborative decision-making. The use of these carefully crafted scenarios allows the researcher to emulate a more realistic team collaborative problem-solving setting. The undeniable result is a more comprehensive understanding of the collaboration process in the intended domain. Because reliance on effective collaboration tools for distributed teams has become critical to the military's efforts in mission planning and decision-making, it is imperative advancing technology enhance the collaborative effort rather than impede it. Continued research into these macrocognitive processes provides a key to improving network-centric operations and efficient warfighter initiatives.

References

Burge, J. (2007). *Knowledge Elicitation Tool Classification*, http://web.cs.wpi.edu/~jburge/thesis/kematrix.html# Toc417957390, accessed 16 January 2007.

Cooke, N. and Kiekel, P. (2006). 'Automatic Tagging of Macrocognitive Collaborative Process through Communication Analysis', *Collaboration and Knowledge Management WorkshopProceedings.* (Arlington, VA: Office of Naval Research, Human Systems Department).

Garstka, J. (2006). *Network Centric Warfare: An Overview of Emerging Theory* <http://www.mors.org/publications/phalanx/dec00/feature.htm>, accessed 10 January 2007.

Handley, H. and Heacox, N. (2006). 'Modeling and Simulation of Cultural Differences in Human Decision Making', *Collaboration and Knowledge Management Workshop Proceedings.* (Arlington, VA: Office of Naval Research, Human Systems Department).

Heuer, R. (1999). *Psychology of Intelligence Analysis.* (Washington, DC: Central Intelligence Agency, Center for the Study of Intelligence).

Keel, P. (2006). 'EWall: Electronic Card Wall: Computation Support for Individual and Collaborative Sense-Making Activities', *Collaboration and Knowledge Management Workshop Proceedings.* (Arlington, VA: Office of Naval Research, Human Systems Department).

Kirsch, D. (2005). 'Passing the Bubble: Cognitive Efficiency of Augmented Video for Collaborative Transfer of Situational Understanding', *Collaboration and KnowledgeManagement Workshop Proceedings.* (Arlington, VA: Office of Naval Research, Human Systems Department).

Klein, G. (1998). *Sources of Power: How People Make Decisions* (Cambridge: The MIT Press).

Klein, G., Calderwood, R. and Macgregor, D. (1989). 'Critical Decision Method for Eliciting Knowledge', *IEEE Transaction on System, Man, and Cybernetics* 19, 3.

Klein, G., Ross, K., Moon, B., Klein, D., Hoffman, R. and Hollnagel, E. (2003). 'Macrocognition', *Human Centered Computing* 18(3), 81–85.

Klein Associates. (2005). *Putting Cognitive Task Analysis for Work: Understanding Users, Customers, and Experts* (Fairborn, OH: Klein Associates, Inc.).

Klugiewicz, G. and Manreal, G. (2006). 'Simulation Training: Getting off on the Right Foot', *Law Officer Magazine*. <www.policeone.com/writers/columnists.lom.articles/134990/>, accessed 27 December 2006.

Noble, D., Christie, C., Yeargain, J., Lakind, S., Richardson, J. and Shaker, S. (2004). 'Cognitive Guidelines to Support Effective Collaboration', *Collaboration and Knowledge Management Workshop Proceedings*. (Arlington, VA: Office of Naval Research, Human Systems Department).

Phillips, J., McCloskey, M., McDermott, P., Wiggins, S., Battaglia, D., Thordsen, M. and Klein, G. (2001). *Decision-Centered MOUT Training for Small Unit Leaders* (Fairborn, OH: Klein Associates, Inc.).

Rentsch, J. (2006). 'Developing Shared Understanding through Knowledge Management: Collaboration and Meaning Analysis Process (C-MAP)', *Collaboration and Knowledge Management Workshop Proceedings* (Arlington, VA: Office of Naval Research, Human Systems Department).

Stasser, G., Stewart, D.D. and Wittenbaum, G. M. (1995). 'Expert Roles and Information Exchange during Discussion: The Importance of Knowing Who Knows What', *Journal of Experimental Social Psychology* 31, 244–265.

Warner, N., Letsky, M. and Cowen, M. (2005). 'Cognitive Model of Team Collaboration: Macro-Cognitive Focus', *Proceedings of the Human Factors and Ergonomics Society 49th Annual Meeting*, pp. 269–273. (Orlando, FL: Human Factors and Ergonomics Society).

Warner, N.W., Wroblewski, E. and Shuck, M. (2004). 'Noncombatant Evacuation Operation: Red Cross Rescue Scenario', *Collaboration and Knowledge Management Workshop*. (San Diego, CA: Office of Naval Research).

Wroblewski, E., Warner, N., St. John, M., Smallman, H. and Voigt, B. (2007). *Special Operations Reconnaissance Scenario: Intelligence Analysis, Mission Planning, and Mission Execution* (Forthcoming).

Zsambok, C. and Klein, G. (1997). *Naturalistic Decision Making*. (Mahwah, NJ: Lawrence Erlbaum Associates).

Chapter 23

Macrocognition Research: Challenges and Opportunities on the Road Ahead

Stephen M. Fiore, C.A.P. Smith and Michael P. Letsky

This volume has presented a series of chapters describing multidisciplinary efforts investigating collaboration in dynamic environments, all in support of understanding complex collaborative problem-solving. As the many chapters in this volume illustrate, multidisciplinary research blending theory and methods from the cognitive, organizational and information sciences has come a long way in helping us understand teams. This body of research has substantially influenced the study of collaborative activity and we are moving towards new theories that transcend disciplinary boundaries. *Macrocognition* is one such theory in that it is being developed to help us understand teams in more complex environments, their interactions with technologies and the measurement of these factors.

For the conclusion to this volume we both summarize these chapters as well as discuss how these contributions can be developed to support this burgeoning field of macrocognition. First we summarize the theoretical and methodological findings as presented in this volume. We follow this with discussion of how these findings can make a difference to our support of complex collaborative problem-solving. We conclude with a discussion of some of the significant challenges facing multidisciplinary research in macrocognition. Our goal is to highlight some of the important issues emerging in this research so as to stimulate thinking on, and to encourage discussion of, future research for macrocognition.

Summarizing theoretical and methodological developments in macrocognition research

The authors of the chapters in this book represent a wide variety of scientific disciplines and we see represented a number of different approaches to the study of teams – ranging from cognitive science, to human factors to industrial/organizational psychology, to organizational behavior and computer supported collaborative work. As such, each chapter provides insights regarding the way that macrocognition is studied within those separate disciplines. While the backgrounds of the authors vary widely, each is attempting to answer how teams are able to maximize their effectiveness. Nonetheless, each author in this book has a somewhat different perspective and in this section we present a synopsis of the contributions to this volume. We do not attempt to be exhaustive in these summaries, rather we choose to highlight what are some of the more critical contributions from this diverse body of research. By highlighting developments we illustrate not only the varied nature of this research, but we also try to capture the theoretical implications of the findings and the tools being developed.

First, what we see in the discussion of the CKI model presented by Warner and Letsky (Chapter 2), is an important stepping off point for macrocognition research. Via the development and test of the initial model along with representative scenarios to examine macrocognitive processes, that is, collaborative activity as manifest in unique and time-stressed problem-solving environments, this research has illustrated the complexity inherent in such tasks. With findings ranging from the domain dependency of macrocognitive processes to their dynamic and developing nature, these initial studies provide an important foundation for models of macrocognition.

McComb (Chapter 3) reviewed the relationship between shared mental models and team performance and discussed how a number of varied relationships have been found. Reporting on what is essentially a common theme throughout much of this volume, she notes the complex relationship between team performance and concepts addressing the sharedness of knowledge within a team. For example, she reports that Mathieu and colleagues (2005) found that mental model accuracy moderates the relationship between teamwork shared mental models and team processes. Additionally, several mediated relationships have been reported. Specifically, team processes have been shown to mediate the relationship between shared mental models and team performance (Marks *et al.* 2002; Mathieu *et al.* 2000); team mental model accuracy partially mediates the relationship between member ability and team performance (Edwards *et al.* 2006); teamwork schema mediates the relationships between percentage of experienced team members and member growth, percentage of experienced team members and team viability and percentage of team members recruited and team viability (Rentsch and Klimoski 2001); and mutually shared cognition mediates the relationship between team learning behavior and team performance (Van den Bossche *et al.* 2006). McComb also found studies reporting that multiple mental models can exist simultaneously, and that the content of these mental models shifts over time. Further, while shared beliefs may exist within groups, these shared beliefs vary across groups within the same organization; mental models also vary between managers/experts and team members when they are confronted with unstructured tasks. From this we see how shared mental models, common ground and cognitive similarity appear to be related constructs in that all have been shown to influence macrocognitive processes in teams; yet we recognize that an important issues here has to do with complementary knowledge rather than simply or only similar knowledge. To help distinguish the differences, if any, McComb offers several measurement techniques to assess the content and structure of knowledge held by team members. These techniques include structured observation, repertory grids and linguistic analysis. McComb's prescriptions provide a roadmap for further studies of the complex relationships between team performance and the degree of cognitive homogeneity.

Clearly, this issue of measurement is one of the primary challenges to the study of macrocognition in teams – that is, finding ways to reliably measure what team members are thinking as they interact. The classic approach of cognitive science is to focus on individual-level constructs such as perception, attention and memory. However, in the context of teams, it becomes difficult to extend these individual-level constructs to useful team-level metrics of process and performance. To help address this, Cooke, Gorman and Kiekel (Chapter 4) consider macrocognition from a human factors perspective and focus on communication analyses. They distinguish between team cognition research that is focused 'in the head' (ITH) and research that seeks to measure 'between the heads' (BTH) processes. Cooke *et al.* argue that an important key to understanding BTH macrocognition is the study of *interactions* among team members. They report on their development of tools to analyze team communication flows and patterns and posit that communications is a large part of team cognition. For example, they report regression analyses suggesting that communication flow techniques like ProNet can be used to discriminate high- from low-performing teams. They also report that flow patterns can be used to differentiate co-located from geographically distributed teams. They caution that while there may be general patterns of flow that are indicative of conflict, problem-solving, or knowledge sharing, flow patterns will, nevertheless, be largely specific to a domain and a task within that domain – indicating the importance of additional research to identify the extensibility of such methods. For example, what may be a pattern indicative of good team performance in an aviation task may be indicative of poor situation awareness in a team planning setting. Analyses of communication flow provide essential clues to the cognitive states of the team members and the techniques reported by Cooke *et al.* are useful for analyzing written and spoken communications. But not all communications can be transcribed into text; teams often engage in critical non-verbal communications, for example by gesture, or by placing coordinating representations into the shared environment, leaving us with a need to come up with innovative ways of addressing these measurement needs.

Hayne and Smith (Chapter 5) discuss the study of self-synchronizing behaviors in teams working without verbal or textual communications. They describe the cognitive mechanisms by which teams learn to recognize patterns and share their knowledge without talking. Hayne and Smith propose that individual team members develop pattern templates having a 'core' of discriminating features, and potentially numerous additional 'slot' features learned through experience. The core features of templates are invariant, and are typically known to multiple members of a team. Team members use their common knowledge of core features as the basis for team pattern recognition. Patterns were communicated within the team through the use of the 'stigmergy' strategy of placing stimulating structures into the shared virtual environment. Hayne and Smith found that performance was improved in a pattern-recognition task when teams were trained explicitly to associate the core features of a pattern into a single chunk, and to assign a common label to the pattern within the team. Placement of artifacts representing chunk labels into the shared environment was all the interaction required for effective team performance. The findings by Hayne and Smith are significant because they suggest a cognitive mechanism by which common ground, shared mental models and cognitive similarity are created. Over time, team members build templates through experience. The templates consist of core features that help to discriminate between different interpretations of the current situation, and slot items that serve to facilitate response selection. Hayne and

Smith find that high-performing teams develop a set of commonly held templates for their task domain and their assigned roles. Shared template labels promote rapid communications within a team through the use of stimulating structures and coordinating representations. These templates appear to be an important bridge between individual-level cognition and team macrocognitive processes.

Another perspective on the consistent issue of cognitive similarity with teams comes from the chapter by Carroll (Chapter 6) who approaches macrocognition from the perspective of Computer Supported Collaborative Work (CSCW). He proposes that macrocognition is greatly affected by the degree of 'common ground' within a team. Carroll argues that larger degrees of common ground enable improved communication, coordination and collaboration within teams. Carroll develops a conceptual model of common ground, and present techniques that researchers can use to manipulate and measure common ground. Nonetheless, he cautions that building common ground within a team is a complex process, and that requirements for support of the common ground process may change depending on the state of the activity and the nature of the task.

In their chapter on agents, Deshmukh, McComb and Wernz (Chapter 7) focus on the important emerging issue of augmenting team cognition through the implementation of intelligent agents. By examining how agent technology is able to address macrocognitive processes, they show how it may be possible to enhance a number of the foundational drivers of team collaboration. Here we come to understand the state-of-the-art regarding human–agent interaction but, importantly, we see areas needing development. From understanding the subtleties of sociotechnical structure when agents are team members, to the more general issues surrounding agent capabilities in monitoring interaction in service of dynamic aiding, we come to a fuller appreciation of the theoretical and technical factors that need to be addressed when considering agent aids to teamwork.

The chapter by Rentsch, Delise and Hutchison (Chapter 8) approaches the topic of macrocognition from the perspective of industrial–organizational psychology. They work with the analogous concept of 'cognitive similarity', defined as 'forms of related meanings or understandings attributed to and utilized to make sense of and interpret internal and external events including affect, behavior, and thoughts of self and others that exist amongst individuals'. Rentsch *et al.* propose that effective decision-making teams must develop a configuration of cognitive similarity that includes: (1) cognitive similarity with respect to communicating with team members and, (2) cognitive similarity related to the task. Cognitive similarity among team members with regard to communication methods should lead to increased perspective-taking, which should lead to the development and use of meaningful schema-enriched communications and knowledge objects. Schema-enriched communication and the use of knowledge objects should promote knowledge interoperability and cognitive similarity regarding the task. Further, Rentsch *et al.* (Chapter 8), provide a cogent description of how knowledge objects foster collaboration. They explicate how external representations (e.g., graphics/icons) serve to not only clarify meaning for the team member creating such objects, but also provide the means through which complicated concepts can be understand so as to support knowledge interoperability. Thus, Rentsch *et al.* find that high-performing teams are able to effectively utilize knowledge objects and that they tend to have some degree of cognitive similarity among members, facilitating effective macrocognitive processes such as consensus-building and self-synchronization. But how much cognitive similarity is required? Too much similarity of experience within a team is inefficient; but not enough similarity leads to breakdowns in communications and expectations. Evidently there are other factors that influence the degree of cognitive similarity required for effective teams.

Fiore, Rosen, Salas, Burke, and Jentsch (Chapter 9) describe a more dynamic conceptualization of macrocognition by attending to the internalization and externalization of cognitive processes. Further, by addressing definitional issues associated with macrocognition terms, and linking this to the iterative nature of the collaborative process, they are able to provide a succinct representation of complex problem-solving. Pointing to the thematic issues arising out of their framework, Fiore *et al.* suggest a number of research issues for future macrocognition research.

Kirsch (Chapter 10) describes investigations of the cognitive efficiency of mediating, coordinating, and attention focusing artifacts from a more traditional cognitive science perspective. His study manipulated conditions under which a variety of presentation types were created, presented and distributed to team members. Kirsch found that presentations are most effective when given live and either face-to-face, or with audio and video of sufficient quality that remote viewers can easily talk with the speaker, and see his or her gestures on the media they are discussing. When remote audiences lack two-way audio they get less from a presentation. Kirsch concludes that the implication for sharing situation awareness among distributed members is that transfer is best when team mates have two way audio, when they ask questions, and when they have clear view of gestures and annotation.

In short, the theoretical component of this volume contains a number of substantive chapters theorizing about team macrocognitive processes, and their relationships to team performance. These are written from varied perspectives grounded in separate disciplines, nonetheless, a common thread emerges from these disparate perspectives: teams appear to need some degree of congruency in their knowledge and experience in order to be effective. Further, it is possible to greatly reduce the amount of effort dedicated to communication and coordination by using stimulating structures or coordinating representations. These findings are clearly interdependent: in order for the structures to stimulate coordination, they must first have shared meaning to the users. A congruency of knowledge and experience within a team can lead to the team members having similar interpretations of decision contexts. We turn next to a summary of the equally diverse discussion of the development of tools and methods arising out of research in macrocognition.

How findings can make a difference

What one should understand from the varied research in macrocognition is that there are multiple ways these results can make an impact both epistemologically and ecologically. In particular, the value from these findings comes from the programmatic ways in which context has been integrated into the research. Context has as its Latin root, *contextus*, 'a joining together', which, in turn, was derived from *contexere*, 'to weave together', with *com* meaning 'together' and *texure*, 'to weave'. Considering this etymology in regards to macrocognition, we can see that a particular context helps us to weave together our understanding of events in order to form a model of the world with which we are interacting at any given moment in time. Attending to context has a long history in the social sciences, primarily originating with ecological psychology (e.g., Gibson 1966) and even as far back as Bartlett (1932) and Dewey (1902). But most forcefully, researchers in ecological psychology argued that, in order to truly understand human behavior, we must understand its relation to the environment. The main premise was that humans are inextricably linked with a larger system. From this, if we are to adequately understand humans within this 'system' (i.e., the environment), the *context*, must always be part of the analysis. Others have similarly argued that ecological factors must be part of psychological research if we are to truly understand humans in context. Specifically, research in human behavior should be pursued in terms of *epistemological relevance* and *ecological salience* – that is, the degree to which the experimental approach relies on concepts from existing theories, *and* the degree to which the materials or tasks of study pertain to what is actually perceived or done in the context in which the cognition is occurring (Hoffman and Deffenbacher 1993).

What we see in macrocognition research is this type of careful attention to both task context and environmental context. Essentially, contextual elements represent a critical factor for understanding human performance and the contributions to this volume show how attending to context with respect to design and/or methods can have tremendous value. Because of this, it is possible to envision a potential for impact on both *understanding* and on *use* (see Stokes 1997). Specifically, by being both epistemologically *and* ecologically grounded, research can simultaneously be on a 'quest for fundamental understanding', that is, following an epistemological path, while also attending to the eventual 'use' of that knowledge, that is, recognize the ecological path. In light of these meta-scientific issues, in this section we briefly describe the potential for impact on both theory and practice related to macrocognition.

Epistemologically, what we see across this research is a clearer explication of the collaborative process as it unfolds in context. First we see how understanding evolves within the team, whether it be through convergence of mental models (e.g., McComb, Chapter 3) or through processes involving meaning transfer and the use of knowledge objects (e.g., Rentsch *et al.* Chapter 8). These and related processes require a close coupling of perceptual and conceptual processes as teams work to integrate their visual and verbal input. From the theoretical standpoint, this research is making important strides in helping us understand such processes at varied levels of complexity. Second we come to a better understanding of how teams deal with uncertainty. Whether it be through pattern recognition processes (e.g., Hayne and Smith, Chapter 5), or through more effective communication (Cooke *et al.* Chapter 4), we have gained a richer sense of how teams interact to reduce environmental uncertainty associated with both data and their team mates. Further, we see how they use similar processes to work towards a determination of plausibility – that is, teams engage in a more critical evaluation to determine the utility and realizability of their potential solutions. Finally, an important theoretical issue that similarly emerges from these chapters is an articulation of the interplay between what Letsky *et al.* (2007) described as the internalization and externalization of cognition. As described in Warner *et al.* (Chapter 2) and Fiore *et al.* (Chapter 9), we are coming to a better understanding of how collaboration relies upon internal processes such as mental model development while concomitantly interacting through external processes such as sharing unique knowledge. These macrocognitive processes support interpretations and interaction and are at the core of collaborative problem-solving.

In addition to supporting a richer understanding of the theoretical issues involved in collaboration, the findings reported here can similarly make a significant impact on practice. In particular, when macrocognition theory is wedded with technology development we see the potential for a very real ecological impact on the collaboration process. Specifically, the chapters presented in this volume illustrate how this careful integration of theory with technology can support the development and employment of sociotechnical collaboration systems. Large-scale systems such as EWall (Keel, Chapter, 11) show how a broad range of macrocognition processes can be supported when carefully developed based upon a strong foundation of collaboration. Similarly, Carroll's model to support activity awareness (Chapter 6) shows how theory can better inform the development of collaboration tools focused on shared references related to the group work. Similarly, tools such as DCODE (Fleming, Chapter 12) JIGSAW (Smallman, Chapter 17), and SLATE (St. John and Smallman, Chapter 17) are important developments in the melding of macrocognitive theory with technology in that they elegantly illustrate how internalized and externalized cognitive processes can be supported. Across these we see a sophisticated blend of techniques relying on graphical icons or perceptual cueing, to support attentional, decision and interaction processes while scaffolding a number of macrocognitive processes.

Other tools discussed in this volume broaden the lens of macrocognition research by shifting their level of analysis. For example, by incorporating an element of team *development,* Noble's CAT tool (Chapter 19) helps teams monitor and diagnose cognitive problems they might be experiencing. Here we see how the evolving perspective of team members along with their developing experience, can be relied upon to more generally improve their problem-solving processes. The I-DecS approach described by Handley and Heacox (Chapter 13) broadens macrocognition theorizing to the level of culture and shows how the results of this research can support collaborative processing when heterogeneity is a core team characteristic. Along these lines, while stepping up a level to team leadership, CENTER (Hess, Freeman and Coovert, Chapter 14) still draws on macrocognition theory to facilitate leader monitoring and decision-making in service of collaborative critical thinking. Other developments such as IMAGES (Weil *et al.* Chapter 16), CORE (Wroblewski and Warner, Chapter 20), or the Bordetsky and Hutchins (Chapter 21) use of the GIG infrastructure for team experimentation, provide the research community with powerful examples of how technology can be developed to support ecologically valid research in macrocognition. Importantly, CORE also provides a sophisticated suite of tools devised to assess macrocognition. This includes incorporation of techniques to capture externalization of cognition (e.g., video/audio) as well as careful collection of system interaction (e.g., key strokes). Further, the CORE testbed is working towards not only integration of automated assessment of, for example, communication, but also a suite of performance support tools that may be able to guide collaboration through the stages of macrocognition. When coupled with the principled approach to scenario design for naturalistic research (Wroblewski and Warner, Chapter 22), a tremendous array of contextually rich research becomes possible. In particular, as described in Chapter 22, the manipulations and task variations that are possible, allow researchers to assess a important level of granularity while still carefully controlling experimental factors.

In addition to showing how macrocognition can make a difference to both team research and team performance in operational settings, the developments discussed in this volume take us closer to real-time diagnosis of team performance. For example, technologies based upon Cooke *et al.*'s (Chapter 4) communication coding protocols or techniques relying on discourse analysis in support of diagnosing effective and ineffective situation awareness (Foltz *et al.* Chapter 15), may be automatable in not only the laboratory but also in the field. Such developments are important because they provide the possibility of real-time corrective action. For example, algorithms based upon dynamical systems analysis of communication may be implemented in the field in support of diagnosing when process errors occur and in making recommendations for correction. Similarly, as described by Deshmukh *et al.* (Chapter 7), intelligent agents can be developed to, among many macrocognitive processes, analyze discourse. As such, they may be able to pinpoint failures in situation awareness and either initiate corrective actions or flag team members to make corrections.

In sum, the extensive variety of research taking place in the field of macrocognition can make an impact not only on our understanding of collaboration, but also on the practice of collaboration. Nonetheless, despite these gains made in theory and the potential of the developing technologies, there is still much that needs to be done. In the remainder of this chapter we discuss some of the issues that need to be addressed when pursuing complex multidisciplinary research as well as future directions for the study of macrocognition.

Research gaps and challenges for the future

In this final section we describe some of the challenges and future directions for macrocognition research. This is based upon not only some of the current gaps in our understanding, but also on a larger need to better meld macrocognition with relevant theories and methods from other disciplines. First we discuss the difficulties emerging from multidisciplinary scientific collaboration and the significance of carefully defining terminology. Second we discuss potential next steps for improving the measurement of macrocognition. Third we discuss perhaps the greatest challenge, how to develop multilevel macrocognition theory to identify the causal mechanisms of effective team process and provide explanatory power to the field.

Significance of terminology

Within the policy and scientific communities we see an increasing recognition that the drive for *interdisciplinarity* is creating its own problems of coordination across the sciences. In particular, the large-scale scientific efforts emerging in academia and the military are cutting across disciplinary boundaries in ways never before realized. This, in turn, requires that researchers deal with the categories of thinking that have developed within disciplines, that is, the unique languages created to describe sometimes similar terms or through the use of similar terms to describe different concepts. Macrocognition research exemplifies this problem in that, in bringing together researchers from a variety of disciplines, their idiosyncratic scientific perspectives create challenges to the practice of science. At a more meta-scientific level, such problems can be addressed through either etymological or ontological means and we offer these up for consideration in support of improving future research in macrocognition.

With regard to etymology, researchers can develop stronger foundations for collaboration by developing 'insights into the current usage of a word … gained from a full knowledge of the word's history… [and a] better understanding of language generally can be achieved from knowing how words are related to other words' (Merriam-Webster 1986, p. xii). Such scholarly pursuits are neither for semantic debate nor merely an academic exercise. Instead, the goal of such activity is to illustrate how it is that usages have developed in support of better redressing divergent definitions. The CKI program has recently been addressing this within the area of macrocognition by developing more refined definitions of concepts applicable to collaborative problem-solving (Letsky *et al.* 2007).

Should etymological analysis not be sufficient, what may be necessary is the development of a more formal ontology relating an array of theoretical concepts. Ontology development is a mature epistemological approach to analyzing and categorizing our world. Only recently has it been considered as a means to support more practical issues related to interdisciplinarity in science. In particular, with the advent of more sophisticated technologies arising from the information sciences, applied ontology is able to address problems such as polysemy as it exists within and across the sciences. Here we have instances of a term having multiple meanings to capture a variety of phenomenon or we have the same underlying concept associated with it different words. Terminology issues are being addressed in applied ontology development to deal with, for example, such problems arising from medical polysemy. Here the semantic content that exists in either databases, or in the scientific literature, can be analyzed to create an ontological map of a concept or concepts, and thus facilitate more efficient research and development (e.g., Pisanellia, Gangemia, Battagliab and Catenacci 2004).

When considering techniques such as applied etymological or ontological analysis, we see a possibility for the development of methods and tools high in both utility and efficiency. For example, applications such as the 'semantic web', where information on the Internet is expressible in both natural language and in formats understandable by technologies such as intelligent agents, illustrate the utility of careful consideration of the *meaning* behind content. These technologies show how it may be feasible to adopt methods from applied ontology to provide the research community with a means to analyze conceptual relations at varied levels of granularity. In particular, both categorical relations and conceptual relations can be made more apparent when the focus is on both deeper level content and surface level terms. Such techniques will similarly be more sensitive to the subtle variations in meanings derived within differing disciplines. The collective impact of such tools is a clearer conceptualization of differing terms and their varied relations. Considering such approaches in the area of macrocognition may facilitate a more sophisticated integration of concepts from multiple disciplines as well as support coordination across the sciences.

Measuring macrocognition in the 'head' and in the 'world'

Another significant challenge and future direction is the need for additional measurement techniques including offline measures of products that is, knowledge related to problem-solving, as well as online measure of process. More generally, developing such approaches is related to an important theoretical issue of where macrocognition resides – in the minds of the individual team members, as an emergent property visible via the interactions of the teams, or some combination of the two. Essentially, as macrocognition theory evolves its own conceptualization of internalized and externalized cognition, the research community needs to improve its diagnostic capability for capturing such factors. As illustrated in Table 23.1, such measures can be broadly conceptualized as designed to assess macrocognition online or offline (i.e., during performance or before/after performance), and assess internalized or externalized macrocognition (i.e., macrocognition in the head or observable in some way). Along these lines, we next discuss some potential directions for macrocognition measurement.

With regard to measuring online internalized mental processes, that is, macrocognition *in the head*, this can be inferred via the nature of the interaction one has with the environment through his/her task environment and measured by capturing multiple levels of this human–system interaction. Importantly, if viewing certain macrocognitive processes as interaction at the system level, automated performance measurement techniques may be possible. Such techniques include dynamically capturing not only standard data such as key strokes and mouse movement on screens, but also eye-tracking to determine fixation locations and

Table 23.1 Conceptualizing macrocognition measures

		Macrocognition	
		Internalized	Externalized
Form of measurement	Online		
	Offline		

durations as well as gaze direction (e.g., looking at teammates). The confluence of these measures could allow for inferences about the processes engaged during, for example, iterative information collection. Research can similarly examine less intrusive techniques for gauging internal macrocognitive processes such as intuitive decision-making. This may be better understood via physiological measures such as skin conductance. For example, galvanic skin response (GSR) is argued to be diagnostic of what has been referred to as preconceptual phases of decision-making along the lines of Damasio's (1996) *somatic markers* (see Bierman, Destrebecqz and Cleeremans, 2005; Dunn, Dalgleish and Lawrence 2006). Here, non-verbal, and perhaps non-conscious stages of decision-making, are thought to be manifest through low-level physiological indicators measurable via unobtrusive measures such as GSR. Essentially, physiological markers can be an indication of cognitive processes. Given that such measures are used to measure workload, a measure typically tied to cognitive processing *effort*, it may be feasible to use such measures to converge on macrocognitive processes. For example, to the degree there are physiological measures correlated with measures of macrocognitive processing, such measures can be used either independently or in addition to these other measures. This could create not only increases in diagnosticity of macrocognition, it would increase the robustness of systems designed to measure and/or monitor macrocognition.

Furthermore, research coming out of DARPA and ONR is successfully integrating multidisciplinary approaches to understanding human cognition. Via utilization of tools and methods drawing from basic cognitive science, operational neuroscience and human factors psychology, this program is making great strides in developing innovative performance measures (see Schmorrow, Stanney and Reeves 2006). For example, research has shown how neurophysiological measures can diagnose the development of skill acquisition as training proceeds. Our point here is that techniques are increasing in diagnostic sensitivity. Given that skill acquisition is essentially a measure of knowledge acquisition, and because an important component of macrocognition is knowledge building, in the future it may be possible to ascertain the degree to which knowledge building, at a discrete level, is occurring dynamically during the collaboration process.

Further, using non-invasive measurement techniques, research finds that brain activity can be visibly differentiated when engaged in complex cognitive processes in the lab and the field (Ciesielski and French 1989; Deeny *et al.* 2003; Kerick *et al.* 2004). Other possibilities include integrating non-invasive physiological techniques such as eye movement metrics (e.g., micro saccades) with neurophysiological measures such as EEG (McDowell *et al.* 1997). Such techniques may be adaptable to macrocognition research to the degree they can be both automated and more portable. By automated we mean physiologically and behaviorally derived assessment criteria that can be captured in real-time from the operator's performance and used to feed back into a system. There is promise in that this line of research is making strides in automating and integrating previously subjective measures of performance and workload derived directly from human operators (e.g., self-assessments of workload and performance) with objective physiological measures of process (e.g., GSR, eye-tracking, fNIR), and objective measures of performance (e.g., task proficiency). Because they have the potential to measure the interaction between cognitive processes and cognitive and behavioral products in examination of human and system interaction in a more diagnostic manner (see Nicholson *et al.* 2006), they may be well suited for research in collaborative problem-solving. Although the appropriateness of such techniques for macrocognition is not necessarily clear at this time, they do provide tantalizing evidence that previously un-measurable dynamic processes, can now be plausibly measured online.

With regard to offline techniques for assessing internalized mental process, a variety of techniques have been developed to assess knowledge organization, that is, the degree to which elements of knowledge are interconnected and integrated within meaningful networks in long-term memory (Glaser 1989; Jonassen, Beissner and Yacci 1993). These consist of quantitative and qualitative methods, each presenting unique advantages and disadvantages in assessment, and which may elicit distinct aspects of knowledge organization (e.g., Dorsey, Campbell, Foster and Miles 1999). Common techniques include concept mapping (e.g., Dorsey *et al.* 1999), card sorting (e.g., Fiore, Cuevas, Scielzo and Salas 2002), and Pathfinder analysis (Schvaneveldt 1990). Responses are often compared to a referent (or expert) model of the domain, permitting diagnosis of organization (e.g., in the case of understanding expert knowledge), or misconceptions (e.g., in the case of training research); in both cases, the goal is to assess how critical task-relevant concepts are related. Specifically, what is of interest is integrated knowledge, that is, interconnected knowledge consisting of facts and integrated via, for example, functional relationships (e.g., Bassok and Holyoak 1993).

Methods include techniques such as concept maps, card sorting, similarity ratings and even more advanced contextualized technologies such as EWall (Keel 2007). Further, recent developments in automated card sorting and concept mapping techniques have been very helpful to both data collection and data analysis (see Harper *et al.* 2003: Hoeft *et al.* 2003). Concept maps require the trainee to visually represent task-relevant terms in an interconnected visually represented format. With this, they attempt to identify both type of relations (e.g., functional), and similarity of items (e.g., distance). Studies suggest that concepts maps are a reliable indicator of knowledge acquisition in tasks as varied as biological science and software applications (see Dorsey, Campbell, Foster and Miles 1999; McClure, Sonak and Suen 1999). Card sorts are an additional method requiring trainees to indicate the degree to which they believe concepts are related. Participants are presented with cards on which concepts are typed and instructed to group these behaviors into as many or as few categories as they desire. Although a somewhat limited method, because trainees are forced to group together items rather rigidly, studies suggest that card sort data may be used to ascertain the degree to which one accurately views conceptual relations (see Fiore *et al.* 2002; Jones, Carter and Rua 2000; Maiden and Hare 1998). Similarity ratings based upon paired comparisons between key concepts are used to generate a quantitative and graphical representation of organized knowledge. A consistent body of research shows that such techniques capture critical relations among concepts and can reliably differentiate between experts and novices (see Johnson, Goldsmith and Teague 1995; Wyman and Randel 1998). More recent developments include techniques that use computational-based contextualized tools to support and analyze collaboration. For example, EWall is software developed as a tool for visual analytics, supporting collaborators who interact at a distance (see Keel 2007). EWall utilizes knowledge organization tools (e.g., electronic cards that contain a variety of information formats) to infer the underlying representations users maintain and develop while engaged in tasks required complex collaborative processes.

Furthermore, with regard to these forms of offline techniques for assessing internalized mental process, research will need to pursue further innovations in knowledge elicitation methods. For example, integrating techniques such as signal detection theory with more standard methods of knowledge elicitation, may lead to more precise measures of the types of complex knowledge structures typically diagnosed with techniques such as mental model measurement. For example, this combination of approaches can provide an index of 'sensitivity' (see Fiore, Cuevas and Oser 2003; Fiore *et al.* 2002) in determining whether a team or teammate's mental model is structurally similar to an expert model. To illustrate, this method was used in a relational problem-solving assessment where the participants' accuracy in the semantic integration of concepts was determined. Participant accuracy in making such connections was hypothesized to be related to performance as indicated by standard measures of task execution. The scoring algorithm was based on signal detection theory, allowing for proficient diagnosis of critical conceptual relations. Signal detection theory posits that target identification ability (in this study, accurately recognizing relations among concepts) is dependent upon a combinatory process of one's *sensitivity* and *bias* towards detection. Sensitivity refers to the accuracy of one's detection mechanisms, in this case, the degree to which a participant is more likely to appropriately identify conceptual relations. By comparing data from a knowledge elicitation method (e.g., card sort) to an expert's data, a participant's sensitivity to identifying critical conceptual relations was determined. The proportion of 'hits' (i.e., correct conceptual relations) and false alarms (i.e., incorrect conceptual relations) were used to calculate the measure of sensitivity. By using this as a form of relational problem-solving which required the semantic integration of concepts, it was possible to document how sensitivity to critical conceptual relations in a synthetic task environment was directly related to measures of task performance. With respect to more complex task environments, these findings are consistent with the theme of problem-solving involving accurate interconnectivity or integration of concepts (e.g., Frensch and Funke 1995). Further, this is easily applied as a diagnostic technique for assessing team related problem-solving knowledge when card sorting and/or concept mapping is used (cf. Smith-Jentsch, Campbell, Milanovich and Reynolds 2001).

Additionally, with regard to online measuring of external processes, that is, observable macrocognition, theory and methods from virtual reality may offer fresh insights for measuring macrocognition as 'knowledge in the world'. Recent research has used tracking technologies to examine movement and the efficacy of training in mixed reality environments. This involves the use of, for example, head tracking, allowing researchers to visually represent and compare movements of participants in a virtual environment (see Fidopiastis *et al.* 2006). To the degree any macrocognitive problem-solving environment requires an understanding of more gross levels of movement, this could be an additional consideration for externally exhibited problem-solving behaviors within teams. Additionally, more carefully attending to the creation and use of knowledge objects, and the affordances they provide, may provide important insights to the ways in which macrocognition is manifest 'in the world'. Combining such techniques with virtual reality may be productively pursued in what are called 'mixed-reality' environments. Here the virtual and the real are integrated in such a way that real and artificial objects become part of the interaction (see Malo, Stapleton and Hughes 2004).

Advancing theories and models of macrocogntion

Finally, perhaps the greatest challenge and an important next step for the field is the development of more sophisticated models and theories of macrocognition. Building a comprehensive and coherent theory for macrocognition requires that we understand the many factors and their interrelationships that emerge when individuals and teams attempt to make decisions and solve complex problems. In order to address this need, it has been recommended that members of the operational and analytic communities cooperate and bring to bear theory and methods from a variety of scientific disciplines, including complex systems, organizational theory and cognitive science (Alberts, Garstka and Stein 1999). In particular, the complexity inherent in operational environments is argued to consist of a network of non-linear interactions within an open system and which produce a form of self-organization and emergence (cf. Liljenström and Svedin 2005). From this, we see how we can approach complex systems in the physical, biological or social world at different levels of analysis. For example, in the biological sciences, microscopic can refer to the molecular level and macroscopic to the organ level (see Liljenström and Svedin 2005). From the organizational sciences, microscopic refers to the behavior of individuals as they work alone and macroscopic describes the behavior of entire organizations (Wagner and Hollenbeck 2004). By adapting the aforementioned distinctions, macrocognition research can mature in its explication of the varying levels of cognition and how they are coupled. As it stands, we currently have *micro*cognition, defined as individual level cognition resulting from the encoding, retrieval and use of information, and *macro*cognition, defined as the internalized and externalized high-level mental processes employed by teams to create new knowledge during complex, one-of-a-kind, collaborative problem-solving. More importantly, though, we need to ask whether there are additional levels of analysis to consider: levels above, below, or between. If so, complex systems theory may help the field better understand the intricacies arising from factors interacting within and across levels.

An important additional gap that needs to be addressed is the development of plausible computational models of macrocognition. More standard cognitive processes, referred to in the present context as microcognition, are quite adequately modeled with techniques such as ACT-R (see Schoelles, Neth, Myers and Gray 2006). But computational models of collaborative problem-solving have to not only address a complex problem representation, but do so while taking into account team interaction. Further, computational models, while able to access a problem-solvers domain knowledge so as to retrieve problem relevant information, would have to do so while taking into account not only the varied knowledge across team members, but also address the sharing processes that occur (and do not occur) during collaborative problem-solving. While these and related problems such as computational modeling of reasoning strategies (Paritosh and Forbus 2004) are not intractable, they do add a level of complexity to the modeling process that needs to be overcome if macrocognition is to be computationally modeled.

Such distinctions are important for a number of reasons. First, through empirical analysis they allow both the research and the operational communities to more fully identify boundary conditions related to the emergence and effective use of cognition in complex environments. Second, refining and/or augmenting these distinctions will allow us to consider macrocognition independent of the context in which it might occur. Specifically, researchers could better examine macrocognition, not only in the field, but also in appropriately designed laboratory studies – for example, within simulations that reify the contextual conditions necessary to mimic dynamic and complex collaboration across varying levels of analysis (cf. Hoffman and Deffenbacher 1993). Third, clearly identifying these levels of cognition emerging in complex collaborative activity can more fully support the development of a truly multilevel theoretical approach to this research, (e.g., individuals, groups, and organizations), helping us to better specify how we are conceptualizing construct(s) that may cut across levels (see Dansereau and Yamarino 2002). Finally, computational models will allow us to predict performance outcomes more precisely and improve collaborative systems design.

An example of a macrocognitive theory that spans levels of analysis is found in the chapter by Hayne and Smith (Ch 5, this volume). Hayne and Smith describe Template Theory, a micro-level explanation of the way that individuals encode and recall knowledge gained by experience. By extending this theory to groups, and providing empirical support for the implications of this theory to macrocognitive constructs (e.g., shared meaning, coordinating representations, decision context), they illustrate how macrocognition research can explore multiple levels. Clearly, more work is required to expand the available models for cross-level cognitive research. From all of this what we would gain is substantial improvement in our explanatory power by better diagnosing causal factors associated with a given phenomenon of interest. For example, Hackman (2003) showed how shifting focus to a higher or a lower level led to new insights into causal mechanisms in shaping team process and performance. He argues that by bracketing the main phenomenon via a level above and a level below, greater explanation is possible, stating that the 'explanatory power of bracketing lies in crossing levels of analysis, not blurring them' (p. 919). Essentially, what the field needs to develop is theory that captures not only micro- and macro-level issues, but also the interrelations among these levels. By moving the analytical lens either one level up, or one level down, we may be able to provide both explanatory and causal power.

Clearly model development for macrocognition is still in the early stages of development. Nonetheless, simultaneously researching the factors influencing collaboration processes, while carefully considering the results coming from the varied studies looking at macrocognition, is taking us towards both an improved conceptual understanding of problem-solving, as well as a more sophisticated functional model of macrocognition. As our models of macrocognition mature, research will need to better address how the intricate relationship among macrocognitive processes can be expressed, and potentially modeled computationally. This includes not only modeling the parallel, interdependent and iterative nature of these processes as they unfold in the context of collaboration, but also a clearer understanding of how these processes drive specific problem-solving outcomes (e.g., solution viability).

In short, our goal in moving towards multilevel and computational theories is to enable the development of precise research propositions that may provide more explanatory power with regard to causal mechanisms underlying collaboration. These will help us address the important questions that have arisen out of the collective efforts articulated in this volume. Specifically, across a number of these chapters, we see a set of fundamental issues needing further research. First, we need to more fully examine the development of understanding of knowledge within the team, including the complex interplay between the perceptual *and* conceptual in collaborative problem-solving. Second, we need to come to better understand the iterative nature of internalization *and* externalization of knowledge. Here the research must assess how, for example, *interpretation* and *interaction* within teams supports the comprehension of task elements. Finally, we need to gain a clearer understanding of the evolution of team interaction from uncertainty reduction *to* determination of plausibility, that is, how macrocognitive processes drive information interrogation early in the problem-solving, but evolve more to solution evaluation later in the collaboration. A well integrated multilevel theory of macrocognition, taking into account factors at the individual and the team, and the team of teams, level, will provide an important foundation for these research issues.

Conclusions

This volume was conceived in order to address a need in the team cognition literature, a need to provide an avenue through which researchers from diverse fields can present their theories, methodologies and tools related to macrocognition in a unified volume. Here at the conclusion of this volume we hope the reader recognizes this need for, and the importance of, the concept of *macro*cognition. In particular, macrocognition represents a focal area to unite different disciplines in the study of collaborative problem-solving – taking us towards what is truly *interdisciplinarity* in research. With interdisciplinarity, we see an integration of varied disciplines creating a unified outcome. Specifically, researchers in this volume have shown how to integrate techniques, tools, perspectives, concepts and theories and illustrate how collaboration as a practice of science can be successfully implemented. From this emerges the value added of the macrocognition construct; that is, progress in science typically occurs when disciplines begin to adopt and adapt theoretical and empirically derived principles from tangential disciplines (Dunbar 1997). This field is progressing in important ways because constructs coming out of the cognitive, organizational and computational sciences are being adopted to aid our understanding of collaborative problem-solving. Indeed, this is helping the field of team cognition by providing an important new concept for collaborative problem-solving research – one that encompasses the myriad of processes engaged as teams interact to solve complex one-of-a-kind problems.

In sum, the more we understand, for example, the interplay between internalization and externalization of cognition and the mechanisms that allow team members to co-construct knowledge, the better we will be able to offer guidelines for the design of collaborative systems. Similarly, the more we understand measurement of process and performance, in due time, we may be able to develop online methods of remediation. Essentially, the deeper the thinking, and the better specified the research, the more likely it is that we can develop and offer accurate principles and guidelines as well as effective tools and interventions that facilitate macrocognition. We hope this volume takes us closer and motivates others to continue to investigate the many factors at play when teams collaborate to solve complex problems in naturalistic environments.

Acknowledgments

The views, opinions and findings contained in this article are the authors and should not be construed as official or as reflecting the views of the Department of Defense or the University of Central Florida. Writing this chapter was partially supported by Grant N000140610118 from the Office of Naval Research awarded to S. M. Fiore, S. Burke, F. Jentsch and E. Salas, University of Central Florida.

References

Alberts, D.S., Garstka, J. and Stein, F. (1999). *Network Centric Warfare: Developing and Leveraging Information Superiority.* Command and Control Research Program, OSD NII. (Washington DC: DoD/CCRP).

Bartlett, F.C. (1932). *Remembering*. (Cambridge: Cambridge University Press).

Bassok, M. and Holyoak, K.J. (1993). 'Pragmatic Knowledge and Conceptual Structure: Determinants of Transfer Between Quantitative Domains', in D.K. Detterman and R.J. Sternberg, (eds), *Transfer on Trial: Intelligence, Cognition, and Instruction*, pp. 68–98. (Norwood, NJ: Ablex).

Bierman, D., Destrebecqz, A. and Cleeremans, A. (2005). 'Intuitive Decision Making in Complex Situations: Somatic Markers in an Implicit Artificial Grammar Learning Task', *Cognitive, Affective, and Behavioral Neuroscience* 5(3), 297–305.

Ciesielski, K.T. and French, C N. (1989). 'Event-Related Potentials Before and After Training: Chronometry and Lateralisation of Visual N1 and N2', *Biological Psychology* 28, 227–238.

Damasio, A.R. (1996). 'The Somatic Marker Hypothesis and the Possible Functions of the Prefrontal Cortex', *Philosophical Transactions of the Royal Society of London (series B)* 351(1346), 1413–1420.

Dansereau, F. and Yamarino, F. (eds) (2002). *Research in Multi-level Issues*. (Oxford: Elsevier Science Ltd).

Deeny, S.P., Hillman, C.H., Janelle, C.M. and Hatfield, B.D. (2003). 'Cortico-cortical Communication and Superior Performance in Skilled Marksmen: An EEG Coherence Analysis', *Journal of Sport and Exercise Psychology* 25, 188–204.

Dewey, J. (1902). *The Child and the Curriculum.* (Chicago, IL: University of Chicago Press).

Dorsey, D.W., Campbell, G.E., Foster, L.L. and Miles, D.E. (1999). 'Assessing Knowledge Structures: Relations with Experience and Posttraining Performance', *Human Performance* 12, 31–57.

Dunbar, K. (1997). 'Conceptual Change in Science', in T.B. Ward, S.M. Smith and J. Vaid (eds), *Creative Thought: An Investigation of Conceptual Structures and Processes*, pp. 461–494. (Washington, DC: American Psychological Association).

Dunn, B.D., Dalgleish, T. and Lawrence, A.D. (2006). 'The Somatic Marker Hypothesis: A Critical Evaluation', *Neuroscience and Biobehavioral Reviews* 30, 239–271.

Edwards, B.D., Day, E.A., Arthur, W., Jr. and Bell, S.T. (2006). 'Relationships Among Team Ability Composition, Team Mental Models, and Team Performance', *Journal of Applied Psychology* 91, 727.

Fidopiastis, C.M., Stapleton, C.B., Whiteside, J.D., Hughes, C.E., Fiore, S.M., Martin, G.A., Rolland, J.P. and Smith E.M. (2006). 'Human Experience Modeler: Context Driven Cognitive Retraining to Facilitate Transfer of Training', *Cyberpsychology and Behavior: Special Issue: International Workshop on Virtual Reality 9(2),* 183–187.

Fiore, S.M., Cuevas, H.M. and Oser, R.L. (2003). 'A Picture is Worth a Thousand Connections: The Facilitative Effects of Diagrams on Task Performance and Mental Model Development', *Computers in Human Behavior* 19, 185–199.

Fiore, S.M., Cuevas, H.M., Scielzo, S. and Salas, E. (2002). 'Training Individuals for Distributed Teams: Problem Solving Assessment for Distributed Mission Research', *Computers in Human Behavior* 18, 125–140.

Frensch, P.A. and Funke, J. (eds) (1995). *Complex Problem Solving: The European Perspective*. (Hillsdale, NJ: LEA).

Gibson, J.J. (1966). *The Ecological Approach to Visual Perception.* (Boston, MA: Houghton Mifflin).

Glaser, R. (1989). 'Expertise in Learning: How do we think about Instructional Processes now that we have Discovered Knowledge Structure?', in D. Klahr and D. Kotosfky (eds), *Complex Information Processing: The Impact of Herbert A. Simon*, pp. 269–282. (Hillsdale, NJ: LEA).

Hackman, J.R. (2003). 'Learning more from Crossing Levels: Evidence from Airplanes, Orchestras, and Hospitals', *Journal of Organizational Behavior* 24, 1–18.

Harper, M.E., Jentsch, F.G., Berry, D., Lau, H.D., Bowers, C. and Salas, E. (2003). 'TPL-KATS-Card Sort: A Tool for Assessing Structural Knowledge', *Behavior Research Methods, Instruments, and Computers* 35(4), 577–584.

Hoeft, R.M., Jentsch, F.G., Harper, M.E., Evans, A.W., Bowers, C.A. and Salas, E. (2003). 'TPL-KATS – Concept Map: A Computerized Knowledge Assessment Tool', *Computers in Human Behavior* 19(6), 653–657.

Hoffman, R.R. and Deffenbacher, K.A. (1993). 'An Analysis of the Relations of Basic and Applied Science', *Ecological Psychology* 5, 315–352.

Johnson, P.J., Goldsmith, T.E. and Teague, K.W. (1995). 'Similarity, Structure, and Knowledge: A Representational Approach to Assessment', in P.D. Nichols, S.F. Chipman and R.L. Brennan (eds), *Cognitively Diagnostic Assessment*. pp. 221–249. (Hillsdale, NJ: LEA).

Jonassen, D.H., Beissner, K. and Yacci, M. (1993). *Structural Knowledge: Techniques for Representing, Conveying, and Acquiring Structural Knowledge*. (Hillsdale, NJ: LEA).

Jones, M.G., Carter, G. and Rua, M.J. (2000). 'Exploring the Development of Conceptual Ecologies: Communities of Concepts Related to Convection and Heat', *Journal of Research in Science Teaching* 37, 139–159.

Keel, P. (2007). 'EWall: A Visual Analytics Environment for Collaborative Sense-making', *Information Visualization* 6, 48–63.

Kerick, S.E., Douglass, L.W. and Hatfield, B.D. (2004). 'Cerebral Cortical Adaptations Associated with Visuomotor Practice', *Medicine and Science in Sports and Exercise* 36, 118–129.

Letsky, M., Warner, N., Fiore, S.M., Rosen, M.A. and Salas, E. (2007). 'Macrocognition in Complex Team Problem Solving', *Proceedings of the 11th International Command and Control Research and Technology Symposium (ICCRTS)*. (Washington, DC: US Department of Defense Command and Control Reearch Program).

Liljenström, H. and Svedin, U. (eds) (2005). *Micro – Meso – Macro: Addressing Complex Systems Couplings*. (London: World Scientific Publications Company).

Maiden, N.A.M. and Hare, M. (1998). 'Problem Domain Categories in Requirements Engineering', *International Journal of Human–Computer Studies* 49, 281–304.

Malo, S., Stapleton, C.B. and Hughes, C.E. (2004). 'Going beyond Reality: Creating Extreme Multi-Modal Mixed Reality for Training Simulation'. Paper presented at the Interservice/Industry Training, Simulation, and Education Conference, Orlando, Fl, National Training Association, Arlington, VA.

Marks, M., Sabella, M.J., Burke, C.B. and Zaccaro, S.J. (2002). 'The Impact of Cross-training on Team Effectiveness', *Journal of Applied Psychology* 87, 3–13.

Mathieu, J.E., Heffner, T.S., Goodwin, G.F., Cannon-Bowers, J.A. and Salas, E. (2005). 'Scaling the Quality of Teammates' Mental Models: Equifinality and Normative Comparisons', *Journal of Organizational Behavior* 26, 37–56.

Mathieu, J.E., Heffner, T.S., Goodwin, G.F., Salas, E. and Cannon-Bowers, J.A. (2000). 'The Influence of Shared Mental Models on Team Process and Performance', *Journal of Applied Psychology* 85, 273–283.

McClure, J.R., Sonak, B. and Suen, H.K. (1999). 'Concept Map Assessment of Classroom Learning: Reliability, Validity, and Logistical Practicality', *Journal of Research in Science Teaching* 36, 475–492.

McDowell, S, Whyte, J, and D'Esposito, M. (1997). 'Working Memory Impairments in Traumatic Brain Injury: Evidence from a Dual Task Paradigm', *Neuropsychologia* 35, 1341–1353.

Merriam-Webster's Collegiate Dictionary (9th edn) (1986). (Springfield, MA: Merriam Webster).

Nicholson, D., Stanney, K., Fiore, S.M., Davis, L., Fidopiastis, C., Finkelstein, N. and Arnold, R. (2006). 'An Adaptive System for Improving and Augmenting Human Performance', in D. Schmorrow, K. Stanney and L. Reeves (eds), *Augmented Cognition: Past, Present and Future*, pp. 215–222. (Arlington, VA: Strategic Analysis, Inc.).

Paritosh, P. and Forbus, K. (2004). 'Using Strategies and AND/OR Decomposition for Back of the Envelope Reasoning', in *Proceedings of the 18th International Qualitative Reasoning Workshop, Evanston, Illinois, August.*

Pisanelli, D.M. Gangemi, A. Massimo, B., Catenacci, C. (2004). 'Coping with Medical Polysemy in the Semantic Web: The Role of Ontologies', in M. Fieschi, E. Coiera and Y.-C.J. Li, in *Proceedings of the 11th World Congress on Medical Informatics, Volume 107, Studies in Health Technology and Informatics*, pp. 416–419. (Amsterdam: IOS Press).

Rentsch, J.R. and Klimoski, R. (2001). 'Why do "Great Minds" Think Alike?: Antecedents of Team Member Schema Agreement', *Journal of Organizational Behavior* 22, 107–120.

Schmorrow, D., Stanney, K. and Reeves, L. (eds) (2006). *Augmented Cognition: Past, Present and Future.* (Arlington, VA: Strategic Analysis, Inc.).

Schoelles, M.J., Neth, H., Myers, C.W. and Gray, W.D. (2006). 'Steps Towards Integrated Models of Cognitive Systems: A Levels-of-analysis Approach to Comparing Human Performance to Model Predictions in a Complex Task Environment', in *Proceedings of the 28th Annual Conference of the Cognitive Science Society*, pp. 756–761. (Wheatridge, CO: Cognitive Science Society).

Schvaneveldt, R.W. (Ed.) (1990). *Pathfinder Associative Networks: Studies in Knowledge Organization.* (Norwood, NJ: Ablex Publishing Corp).

Smith-Jentsch, K.A., Campbell, G., Milanovich, D. and Reynolds, A. (2001). 'Measuring Teamwork Mental Models to Support Training Needs Assessment, Development and Evaluation: Two Empirical Studies', *Journal of Organizational Behavior* 22(2), 179–94.

Stokes, D.E. (1997). *Pasteur's Quadrant: Basic Science and Technological Innovation.* (Washington, DC: Brookings Institution Press).

Van den Bossche, P., Gijselaers, W.H., Segers, M. and Kirschner, P.A. (2006). 'Social and Cognitive Factors Driving Teamwork in Collaborative Learning Environments', *Small Group Research* 37, 490–521.

Wagner, J.A. and Hollenbeck, J.R. (2004). *Organizational Behavior: Securing Competitive Advantage.* (Englewood Cliffs, NJ: Prentice-Hall).

Wyman, B.G. and Randel, J.M. (1998). 'The Relation of Knowledge Organization to Performance of a Complex Cognitive Task', *Applied Cognitive Psychology* 12, 251–264.

Index

Accountability, 129, 341–342
Accuracy, 42–44, 46, 127, 136, 195, 204, 224, 229, 261, 264–265, 315, 359, 374, 409
ACH diagram, 334
Acknowledgment, 99, 129
Activity
  awareness, 95, 157, 342–343
  theory, 8, 87–88, 91–93, 95
Activity system model, 89, 91–93
  activity theory and, 92
ACT-R, 3, 410
Ad hoc
  data collection, 285, 325
  teams, 1, 2, 107, 145, 255, 339
Adjustable autonomy, 114–115
Agent, 3, 6, 107–118, 120–121, 187–188, 195–197, 201–204, 281–283, 288–289, 366–367, 370, 400
  adaptation, 197
  autonomous, 111
  balance, 201
  characteristics, 108, 111, 114, 118–119, 121
  collaboration management, 111
  collaborative, 111
  examples of, 118
  implementations, 189
  intelligent, 400, 404–405
  interaction, 201
  pattern, 202
  popularity, 202
  relevance, 201
  support, 11, 119
  system, 109, 187, 194–197, 204
  see also human-agent teams
'Agent × Implicit meta-matrix' network, 283, 289–290
'Agent × Knowledge' network, 283, 289–290
Agreeableness, 254
Agreement, 5, 9, 21, 42–43, 131, 133, 153, 158, 226, 246, 345–6, 379, 381
Air support, 27–8
Air warfare, 315, 374, 376–377
Ambiguity, 7, 18, 145–147, 159, 225, 244, 246–7, 298–9, 344, 359, 374–375
Analysis
  automated communication, 10, 54–62, 259–272
  collaborative, 3, 324, 326, 334
  multilevel, 253–4
  network, 287
  statistical, 99, 359
  trend, 147–8
Analysis plan generator (APG), 326–329, 332–333
Anchoring, 101, 103, 242
Animations, 166, 168, 178–80
  versus stills, 171–173
Annotations, 100, 166–169, 171–176, 178, 182–183, 185, 306, 309, 311–316, 401
  dynamic vs. static, 180, 181
  foreground, 173, 179
  graphical, 100, 185, 306
  moving, 180–181, 185
  multimedia, 306, 317
  sequence of, 312
  static, 174, 178, 180, 185
Articulatory loop memory, 66
Artificial intelligence, 16, 108, 110
Assessment interactions, 242
Assistance, 25–26, 114–117, 226, 324, 371
Asynchronous collaboration, 264, 306, 308, 318, 351
Attention management, 173, 181
  tools, 81, 306, 318
Augmentation
  see also cognitive augmentation
AutoMap, 282, 288–290
Automation, 62, 281, 290
  communication analysis, 54–59

content analysis, 55
flow analysis, 56–57
Autonomy, 111, 114–115, 118–119
Awareness, 20, 93, 98–99, 147, 155, 157, 167, 202–203, 231, 242, 268
see also situation awareness

'Between the heads' (BTH) approach, 51–53, 399
Brainstorming, 5, 187, 341, 366
collective, 326
collaborative, 193

C3Fire, 264, 266–272
ChainMaster, 56, 284, 291–292, 295
Chunking (pattern recognition), 66–68, 79
Classical information processing theory, 52
Co-located teams
collaboration and, 15
versus distributed teams, 60–61, 129
see also collocated teams
Coalition-wide area network (COWAN), 227
Cognition
individual, 51, 54, 65, 240
team communication and, 53–54
theories of, 7
Cognitive Assessment Tags (CATs), 210–214, 218–220
basic, 212–213
course of action, 212
display, 215–217
impact, 213
importance, 213
information quality, 213
knowledge elicitation approach, 214–215
Cognitive augmentation, 107–110
human-agent teams, 110–112
individual human, 109–110
team macrocognitive processes, 117–120
through agents, 108–112
Cognitive conflict, 37
Cognitive consensus, 36
Cognitive fit, 70, 79
Cognitive processes
collaborative team problem-solving and, 27–28
knowledge construction and, 26
outcome evaluation and revision and, 29
team consensus and, 28
Cognitive resources, 65, 66, 69, 79, 81
Cognitive similarity, 128–129, 135–139, 398, 400–401
configurations, 131
in teams, 130–131
schema-enriched communication and, 134–135
Cognitive task analysis, 109, 113, 144, 331, 387, 391–392
Collaboration, 4, 68, 107, 114, 277, 400
asynchronous, 264, 306, 308, 318, 351
CATS, 339–349
CORE testbed and, 358
distributed, 308
empirical model, 6–7, 16–31, 95, 260–261, 340, 352–353, 374–375, 386
environment, 4
GIG and, 365–366
GIG testbed, 366–373
I-DecS and, 223–224, 232
in the military, 352
intra-team, 322–323
macrocognition and, 10
military, 352
measuring, 375–382
naturalistic decision-making-based scenarios and, 385–386, 389–391
performance monitoring, 116–117
research spectrum, 354–358
research strategies, 353–354
SLATE and, 305, 309
simulation experiment, 266–272
support for, 223–224, 331–334
tool development 351–352
tools, 15–16
understanding, 239
Collaboration Advizor Tool (CAT), 339–349
cognitive framework, 340–344
collecting team data, 345
diagnosis drill-down, 346–347
enablers, 340–343
functions, 344–348
knowledge basis, 344
life cycle, 344–345
prioritized issues, 347
overall team diagnosis, 345–346
recommendations, 348
risks, symptoms, and diagnosis, 343–344

Collaboration and Knowledge Interoperability (CKI) Program, 1–3, 10–11, 223, 321, 360, 390, 398
research questions, 5–6
trusts, 4–5
Collaboration and meaning analysis process (C-MAP), 10, 137–140
Collaboration for ENhanced TEam Reasoning (CENTER), 245–255
fielding, 254–255
managing knowledge, 249
measuring knowledge, 246–247
monitoring knowledge, 247–248
validation study, 249–254
Collaboration scenarios, 385–394
gaming, 388–389
generating, 391–393
multidimensional, 389–390
multidimensional/multidomain, 390–391
one-dimensional, 389
pitfalls, 393
Collaborative Agents for Simulating Teamwork (CAST), 111
Collaborative critical thinking (CCT), 239–255, 403
behaviors defined and applied, 244
CENTER, 245–255
context, 241
theory of, 240–245
Collaborative Operational and Research Environment (CORE), 10, 351–352, 361, 385–386, 390–391, 403–404
testbed, 358–360
tool suite, 360–361
Collaborative team problem solving, 10, 20, 26–27, 44, 209, 223–224, 261, 353, 374, 376
cognitive processes and, 27
JIGSAW, 333
stages, 309, 310
Collocated teams, 96, 205, 277
see also co-located teams
Command authority, 226
Command and control groups, 210
Common ground (in teamwork), 3, 11, 87–103, 135–136, 151, 158, 165, 299, 398, 400
developing, 152
knowledge transfer and, 135–136
see also common ground conceptual model
Common ground conceptual model, 87–103
implementation, 96–98
macro-models of collaboration and, model, 93–94
motivation and strategy for, 89
theoretical context, 88–89
theoretical sources, 89
Communication
distributed, 308
forms of, 53
miscommunication, 277
pattern interpretation, 58–59
referential, 151
Communication analysis
applications, 59–62
approaches, 280–284
automating, 54–59
communication content analysis, 100
communication structure analysis, 100
method comparison, 57–58
pattern interpretation, 58–59
simulation experiment, 266–272
Complex team problem-solving processes, 143–158
Comprehension, 4, 109, 110, 152, 187, 260, 262, 267, 272, 278, 311, 389, 412
Computational systems, 197–189, 191, 192, 194–197, 200, 203–204
Computational opportunities, 193
Computer computational model, 8
Computer mediated communication, 87, 88, 102
Computer-supported cooperative work (CSCW), 87, 88, 91, 92, 95, 102, 400
Conceptual models, definition, 89
Conflict resolution, 91, 196, 209–210, 212, 218
Consensus, 2, 4–5, 20–21, 26, 28, 45, 61, 218–220, 224, 376, 379, 382
building, 209–210, 309
development, 153–156
JIGSAW, 333
Content analysis
automating, 55
communication, 54
Coordinating representations, 80, 329, 399–401, 411
SLATE and, 309, 313–314
Coordination
effective teamwork and, 156–157

Course of action analysis (COA), 209
Crisis action planning process, 227, 339
Critical decision method (CDM), 387, 388
Critical thinking, 5, 24, 153, 155, 158, 353
  collaborative critical thinking (CCT), 239–255, 403
Culture, 199, 210, 223–224, 228, 231–232, 359, 403
differences, 223
interface, 226
national, 224–225
work processes and, 224–226

Decision-making, 187, 225, 279
  biases, 325–326
  CENTER and, 245–246, 255
  CORE, 351–361
  critical thinking and, 239–240
  distributed teams, 239–240
  special operation forces, 306
  TADMUS, 263
  see also naturalistic decision making
Decision-making concepts in a distributed environment (DCODE), 209–211, 213, 215, 217–220, 330, 358, 361, 403
  concept, 209–212, 214
Decisiveness, 225
Delegation, 225
Developmental biology, 8
Discourse theory, 309, 312–313, 316
Distributed cognition, 8, 66, 67, 68, 79, 88, 165, 192
Distributed teams, 127–130, 277, 308
  cognitive similarity, 130–131
  collaboration and, 15
  decision-making and, 239
  face-to-face teams and, 127
  information-sharing, 131–132
  knowledge interoperability, 132
Dominance, 54, 56–60

Ecological psychology, 402
  versus processing theory, 52
Empirical model of team collaboration, 16–31, 223, 260–261, 340, 352–353, 374–375, 386
  components, 19–24
  development, 16–19
  empirical validation, 29–30
  example, 24–25
  validation of, 374
Engestrom's activity system model
  see activity system model
Errors of omission, 277
Event-based scenarios, 388–391
  gaming, 388–389
  multidimensional, 389–390
  multidimensional/multidomain, 390–391
  one-dimensional, 389
EWall, 187–189, 203–205, 314, 367
  benefits, 197–198
  cards, 189–191
  design criteria, 196–197
  interpretation agents, 198–200
  system agent, 194–203
  transformation agents, 200–203
  user interface, 189, 194
  views, 191–193
Expert knowledge, 244, 408
  externalizing, 134–138
  transfer of, 132–134
Expertise recognition, 150, 158
Externalized processes, 7, 19, 21–24, 143, 145, 150, 151, 153, 158, 192

Face-to-face presentation, 165, 172, 174, 180–184, 185
  audio interactivity, 184
  live remote vs. typical, 183–184
  minimal vs. canned, 180–181
FAUCET metrics, 290
Feature matching, 65, 67
Flow analysis, 19, 31, 54, 284, 295, 386
  automation of, 56–57

Global Information Grid (GIG), 365
  collaborative technology and, 365–366
  maritime testbed, 366–373
Grounded transactions, 309, 315–318
Group
  cognition, 9
  outcomes measurement, 101
  process theory, 87
Guidance
  reliance on, 225

HCI theory, 88
Hierarchical shared-plans model, 110
High-dimensional data
  visualizations, 107

Human-agent teams, 110–112
benefits, 120
design, 112–117

Immersion, 167
Implicit meta-matrix, 283, 289–290
'In the head' (ITH) approach, 50–52, 399
Individual cognition, 239
Individual critical thinking, 240, 246
Individual task knowledge, 379
Information
encapsulation, 309, 314–315
flow, 225
processing, 8
space, 192
Information object (IOB), 210, 214–220
Information sharing
among team members, 131–132
Input-process-output framework, 51–52
Instrument of the Measurement and Advancement of Group Environmental Situation awareness (IMAGES), 278, 295, 297–298, 300, 403
Integrative decision space (I-DecS), 224–226
Decision Support System for Coalition Operations (DSSCO) integration, 231
decision support tool, 226–229
usability and utility assessments, 229–231
Intelligence analysis, 2, 5, 10, 24, 30, 35, 80, 108, 113, 118, 243, 325, 343, 352, 360, 385, 391
CENTER, 254
JIGSAW, 321–326
Intelligence cycle
process problems, 322–324
Intelligence quotient (IQ) tests, 280
Interface culture, 226
Internalized processes, 7–10, 18–19
Intuitive decision-making, 154–155

Joint Intelligence Graphical Situation Awareness Web (JIGSAW), 321, 324, 326–334
approach and target domain, 326–327
collaborative intelligence landscapes (I-Scapes), 329–330
concept, 326
design features, 327
hypothesis manager, 328–329
I-Posts, 330–331
I-Scape de-clutter, 331
I-Strips, 330
intelligence symbology, 331
macrocognitive perspective, 332–334
RFI manager, 327–328
Joint intentions concept, 110
Judgment process, 101

Keyword indexing, 55–58
compared to LSA, 57–58
Knowledge, 20
building, 1, 4, 8, 145–150, 309, 333
CENTER and, 246–249
construction, 2, 20, 26, 223, 376, 379
development, 4, 149, 379
network, 282, 288–290
sharing and transfer, 151–152
sharing unique knowledge, 149–150
Knowledge elicitation approach, 214–215
Knowledge interoperability
in teams, 132–134, 150, 333, 379
Knowledge objects, 135–136

Language, 8
common ground, 87
Latent semantics analysis (LSA), 54–58, 262–264, 359
compared to keyword indexing, 57–58
TeamPrints and, 262
LTM knowledge, 71

Macro level processing, 7, 18, 143, 203, 411
Macrocognition, 1, 18
characteristics of, 9
construct, 7
development summary, 397–401
how findings make a difference, 401–404
measures summary, 406–412
military perspective, 3
research gaps and challenges, 404–406
Macrocognitive processes, 21–24, 29–30, 67
complex team problem-solving, 143–158
individual knowledge object development, 35
significant, 376–379
tagging, 61–62

McGrath's group process model, 89–91, 93
stages, 91
Meaning making, 8
Mediated cognition, 8
Mental model
development, 143, 148
see also shared mental models
Mental simulation, 155
Meta-matrix networks, 282–283, 287–290
Meta tags, 210
Metacognition, 8, 18, 20, 240, 241
processes, 203
Metrics, 10
objective, 16
Microcognition, 67
Micro level processing, 7, 18, 143, 203, 411
MIT's Electronic Card Wall software, 209, 356
Monitoring
CCT and, 241–242
Multi-agent teams, 111
Multimedia to share information
current technology, 168
slideshows vs. videos for non-experts, 168–171

National culture, 224–225
Naturalistic decision-making, 2, 3, 95, 211, 351, 391
scenarios, 386–388
versus laboratory research, 353–354
Navy Tactical Decision-Making Under Stress (TADMUS), 263
Negotiation
solution alternatives, 155–156
Network-Centric Operations (NCO), 2, 3, 277–278, 352, 394
Networked technology, 306
limitations, 308–309
Network Text Analysis (NTA), 282
Non-combat Evacuation Operation (NEO) scenario, 211, 218, 356–357
TeamPrints, 264–266

Operational Planning Team, 211
Option generation, 153–156
Outcome
appraisal, 156
evaluation and revision, 224
Organizational culture, 225

Participant Subjective Situation Awareness Questionnaire (PSAQ), 268, 271
'Passing the bubble', 167–168, 173, 174–181
Pathfinder, 56
Pattern recognition, 23, 65–66
discrete elements
chunking, 66
knowledge-building and, 147–148
Pena and Parshall's problem seeking methodology, 193
Perceived workload, 230
Perceptual learning, 71
Perspective-making, 136
Perspective-taking, 136–137
Physiological neural networks, 8
Power distance,
Preconceptual phase, 154
Problem-solving, 187, 309
types of situations, 374–375
Priming, 171
Procedural learning, 71
ProNet, 56–57, 58, 60

Real-time collaborative surfaces, 193
Recognition Primed Decision Making model (RPD), 67, 78
Reconfigurable team augmentation, 114–116
autonomy, 114–115
learning, 115–116
level of support, 117
planning horizon, 115
position in team hierarchy, 116
reactivity, 115
Reductive bias, 144
Risk-taking, 225
Rules
adherence to, 225

Schema-enriched communication, 134–137
depth of meaning, 134
Sense-making, 130, 187–189, 192, 193–197, 200, 201–204, 318, 330
Sensitive measures, 10
Sequential communication content, 54
Sequential communication data, 54
Sequential communication flow, 54
Shared mental models, 52, 130, 154, 261, 398–399
accuracy, 42
antecedents and consequences, 36–37
content, 42

- convergence, 35–36, 38–41
- defined, 35–36
- degree of integration, 42–43
- implications of, 45–46
- implicit learning and, 80
- macrocognition and, 44–45
- measurement of, 41–44
- multi-agent teams and, 110–111
- multidimensionality, 42
- revision, 43
- structure, 42
- terminology, 36
- triangulation, 43–44

Shared problem conceptualization
- development, 150–153
- developing common ground, 152
- knowledge sharing and transfer, 151
- team problem model, 152–153
- visualization of data, 151

Shell for TEAMwork (STEAM), 110
Situated learning, 8
Situational awareness, 165, 259, 277–299
- assessment, Enron example, 284–297
- augmenting video to share, 165–184
- CKI,
- coordinating representations, 309, 313–314
- distributed teams and, 165–184
- IMAGES, 278
- intelligence analysis, 325
- macrocognition and, 279–280, 298–299
- organizational network types,
- predicting, 263–264, 265
- PSAQ, 268, 271
- SAGAT, 267, 271
- simulation experiment, 266–272
- team communication and, 261
- team performance and, 260–261
- through communication assessment, 280–284

Situation Awareness Global Assessment Technique (SAGAT), 267, 271
Shared Lightweight Annotation TEchnology (SLATE), 10, 305–306, 309–319
- characteristics and requirements, 306
- coordinating representations, 309, 313–314
- key innovations, 306–309

Slideshows
- versus video clips, 168–171

SOAR, 3
Social network, 282–283, 287–290
Sociotechnical systems (STS), 112–114
- sociotechnical team structure, 113–114

Solution alternative negotiation, 143
Somatic marker, 154
Spacial annotations, 309, 311–312
Special Operation Forces, 305
- SLATE, 305–306, 309–319

Static communication data
- summary measures, 54

Static content measure, 54
Static flow measures, 54
Stigmergy, 68
Stills, 173, 174–181
- versus animation, 171–173

Storyboarding, 156
Supervision, 225

Tacit knowledge, 171
- *see also* expert knowledge

Tactical network topology (TNT), 366, 368
Team knowledge development, 379
Team mental models, 36
Team negotiation, 23, 28, 35, 119, 153, 155, 158, 202, 353, 360, 378
Team performance
- predicting, 263, 265–265

TeamPrints, 261–262, 268, 271–272
- NEO exploratory analysis, 264–266
- predicting situational awareness, 263–264, 265
- predicting team performance, 263, 264–265

Team problem model, 152–153
TeamWise Collaboration theory, 17
Template Theory, 69, 79, 81
The NoteCards, 193
Time, Interaction, and Performance (TIP) theory, 91
Transactive memory, 309, 310–311
Transformational Teams, 2, 107, 113, 117, 118, 120
Trend analysis, 147

Uncertainty awareness, 242–243
Uncertainty reduction (UR)
- about information, 147–148
- about teammates, 148–150

Unmanned aerial vehicles (UAV), 368
Unmanned autonomous vehicle (UAV) videos, 166, 168

Unmanned ground vehicles (UGV), 368
Unmanned surface vehicles, 368–369

Video annotation, 172, 173–174
Video clips, 173, 174–181
  versus slideshows, 168–171
Virtual teams, 277
Visiospatial cognitive resources, 66
Visualization of data, 151
Vocabulary schema, 152

Workload, 3, 58, 109, 114, 342, 374, 407
  assessment, 230
  cognitive, 212, 375
  perceived, 230, 231, 268, 271–272
  subjective, 357, 359